**$37.00**



*Illustrated Guide to the*

# INTERNATIONAL PLUMBING & FUEL GAS CODES

*Howard C. Massey*

**Craftsman Book Company**
6058 Corte del Cedro / P.O. Box 6500 / Carlsbad, CA 92018

# Acknowledgements

Portions of this work are reproduced from the 2003 edition of the *International Plumbing Code®*, and the *International Fuel Gas Code®*, copyrights 2003, International Code Council, Inc., Falls Church, Virginia. Reproduced with permission.


Although the International Code Council, Inc. has granted permission for the reproduction of tables contained within these two codes, Craftsman Book Company indemnifies them from any claim arising from this reproduction.



**Library of Congress Cataloging-in-Publication Data**

Massey, Howard C.
Illustrated guide to the international plumbing & fuel gas codes / by Howard C. Massey.
p. cm.
Includes index.
ISBN 1-57218-120-6
1. Plumbing--Standards--Handbooks, manuals, etc. 2. Pipe fitting--Standards--Handbooks, manuals, etc. I. Title: International plumbing & fuel gas codes II. Title.
TH6125.M3597 2003
696'.1--dc21 2003055332


Layout design by Ian McCauley

# Contents

Chapter 1

# Let's Talk About the Plumbing Code

Everyone who designs or installs plumbing systems knows how important it is to follow the plumbing code exactly. Even a small mistake can keep a plan or job from being approved, causing expensive delays. However, there's no step-by-step explanation to guide you through following the plumbing code.

Back when I was a plumbing contractor, I had the experience of having my work turned down by the authorities. Probably, if you've been in the business for any length of time, you too have had a job held up by a plumbing inspector for some minor discrepancy that you felt didn't amount to much. No matter what your viewpoint, the plumbing code is a fact of life that every plumbing professional has to deal with. My purpose in writing this book is to make it as simple and painless as possible for you to keep up with the code. Whether you're a plumbing engineer, designer, contractor or a plumber, I believe you'll find the information you're looking for right here.

Every plumbing professional is expected to know the code. You're not going to get paid until the plumbing you've designed or installed passes inspection. So, you must know the code and follow it to the letter. But trying to learn plumbing from a code book can be like finding your way through a maze of blind alleys. Plumbing codes are complex regulations that were written for enforcement purposes, not as a guidebook for plumbers. Yet, lack of knowledge and failure to understand minimum code requirements have left many a would-be plumbing professional discouraged, confused and frustrated.

This book will help you understand the important design and installation principles you need to do professional-quality plumbing jobs. What you learn here should be applicable nearly anywhere in the U.S., regardless of the code used in your particular area. Reading and understanding what is written here should be much easier than reading and understanding the code. However, this book doesn't claim to be the plumbing code. You'll still need to refer to your local code from time to time. But this book can be very handy as an answer book for code problems. For example, if you're having a hard time figuring out the rules for a certain type of vent, go to the chapter or section in this book on vents. You'll find a clearer explanation than you'll find in the code, and probably a drawing or two. You'll get a background in what the code demands, so you can save time and avoid expensive mistakes.

## A Word About the Code Itself

The plumbing code I'm referring to in this book is the 2003 *International Plumbing Code (IPC)*. You'll need a copy of whichever version of this code that's being enforced in the community where you do business. Some building and zoning departments sell the code they enforce. Bookstores specializing in construction books definitely sell the *IPC*. If you have trouble finding it, contact the publisher at:

(ICC®) International Code Council®, Inc.
5203 Leesburg Pike, Suite 600
Falls Church, VA 22041-3401
800-786-4452
*www.iccsafe.org*

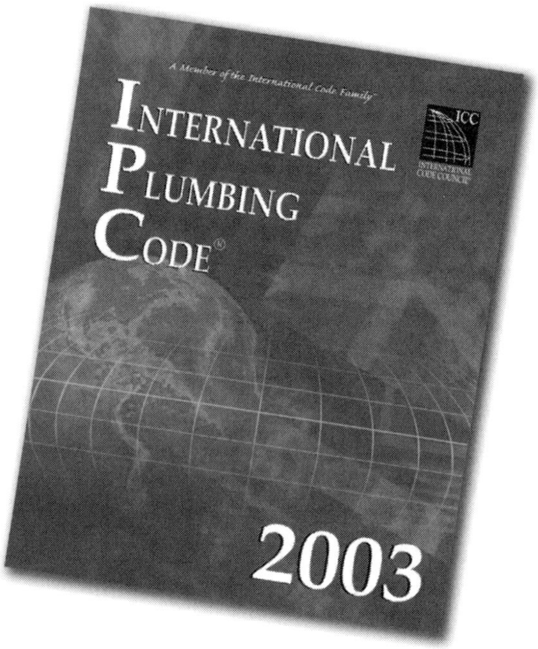


Most communities don't have the resources and time to create their own plumbing code from scratch. So they adopt one of the model codes published by one of the national code-writing organizations. But before a community adopts a code, the local jurisdiction may change selected sections, delete others, or even add material that they think is necessary to meet their particular needs. But once adopted, the code (with changes) becomes a regulation that's enforced like a law. That code remains in effect as adopted until it's amended or replaced by the adoption of another code.

Every three years the *IPC* is re-published as a recommendation and made available to the various building departments. Each local area then decides if it will adopt the revisions as regulations for their particular community. Most adopt the revisions routinely, but again, some make changes. And even if you know which code your community follows, it's not safe to assume it has adopted the most recent version. Many jurisdictions, especially smaller ones, are still enforcing older versions of the code. You can see why it's essential that you have a copy of the code *that's current in the city or county you work in.*

Let's say you've been doing quality plumbing work for 10 years and finally have your master's license on your office wall. Your head is bursting with knowledge about the plumbing code. Still expect that you're going to have an occasional dispute with your plumbing inspector. Perhaps a word of advice is in order here. I've been on your side. I was a plumbing contractor for 14 years. I've also worked on the inspector's side for 15 years and have probably heard most of the arguments. Take it from me — you're not going to win many direct confrontations with an inspector.

If a dispute is minor, it's best to modify your work according to the inspector's wish. He can make your life pretty unpleasant on this and/or future jobs, as he has the power of the building department behind him to enforce the code. Most inspectors would, of course, prefer to have your voluntary cooperation. It's wise to adopt a positive attitude in your relationships with the building department. When taking out a permit, if there's a question about how to do a particular installation, ask for an interpretation from the plumbing chief, the plans examiner or the field inspector. They can tell you what type of installation they'll accept. This helps keep disputes to a minimum.

You'll always have one thing in your favor. The inspector didn't write the code and can only enforce the code as it's written. He can require anything the code demands, but that's all! If he insists on something that you feel isn't a code requirement, you can challenge him. Obviously you'll need a copy of the current code to do this. Ask the inspector, diplomatically, to cite the specific section involved so you can understand your error. Although he may get a little upset, he should comply. Then, in your code book, read the targeted section. If you still feel the code doesn't support the inspector's view, you're in a better position to try to win him over.

Remember that inspectors make their living by knowing the code. Although they can be wrong, they don't like to admit it. If the two of you can't come to an agreement on what's required, and if you can't persuade him to change his mind, there's a good appeal process available. We'll discuss it later in this chapter.

Plumbing inspectors know they can only enforce what the code requires. The national code-writing organization did leave some room for interpretation. This gives code officials room to negotiate, if it seems in the best interest of all concerned. Section 104.2 of the IPC says, in part:

> *The code official shall have authority as necessary in the interest of public health, safety and general welfare to adopt and promulgate rules and regulations to interpret and implement the provisions of this code to secure the intent thereof and to designate requirements applicable because of local climate or other conditions.*

Simply put, this means that the building department has enormous discretionary power to interpret the code.

Sometimes you may face such a tough code issue that there's no way to work it out with an inspector. Before you start appeal procedures, request a meeting with the Chief Plumbing Inspector. Offer to meet when the field inspector can be present. Prepare your case very carefully. Try to show that the code doesn't really make it clear what you're being asked to do. Have a good alternate plan ready and see if you can negotiate a settlement. If your plan doesn't adversely affect public health, safety and general welfare, it's likely that you'll get a satisfactory agreement.

Before you go to this trouble, be certain that it wouldn't be cheaper and easier to just let it go and do what the inspector wants, right or wrong. In any dispute, there's the possibility you could lose. Then you've spent the time and money, and you still have to make the change. Plus, you know the inspector's going to bring his magnifying glass when he comes to inspect your next job.

Obviously, you shouldn't expect a dispute to be waived by a code official if it's a serious violation of the code. Code protection is too valuable to mess with. All code officials realize the importance of the code they administer, and they're not going to stray much from it.

## Basic Principles of the Plumbing Code

Plumbing codes are written and enforced to protect the health of the people who use plumbing systems. Every code requirement listed can be traced directly to basic principles. They're the foundation the plumbing code is based on. Many of these principles go back to the Hoover Code (1924) that is the basis for most model codes used today. These principles are time-tested and clearly show the need for designing and installing a safe and sanitary plumbing system. Some details of plumbing construction may vary, but the basic sanitary and safety principles are the same everywhere. Here are the basic concepts the plumbing code is built on.

- *Potable water:* Every structure equipped with plumbing fixtures and intended for human use must be supplied with a source of potable water. The water supply must be protected against unsafe water sources and must not be subject to backflow.
- *Adequate water:* The plumbing system must provide enough potable water to allow the plumbing fixtures, devices and accessories to function properly without undue noise under normal intended use.

- *Hot water:* In occupied structures, hot water must be supplied to all plumbing fixtures and equipment intended for bathing, washing, culinary use, cleansing, laundry or building maintenance.
- *Water conservation:* Water must be conserved while maintaining proper performance and cleaning of plumbing fixtures.
- *Water heater safety devices:* All storage water heater devices must be designed and installed to guard against overheating or explosion.
- *Public sewer:* In places where a public sewer is available, it must connect to every building (intended for human use) that plumbing fixtures are installed in.
- *Minimum plumbing fixtures required:* Plumbing fixtures must be provided for the type of occupancy of a building and in the minimum number required by the code. For example, every single-family dwelling unit must have at least one water closet, one lavatory, one bathtub or shower, one kitchen sink and one clothes washer connection.
- *Drainage systems:* A drainage system must be designed and installed to keep solid deposits from clogging and fouling piping. Adequate cleanouts must be provided and located so all parts of the system can be readily cleaned.
- *Material durability:* Plumbing fixtures must be manufactured of approved materials, with smooth impervious surfaces, free from defects and concealed fouling surfaces. All porcelain surfaces must be acid-resistant and bear the manufacturer's name. Piping materials must be durable and free from defects. They must be designed and constructed to provide satisfactory service for a reasonable life expectancy. The manufacturer's mark and quality of material must be visible.
- *Plumbing fixture traps:* Fixtures connected directly to a building drainage system must be equipped with a liquid seal trap.
- *Protecting trap seals:* All drainage systems must be designed to provide enough air in the pipes to prevent siphonage, aspiration or back pressure from destroying trap seals under ordinary use and conditions.
- *Vent terminals:* All open vent pipes must be installed to extend outdoors and terminate to the open air.
- *Testing a plumbing system:* A plumbing system must be tested in accordance with code requirements so that all leaks and defects in the installation or materials are found.
- *Excluding certain substances harmful to a building drainage or sewage system:* Any substance that will clog plumbing pipes, produce explosive mixtures, damage drainage systems, or interfere with the normal operation of sewage treatment processes must be excluded from a building drainage system.
- *Prevent contamination:* Proper protection must be provided against sewage backflow, flooding, fouling and stoppage to prevent the contamination of water, sterile goods, distillation, processing, cooling or storage of foods or ice.
- *Light and ventilation:* Washrooms and toilet rooms must be properly lighted and ventilated.
- *Private sewage disposal system:* When a public sewer isn't available, sewage from a building must be provided for by an approved private sewage treatment and disposal system.
- *Prevent sewer backflow:* A plumbing drainage system subject to sewage backflow must be protected by approved means to prevent overflow into a building.
- *Plumbing system maintenance:* All plumbing systems, materials and accessories, both existing and new, must be maintained in a safe and serviceable condition.
- *Fixtures must be accessible:* Plumbing fixtures must be installed to allow easy access for their intended use and cleaning the fixtures and the areas around them.

- *Structural safety:* When installing or repairing any part of a plumbing system, building code requirements must be followed to preserve the strength of structural members and to prevent damage to all surfaces.
- *Protect ground and surface water:* Sewage or other waste from a plumbing system must not be discharged into the ground or into any waterway unless it has been treated by an approved form of treatment.

## Plumbing Code Terminology

Fortunately, most words in the plumbing code mean just what they would in everyday conversation. If you can't understand the meaning of a code term, look it up in a good up-to-date dictionary. But sometimes the English language doesn't quite express the meaning the plumbing code wants. So the plumbing code has assigned special meanings to key terms. To eliminate misunderstandings, the authors of the code include definitions of these terms. Let's look at some examples.

The words *shall* and *may* appear in every plumbing code. We have no problem understanding these words in daily conversation. But in the plumbing code, they have a specialized meaning. *Shall* means you must. It demands that you comply. Take the example of preventing cross-connection in a potable water supply system. The code says, "Cross-connections between a private water supply and a potable public supply shall be prohibited." That means you don't have a choice. As used in the code, *may* is a permissive term. It means allowable or optional, but not required. Here's an example: "Two fixtures set back-to-back may be served by a single vertical drainage pipe." So, a single vertical pipe is permitted with a back-to-back fixture arrangement, but not required. Other options that meet code specifications are acceptable.

Two other commonly misused and misunderstood code terms are *building drain* and *building sewer*. Many professionals assume that both terms refer to the same part of the drainage system. Not so. Here's the code definition of these terms.

> *A building drain is the lowest horizontal collection system which receives the discharge from other drainage pipes within the walls of a building and extends 30 inches beyond the building outer wall.*

The code defines a building sewer as:

> *that part of the horizontal piping of a drainage system which extends from the end of the building drain (30 inches beyond the outer wall of the building) and conveys the liquid waste to an approved point of disposal.*

Before you read the rest of this book, you may find it helpful to look at the glossary at the end of the book. It'll be a handy reference for you because it includes code definitions, plus some useful definitions that I've added. There are also plenty of illustrations to help you understand the definitions. If you're new to plumbing work, I don't recommend that you try to memorize these terms now. But whether you're a veteran or a novice in the field, the glossary will aid your understanding.

## Getting to Know the Administrative Process

The first chapter of the plumbing code tells you who your administrators are. The plumbing department administrators of a local building and zoning department (if large enough) are the chief plumbing inspector, the plans examiner(s) and field inspector(s). Small communities may have one or two inspectors doing it all. They interpret and enforce the plumbing code, approve construction plans and inspect plumbing installations.

The administrative authority that has jurisdiction will create the department of plumbing inspection. The executive code official in charge is known as the chief plumbing inspector. He's appointed by the administrative authority of the jurisdiction and can't be removed from office except for cause and after full opportunity to be heard on specific and relevant charges. He's usually there until he retires, so get to know him and (if possible) even to like him.

The chief plumbing inspector:

- appoints assistants he decides are necessary to carry out the functions of the plumbing department.
- adopts and publicizes rules and regulations.
- decides code intent.
- interprets and implements the code.
- receives applications and issues permits.
- examines any premises considered unsafe, insanitary, dangerous or in any way hazardous.
- issues all necessary notices or orders to ensure compliance with the code.

## Approval of Special Plumbing Systems

Sometimes plumbing installations can't meet the provisions of the code exactly. This happens quite frequently where island sinks and combination waste and vent systems are used. Before installing such a system, if it's not clearly spelled out in the code, check with the chief plumbing inspector. He has the authority to approve changes and modifications in the plumbing code. The details of such changes in a plumbing system will be recorded and filed with the plumbing department.

## Can I Use That Material?

New materials and installation designs for plumbing systems are frequently introduced. You must submit such materials or installation methods to the chief plumbing inspector. He may grant approval when he's satisfied that a proposed design is satisfactory and complies with the intent of plumbing code provisions. When there's not enough evidence that a certain new product or method of construction complies with the plumbing code, the chief plumbing inspector can require you to see that tests are made to prove the manufacturer's claims. These tests must be made at no expense to the plumbing department. If there aren't any recognized and accepted test methods, the chief plumbing inspector must approve the testing procedure you use. All tests for alternate materials or methods of installation must be done by an approved agency. The plumbing department must retain all test results for the period specified for public records.

Don't reuse any materials, equipment, or devices in a plumbing system unless they've been reconditioned, properly tested and approved by the plumbing department. A good rule to follow is to buy all your materials, equipment and devices from your local supplier. It will more than likely stock only materials approved for the area where you do business.

## How Do I Make Appeals?

If you feel you're right and the inspector is wrong, how can you get your work approved? You have the right to appeal an inspector's decision, as Section 109.1 says:

> *Any person shall have the right to appeal a decision of the code official to the board of appeals. An application for appeal shall be based on a claim that the true intent of this code or the rules legally adopted thereunder have been incorrectly interpreted, the provisions of this code do not fully apply, or an equally good or better form of construction is proposed. The application shall be filed on a form obtained from the code official within 20 days after the notice was served.*

The board of appeals is made up of five persons appointed by the chief appointing authority. Each member is appointed to a staggered five-year term. This way, four members of the board are always experienced with board procedures. Each board member must have the following professional disciplines:

1. One member must be a registered architect or a person having 10 years' experience as a builder or superintendent of building construction.
2. One member must be a registered mechanical and plumbing engineer, or a mechanical and plumbing contractor with at least 10 years' experience.

3. One member must be a registered structural engineer.
4. A fourth member must be an electrical engineer, or an electrical contractor with at least 10 years' experience.
5. The final member must be a fire protection engineer, or a fire protection contractor having at least 10 years' experience.

A member of the board of appeals must not hear an appeal if that member has any interest in that particular case. There are two appointed alternate members who must be called by the board chairman if a member is absent or disqualified. Alternate members must meet the same requirements as regular board members. They serve for five years, or until a successor has been appointed. The board members must annually select someone to serve as board chairman.

By a vote of three members, the board may approve, change or reverse a plumbing official's decision. Once the board of appeals has resolved the problem, the plumbing official must then take immediate action in accordance with its decision.

If you have a minor dispute that's easily correctable, you're better off to simply make the correction and not confront the inspector. Suppose he says, "That sink arm (sink drain) is 2 inches over the limit set by code. I want it shortened to meet code requirements." I'd strongly recommend that you cut off the 2 inches, rather than argue with an inspector you may be working with for years to come. The main criterion for deciding whether to appeal is if the matter is in "the gray area" of code requirements. Or, if an inspector's changes could cost you big bucks, consider taking it to the board. They'll listen carefully to both sides and make a judgment call. Hopefully, if the call is in your favor, there should be no hard feelings between you and the inspector, because you won fair and square.

Let me share with you my first and only experience before the board of appeals during my plumbing contracting days. This was when cast iron was used in drainage, waste and vent systems. At that time, the code required buildings three stories or higher to use only extra-heavy cast iron pipe and fittings. My plumbers installed, tested and inspected all the below-ground plumbing using the required extra-heavy material. About that time a new material called "centrifugally-spun service weight" appeared on the market. It was advertised as being superior to extra-heavy pipe and fittings. And the material was stocked by my local supplier who told me it was approved for use in multi-story buildings. Without checking with the plumbing department (my big mistake!), I used it for the job. My plumbers had installed it through the fifth floor before the next inspection was called.

When the plumbing chief in Dade County came out and made the inspection, he looked over the work and signed it off as being okay. About three hours later he returned to the job and turned it down. When my foreman called and told me what had happened, I almost had a heart attack. I could see my business going down the drain if I had to tear out all that material and replace it with extra-heavy cast iron pipe. I began to do a little detective work. I discovered that my plumbing department had already approved this new material and planned to incorporate it into the plumbing code as an amendment, two months down the road. I tried discussing this with the chief inspector but got nowhere. He did recommend that I take it to the board of appeals though (I guess he didn't want to stick his neck out). So I did.

In those days, the board of appeals had nine members appointed by the board of county commissioners. The members were construction trade professionals, as is required for the present board, but only one was a plumbing contractor. I presented my case as a plumbing contractor, explaining that the material in question was approved but would not be officially acknowledged for another two months. I explained that the lost labor and material to replace all the installed material could put me out of business. The board considered the evidence and made an exception in my case. They ruled against the plumbing chief and in my favor. The plumbing chief didn't testify against me, which was good. I know that I was one of the lucky ones.

Several years later I closed my business and went to work for that same plumbing department. Of course, the same plumbing chief who turned down my job was now my boss. I worked as a field inspector for several weeks and then there was a vacancy in the plumbing plans examiner's office. The chief bypassed several senior field inspectors and promoted me to this position — a position I held for 15 years. I later became the assistant plumbing chief. The chief and I became very good friends and stayed in touch through the years. So when circumstances warrant, it's okay to go over the head of an inspector!

## What About Permits?

*IPC* Section 106 gives the scoop on when a permit is or isn't required. Section 106.1 says an owner, authorized agent or plumbing contractor must get a permit from the plumbing department to install, alter, repair, replace or remodel any plumbing system which is regulated by the code. Section 106.2 says you don't have to get a permit to simply repair leaks in a plumbing system, clean out obstructions in sewer lines or waste pipes, or to repair faucets or valves. You also don't have to get a permit to remove one water closet and reset another on the same waste opening.

However, once you get a permit, you must begin work within 180 days from the date the permit was issued to you. If you start work on a building and for some reason you have to stop or temporarily abandon the job, you must start again within 180 days of your last inspection. If you wait more than 180 days, you'll have to get a new permit, which will cost one-half the price of the original permit. If the reason for your delay is satisfactory, you may ask for an extension. The plumbing chief may extend the permit for a period of 180 days, but you can never get more than one extension. Don't ever make false statements or misrepresent the facts to get an extension. You may get your permit revoked entirely. Also if you begin work on a job before you get a proper permit, it could cost you the regular fee, plus 100 percent more of that fee as a penalty. It's not worth it.

The city or county you do a job in will set permit fees. These fees, along with other building department fees, are calculated to pay most, if not all, of the costs of the building and zoning department. Your building department is self-supporting and requires little or no money from local taxpayers.

Finally, in all your contract work, it's extremely important that you have correct and current state and local licenses.

### Required Inspections and Testing

A plumbing permit always requires an inspection. It may be a single inspection for replacing a plumbing fixture. Or it may be for a new building requiring several different inspections. For most residential and small commercial buildings, there are three required inspections. These are set in *IPC* Section 107.1.

The permit holder is responsible for notifying the plumbing official about the required inspections as construction progresses. The plumbing inspector has to sign off on the portion of a plumbing system that he approves. If he finds something unsatisfactory, he'll red tag the job and notify the permit holder what he must do to correct the violation.

Here's the order in which new work inspections are made:

1. Ground roughing-in inspection is made *after* you install and properly bed piping in open trenches, but *before* you backfill the trench.
2. Roughed-in inspection for pipes above ground must be made after the roof and framing of the building are completed. All plumbing piping must be in place. For this inspection, you must prepare bathtubs and showers before the wall and ceiling membranes are installed.
3. Final inspection comes after you've installed all plumbing fixtures, appliances, equipment and devices and connected them to the plumbing system. Once the building passes this inspection, it's ready for a certificate of occupancy.

According to *IPC* Section 107.3.1, new plumbing systems and existing systems that have been changed, extended or repaired must be tested to find possible leaks and defects. As permit holder, you must furnish all equipment, materials and labor required for testing a plumbing system. Of course, if your work doesn't pass inspection, you must make the corrections and call for a reinspection. Some building departments charge an additional fee for reinspections.

As far as you're concerned, a building is ready for a certificate of occupancy when the final inspection is made and approved. And you can be proud of the code-perfect plumbing facilities you've installed.

# *Chapter 1 — Questions and Answers*

**1-1 Like any code, the *IPC* was written to provide**

(A) a guidebook for plumbers.

(B) the specifics needed by architects and contractors.

(C) the specifics needed by every design plumber.

(D) a set of regulations for code enforcement purposes.

***Answer: D***

*Code response: Section 101.3*
*This book, page 5*

**1-2 The code official shall have the authority to**

(A) waive structural performance requirements.

(B) overrule decisions of appointing authority.

(C) adopt and publicize rules.

(D) all of the above.

***Answer: C***

*Code response: Section 104.2*
*This book, pages 7 and 10*

**1-3 The code official is appointed by**

(A) the state governor.

(B) the building department.

(C) the state legislature.

(D) the jurisdictional authority.

***Answer: D***

*Code response: Section 103.2*
*This book, page 9*

**1-4 If practical problems make it difficult to carry out any *IPC* provision, the person who has the authority to change a plumbing system is**

(A) the code official.

(B) the head of the health department.

(C) the job architect.

(D) none of these.

***Answer: A***

*Code response: Section 105.1*
*This book, page 10*

**1-5 If there's not enough evidence that a new product complies with the *IPC*, the code official can require the product be**

(A) tested.

(B) checked out for six months.

(C) used in an alternate system.

(D) installed in tandem with an approved product.

***Answer: A***

*Code response: Section 105.3*
*This book, page 10*

**1-6 You may never reuse any materials, equipment, or devices in a plumbing system until or unless**

(A) the plans specify used materials.

(B) they've been approved and tested.

(C) your supervisor signs a release.

(D) you notify the code official.

***Answer: B***

*Code response: Section 105.5*
*This book, page 10*

**1-7 According to the *IPC*, a permit is *not* required to**

(A) remove and install a new shower.

(B) remodel an existing and approved kitchen.

(C) re-rough an existing bathroom.

(D) remove and install a new water closet.

***Answer: D***

*Code response: Section 106.2*
*This book, page 12*

**1-8 It's the responsibility of the ________ to schedule each portion of plumbing installation before it's covered or concealed.**

(A) general contractor

(B) permit holder

(C) plumbing inspector

(D) job engineer

***Answer: B***

*Code response: Section 107.1*
*This book, page 12*

**1-9 According to the *IPC*, no part of a drainage system may be covered or concealed until**

(A) it has passed a 50 pounds of pressure test.

(B) the owner has approved it.

(C) it has been inspected and approved.

(D) all defects have been corrected.

***Answer: C***

*Code response: Section 107.1 (Subhead 1)*
*This book, page 12*

**1-10 A building is ready for a certificate of occupancy when**

(A) the power and water are turned on.

(B) the health department gives approval.

(C) all plumbing fixture installations are properly installed.

(D) the final inspection is made and approved.

***Answer: D***

*Code response: Section 107.1 (Subhead 3)*
*This book, page 13*

◆ *Chapter 2* ◆

# How Plumbing Systems Are Regulated

*IPC* Section 301 gives you an overall view of code requirements for general plumbing installations. In this section, the *IPC* gives regulations for plumbing installations that aren't discussed in other chapters of the code. I'll discuss and illustrate specific parts of Section 301 throughout this book.

## General Regulations

General regulations in the plumbing code are broad principles that explain what's required when you design, install and maintain plumbing systems. We'll be looking at general regulations for such topics as:

- proper disposal of waste products.
- excluding harmful materials from a sewer system.
- connections to water supplies.
- plumbing materials labeling.
- rodent-proofing plumbing systems.
- pipe protection.
- trenching, excavation and backfill.
- structural safety.
- piping support.
- floodproofing.
- washroom and toilet room requirements.
- toilet facilities for workers.
- tests and inspections.

Your local plumbing code will require that you connect, directly or indirectly, to an approved water source, every plumbing fixture, appliance or device that requires or uses water to operate properly. As you know, the code also insists that you use accepted engineering practices when you design and install a plumbing system. I'll cover all the regulations that are important to you.

### *Disposing of Liquid Waste*

All plumbing fixtures, drains, appliances and accessories used to receive or discharge liquid wastes must drain into an approved drainage system that meets code requirements. If a public sewer is available, a building sewer must connect to it. If no public sewer is available, the building sewer may connect to an approved private sewage disposal system, usually a septic tank and drainfield.

### *Storm Drainage Systems*

Storm drainage systems carry rainwater to an approved disposal point. Storm drainage includes roof drains, area drains, gutters, downspouts, leaders, building storm drains and storm sewers, catch basins and soakage pits. Storm water can't flow into a building's sanitary drainage system unless the plumbing official approves it — and that's not likely!

## Unacceptable Substances

There are many substances that are considered harmful to building drainage systems, public sewers, or treatment plants. Some of them are grease, flammable waste, sand, plaster, lint, hair, ground glass, ashes, cinders, rags and oil. These, and any other insoluble substance, should never be deposited, by any means, into such a system.

Industrial waste can also be a harmful substance. Your local plumbing department will be the authority on how best to treat waste products from manufacturing or industrial operations. Nearly every form of liquid waste can be treated to eliminate contaminants.

## Plumbing Materials Labeling

All pipes, fittings, devices, appliances and plumbing fixtures must include the manufacturer's name or mark. Pipes and fittings must be stamped with the weight and quality of the product. Plumbing products must be listed or labeled by an approved listing agency, such as American National Standard Institute (ANSI), or American Society for Testing Materials (ASTM). See the Glossary for a complete list of agencies.

## Rodent-Proofing Plumbing Systems

Design and install plumbing systems to keep rodents from entering a building. Here are specific preventive measures that are required:

- Design and install drain inlets having strainer plates so no opening is more than $1/2$ inch.
- Design and construct water meter boxes so rodents can't use them to follow the water service pipe into a building.
- Use approved metal collars to seal any openings made in walls, floors or ceilings.

## Protecting Pipes from Breaking, Corrosion and Freezing

When you design or install piping, make certain it's protected so that no structural strain can be transmitted to the piping system. Use protective sleeves, wrapping or some other code-approved method to protect pipes that pass through concrete or cinder block walls and floors or any other corrosive material. Install code-approved sleeves to protect pipes passing through or under walls. Always provide a relieving arch or pipe sleeve two pipe sizes greater than the pipe itself for pipes passing under a footing or through a foundation wall.

When you install piping materials (other than cast iron or galvanized steel) through drilled holes or notches in studs, joists, rafters or similar members that are less than $1^1/2$ inches from the nearest edge of a framing member, protect the pipe with shield plates (steel plates, or stud guards). If you're installing piping along alleys, driveways, parking garages or other locations exposed to damage, recess it into the wall, or protect it in ways approved by your local plumbing official.

When you design or install plumbing systems in areas subject to freezing temperatures, you must follow these code requirements.

- Don't install water, soil or waste pipes in outside walls, attics, crawl spaces, or outside a building unless you provide insulation or heat in accordance with your local plumbing code.
- Install water service pipe at least 12 inches below grade and no less than 6 inches below the frost line.
- The 2003 *IPC* doesn't regulate building sewers that connect to private sewage or to public disposal systems. Check out minimum depth requirements with your local plumbing official.

## Waterproofing Openings

Make vent pipes that penetrate a building roof watertight by using lead, plastic, aluminum, copper, galvanized steel or other approved flashings or flashing materials. Always make sure pipes that penetrate exterior walls are watertight.

## Trenching, Excavation and Backfill

Where a trench bottom forms a bed for pipe, make sure the pipe barrel between joints is in firm contact with solid ground. Provide bell holes, hub holes and coupling holes where the pipe is joined. You can't use blocks to support the pipe to grade. If a trench is overexcavated, be sure to backfill the trench with sand or fine gravel you place in 6-inch layers to the installation level of the bottom of the pipe.

If you hit rock when installing pipe, be sure to observe the following code requirements. First, remove the rock to a depth of 3 inches below the installation level of the bottom of the pipe. Use fine materials for the backfill and tamp them into place up to the installation level of the pipe to provide uniform load-bearing support for the barrel of the pipe between joints.

When backfilling a trench containing a pipe, use loose earth that's free of rocks, broken concrete, frozen chunks or other rubble. Place the loose earth in 6-inch layers and tamp it in place until the top of the pipe is covered by at least 12 inches. Then bring the backfill up evenly on each side of the pipe so it will stay aligned. Compact the backfill under and beside the pipe to support the pipe evenly.

Some approved piping materials for outside installation are considered fragile and require special installation procedures. These include plastic, vitrified clay pipe, bituminized fiber pipe and concrete pipe. Follow the manufacturer's installation instructions if they're more restrictive than your local code.

Under certain conditions you may have to install pipe by tunneling or jacking, or a combination of both. If so, be very careful to protect the pipe from damage during installation or from subsequent uneven loading. If you use tunnels, you must provide adequate support structures to prevent future settling or caving.

## Structural and Fire Safety

When you install a new plumbing system, or remodel or repair an existing plumbing system, you must be sure you leave the building in a safe structural condition. You'll find the limitations for cutting, notching or boring framing members to receive piping in your local building code.

Some buildings have a required fire-resistance rating. When you install pipes that penetrate fire-resistance floors, walls or ceilings, protect the assemblies in accordance with your local building code. The architect will generally give you the building code requirements in the plans or job specifications.

When you trench for piping that must run parallel to the footings of a building, be sure the trench bottom doesn't extend below the 45-degree bearing plane of the footing or the wall. See Chapter 9 for installation details.

## Supporting Pipes Within a Building

Generally speaking, you must support all piping in accordance with the plumbing code. Here are some specific code requirements:

- Where earthquake loads are applicable, follow your local building code. You may find the necessary piping support design on the blueprints. If not, the local building department should have the information you need.
- Any device you install must support the weight of the piping plus its contents. Use hangers and strapping material made of approved material. These items must not promote or cause galvanic action.
- Attach support devices to the building structure in an approved manner.
- Check local code requirements for piping supports for different piping materials.
- When you install 4-inch or larger pipe that changes direction more than 45-degrees, use rigid sway bracing supports.
- Check Chapter 9 for code-required spacing intervals in horizontal and vertical piping support.

- Allow for pipe expansion and contraction as needed. Use expansion joint fittings that are compatible with the type of piping materials you install.
- Support the base of stacks using concrete, bricks laid in mortar, metal brackets attached securely to the building structure, or other approved methods.
- Where you use parallel water distribution systems, support the pipes following your local code requirements. Where direction changes are necessary, provide supports in accordance with the manufacturer's installation instructions. But don't ever install hot water piping where it can come in contact with cold water piping.

### Flood Proofing Required

For buildings in areas subject to flooding, the building code requires that plumbing systems resist hydrostatic and hydrodynamic loads and stresses, including the effects of buoyancy. The plumbing engineer will note on the plans or specifications how parts of the plumbing system must be protected. Here is a list of the parts you must protect:

- water service pipes
- pump seals where pumps are installed below the base flood elevation in individual water supply systems
- potable water well casing covers that extend less than 1 foot above the base flood elevation
- sanitary drainage system
- storm water drainage system
- manhole covers unless they extend to or above the base flood elevation
- building equipment and other plumbing piping systems

### Washroom and Toilet Room Requirements

The building and mechanical codes specify the requirements for lighting, ventilation and type of finished surfaces in toilet rooms. When you install plumbing fixtures, piping and equipment, be sure they don't interfere with normal operation of windows, doors or other openings. Water closets installed in public toilet rooms must occupy a separate compartment with walls and a door to ensure privacy. There's an exception for a water closet installed in a single-occupancy toilet room with a door that locks. This water closet doesn't have to be enclosed in a separate compartment. In day care and childcare facilities, a toilet room that has two or more water closets may have one water closet that's not enclosed in a compartment.

The general contractor is expected to provide adequate toilet facilities for the number of workers he expects to have at a construction job, and must maintain them in a sanitary condition. It's best to use approved portable toilets.

## Tests and Inspections

When your plumbing job is ready for inspection, here's what you need to do to prepare for the inspection:

### General Test Requirements

1. Call for the appropriate inspection. Be sure to allow enough time for the plumbing inspector to make the inspection. Generally, one day's notice is considered enough.
2. Provide all equipment, material, power and labor necessary for the inspection. The system must withstand the test pressure as prescribed by code.
3. Activate test pressure by water or air. Most often, it'll be by water.

### Drainage Test Requirements

1. Remove cleanouts or plugs to make sure the test pressure is reached in all parts of the system if the inspector requests it.
2. Make a final test of the entire system after the plumbing fixtures are set and their traps filled with water.

3. If you use water to test a system, you can test the system in sections or all at one time. Close the system openings tightly to the highest opening and fill the system to overflow. If the system is no more than one story high, or if you test it in sections, the system must withstand a 10-foot head of water. The system must withstand the test for at least 15 minutes and must be tight at all joints.

4. If you use air to test a system, force air into the system and maintain a constant pressure of 5 pounds per square inch for at least 15 minutes. Again, the system must be tight at all joints.

5. Sometimes a completed drainage and vent system is tested with smoke to see if it complies with the plumbing code. In this case, fill all traps with water. Use a smoke machine to fill the entire system with pungent, thick smoke. When smoke appears at stack openings, close them tightly. Maintain a pressure equal to a 1-inch water column for 15 minutes. Then the final inspection can begin.

6. Test gravity sewers by filling the building sewer with at least a 10-foot head of water. Maintain this pressure for at least 15 minutes.

7. Test interior storm drainage piping the same as a sanitary drainage system.

### *Water Test Requirements*

1. Test a water piping system (in sections or all at once) using a water pressure that's at least equal to the normal working pressure of the public system that the water piping is connected to. Use potable water for the test. If there's no potable water, use air at a pressure of at least 50 pounds per square inch.

2. Inspect backflow prevention devices to make sure they're operable. Follow the manufacturer's installation instructions on how frequently the devices must be tested. If the manufacturer doesn't specify the frequency, IPC Section 312.9.1 requires the testing be annual. Let's hope that in this computer age, the owner will be notified when an inspection is due.

## Condensate Disposal

Condensate is generally a clear water waste that contains insignificant amounts of impurities. This waste is generated by water lifts, expansion tanks, liquid combustion by-products of condensing appliances, equipment and appliances containing evaporators or cooling coils, cooling jackets, sprinkler systems, drip or overflow pans, or similar devices. *IPC* Section 314 requires that condensate be disposed of to an approved plumbing fixture or disposal area, according to the manufacturer's instructions.

In Chapter 8, we'll cover drainpipe materials, sizes and installation methods in detail.

# Chapter 2 — Questions and Answers

**2-1 According to the *IPC*, a plumbing system shall not discharge industrial waste into the public sewer unless ______ determine(s) the waste won't damage the public sewer system.**

(A) the health department head

(B) the job engineer

(C) the code official

(D) all of these

***Answer: C***

*Code response: Section 302.2*
*This book, page 18*

**2-2 According to the *IPC*, all pipe, fittings, appliances and plumbing fixtures must be**

(A) installed by a licensed plumber.

(B) labeled with the manufacturer's name.

(C) job prefabricated.

(D) firmly attached.

***Answer: B***

*Code response: Section 303.1*
*This book, page 18*

**2-3 According to the *IPC*, piping installed in corrosive material must be**

(A) first approved by the code official.

(B) made of copper or copper-plated.

(C) properly protected.

(D) provided with ½-inch clearance from the corrosive material.

***Answer: C***

*Code response: Section 305.1*
*This book, page 18*

**2-4 When installing ³/₄-inch copper water pipe through a hole in a stud that's less than 1¹/₂ inches from the nearest edge of the stud, you must protect the pipe with**

(A) insulation.

(B) metal sleeves.

(C) an approved wrapping.

(D) shield plates.

***Answer: D***

*Code response: Section 305.8*
*This book, page 18*

**2-5 If you run into rock when digging a trench for piping, the *IPC* requires you to remove the rock to at least ______ inches below the bottom of the installed piping.**

(A) 2
(B) 3
(C) 4
(D) 10

***Answer: B***

*Code response: Section 306.2.2*
*This book, page 19*

**2-6 According to the *IPC*, you must install rigid support sway bracing for pipe sizes ______ inches and larger, when the pipe changes direction more than 45 degrees.**

(A) 4
(B) 6
(C) 8
(D) 10

***Answer: A***

*Code response: Table 308.6*
*This book, page 19*

**2-7 When using water to test a drainage and vent system, you must use at least a/an ______ foot head of water.**

(A) 8
(B) 10
(C) 12
(D) 16

***Answer: B***

*Code response: Section 312.2*
*This book, page 21*

**2-8 When using air to test a drainage and vent system, you must keep a pressure of 5 pounds per square inch for ______ minute(s).**

(A) 1
(B) 5
(C) 10
(D) 15

***Answer: D***

*Code response: Section 312.3*
*This book, page 21*

**2-9 According to the *IPC*, when testing a water supply system with water, the system must not leak under a water pressure that's**

(A) determined by the code official and city engineer.

(B) 15 pounds over the working pressure of the public system.

(C) 15 pounds under the working pressure of the public system.

(D) equal to the working pressure of the public system.

***Answer:* D**

*Code response: Section 312.5*
*This book, page 21*

**2-10 To rodent-proof a building, the *IPC* requires that no drain strainer plate opening can exceed**

(A) $^1/_{16}$ inch.

(B) $^1/_8$ inch.

(C) $^1/_4$ inch.

(D) $^1/_2$ inch.

***Answer:* D**

*Code response: Section 304.2*
*This book, page 18*

# Chapter 3

# The Sanitary Drainage System

Drainage and vent systems are among the more difficult areas of the code to understand. Because of this, it's best to look at these systems separately. In this chapter, we'll take an in-depth look at a private sanitary drainage system. Remember, I'm using the new *IPC*. Even if you don't use the *IPC*, this book will help guide you through your own code book. You may be pleasantly surprised by how few differences there are between the two. The *IPC* is a first cousin to the three major codes now used nationwide.

This chapter covers:

- all sanitary drainage pipes within the wall line of a building on private property that receive liquid waste or other waste substances (whether in suspension or in solution).
- the drainage pipes that carry this waste to a public sewer or other approved, private sewage disposal system.

## Isometric Drawings

I can't imagine any professional plumber accurately estimating the materials and labor needed for a drain, waste and vent (DWV) system without drawing a layout. I think you'll agree this is a necessary step to figure out the correct size and number of pipe and fittings. It will also help you figure your labor cost, which is based on the number of joints a plumber can put together in a day.

The DWV arrangements are the heart of a plumbing system. This part of a plumbing system is the most complex, misunderstood, and misinterpreted section of the plumbing code. It's the section professional plumbing designers, mechanical engineers, master plumbers and plumbing inspectors frequently have trouble interpreting. This section generates most of the questions asked, and displays most of the isometric drawings required to complete current plumbing examinations. Several years ago, when I worked as a plans examiner for Metropolitan Dade County, in Miami, the plumbing contractor on the state licensing board approached me to develop isometric drawings for the Florida State Plumbing Contractor's Exam. He confided in me that the state exam had become ineffective because some individuals taking the open book state exams were passing the exam, not because they knew the material, but because they had the ability to quickly look up the correct answers. As a result, they were getting licensed even though they weren't qualified to be plumbing contractors. As a member of the board, he was concerned. When a bookkeeper from a large plumbing supplier (with no plumbing experience) passed the state plumbing contractor's exam, it was the last straw. He didn't believe anyone should be able to pass the plumbing contractor's exam without considerable field experience, as well as knowledge of the plumbing code and its relationship to isometric drawings.

He authorized me to develop isometric drawings to be part of the Florida State Plumbing Contractor's exam. The first time my drawings were used in the exam, the pass rate dropped from 75 percent to less than 25 percent. The drawings were simple, yet tested a person's general knowledge of plumbing. The sample isometric drawing shown in Figure 3-1 is similar to the ones I created for the exam. Is the drawing (A) correct or (B) incorrect? If you decided that (A) correct was not the right answer, you'd be required to make a corrected drawing in the blank area labeled "(B) incorrect." The answer to this question is (B) incorrect. A corrected drawing is shown in Figure 3-2, along with the code response. You can read the full *IPC* response to this drawing in Section 1104.1.

This isn't a complex drawing, but if you don't know plumbing design and layout, you wouldn't be able to answer correctly. No bookkeeper who's a whiz at looking things up in a book is going to be able to find out the answer to this in the allotted time.

Like the plumbing contractor on the Florida State Licensing Board, I place great importance on knowing how to draw and read DWV isometrics. An isometric drawing shows a three-dimensional view of where the pipes in a plumbing system go. To give the isometric drawing an appearance of depth, pipes in an isometric drawing meet at angles of 120 (not 90) degrees.

The code doesn't illustrate or explain isometrics. DWV isometrics are generally required by your building department for commercial plans, but not residential. The isometric drawings for a large commercial building are made by the building's architect. When working on commercial projects, you'll need to know how to read and interpret isometrics so you can install the plumbing as it appears on the approved plans.

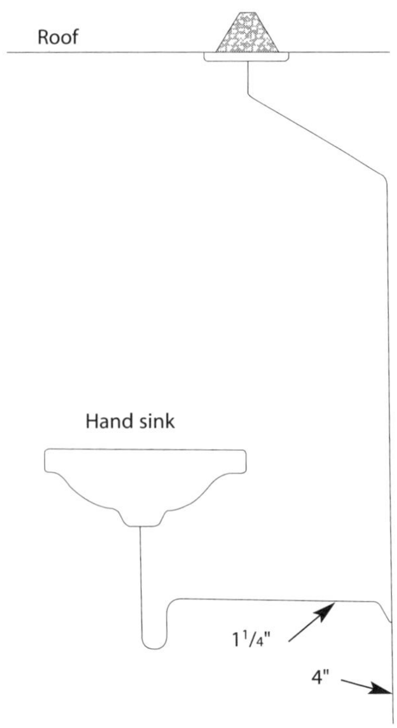


*(A) correct*

*(B) incorrect*

**Figure 3-1**

*According to code, the isometric drawing shown is __. (Pipe size is not a factor.)*

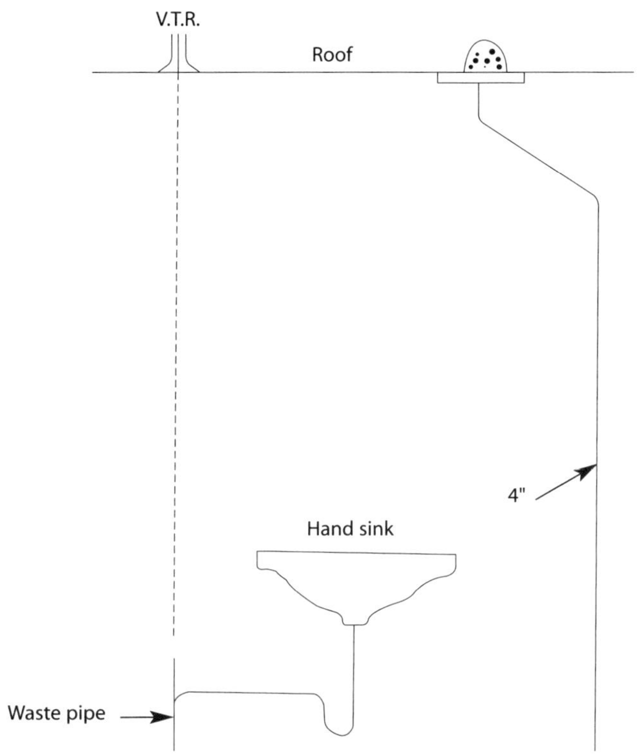


*Answer: (B) Incorrect. The correct drawing is shown above*

**Figure 3-2**

*Note: Code requires that all plumbing fixtures used to receive or discharge liquid wastes shall be connected properly to the drainage system, and that rainwater piping shall not be used as waste pipes.*

Years ago, when I was an apprentice and wanted to learn how to draw plumbing isometrics, I couldn't find any books to help. Fortunately, the plumbing contractor I worked for showed me how to make acceptable isometrics. Having been blessed with some ability to draw, I developed a plan to simplify these drawings.

To begin, there are only three basic pipe angles used to draw a plumbing system — horizontal, vertical and 45-degrees. These three angles are shown in the left half of Figure 3-3. You can use these same angles to make other piping diagrams. When you learn to draw sanitary systems, you also learn to draw all the other piping diagrams used in a plumbing system (storm water systems, gas piping, water piping, etc.) In Figure 3-3, the solid lines indicate the pipes you use for a drainage and waste system. The horizontal broken line is just a constant to make the angles more obvious. The right half of Figure 3-3 shows a simple isometric drawing that has the three basic angles you need to design plumbing piping diagrams for any building.

## Making an Isometric Drawing

The only equipment you need to make an isometric is a sharp No. 2 pencil and a 30-60 right triangle. If you follow the directions and practice the exercises in this chapter, you'll be able to produce your own isometric drawings and understand those made by others.

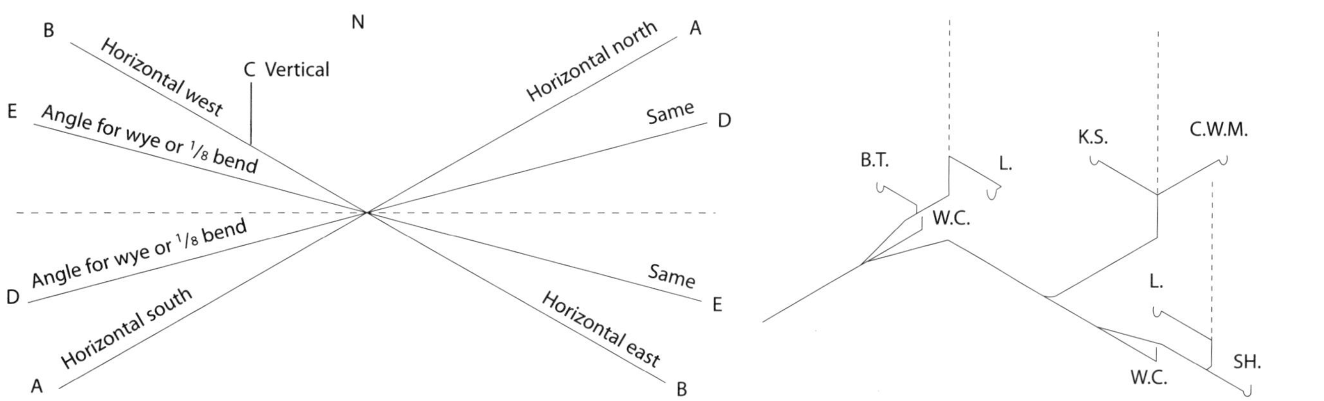


**Figure 3-3**

*Isometric drawing showing various angles*

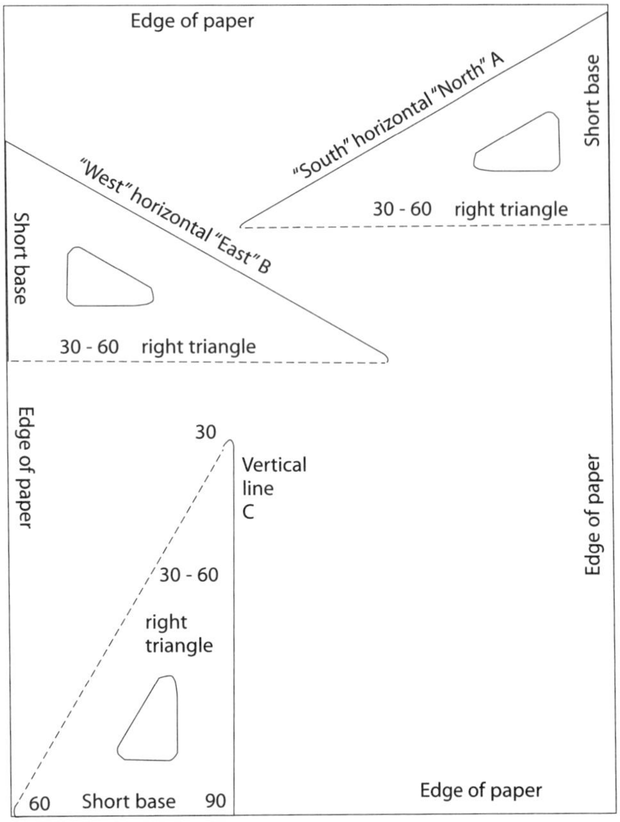


**Figure 3-4**

*Using a 30 - 60 right triangle to draw lines and angles*

First, get a sheet of paper and draw a dot about in the center of the sheet and write the letter "N" at the top of the paper to show the direction of north. Square the short base of your right triangle with the right edge of the paper and draw line A through the center dot. This represents the north-south horizontal pipe. See Figure 3-3. Again using the triangle, square the short base with the left edge of the paper and draw line B through the center dot. This represents the east-west horizontal pipe. To draw the dashed line, put the short side of the triangle at the right or left edge of the paper so the long leg of the right angle of the triangle goes through the center dot. Draw the dashed line along that long leg. To draw line C (the vertical line), square the short base of the triangle with the lower edge of the paper and use the long base as a straight edge. Connect line C with the desired horizontal pipeline. Line C represents the stacks in your isometric. Refer again to Figures 3-3 and 3-4.

To find where to place line E in Figure 3-3, divide the smaller angle between the horizontal line B and the dashed line in half. Then draw line E east and west through the center dot. This shows the change in direction for 45-degree fittings, such as a wye or $^{1}/_{8}$ bend. Use the same method to draw line D north and south.

Take time to practice this process until you get the hang of it. Soon you'll be drawing acceptable plumbing systems. Remember, you can't become proficient in the plumbing trade until you learn how to read and draw isometric layouts correctly.

## Fittings in an Isometric Drawing

The solid lines on isometric drawings represent pipe and fittings for the drainage part of a DWV system. The broken lines represent pipe and fittings for the vent part of a DWV system. The fitting symbols are shown the same way in a drawing regardless of the type of material the pipe and fittings are constructed from. Figures 3-5, 3-6 and 3-7 show isometrics with close-ups of the various pipe fittings shown in the drawings.

## Fixture Abbreviations

There are several common abbreviations used to identify various types of plumbing fixtures in isometric drawings and floor plans. For example, the symbol for a shower and bathtub are exactly the same, so you need to use an abbreviation to differentiate between the two in an isometric drawing. Looking back at the drawing in Figure 3-3, you'll notice that I've used the fixture abbreviations "SH." and "B.T." to indicate the locations of the shower and the bathtub in the isometric of a two-bathroom house.

Abbreviations may differ slightly depending on the illustrator. Some architects and engineers use the letter "L" to designate a lavatory, while others may use "Lav." Figure 3-8 shows the abbreviations I'll use to identify plumbing fixtures in isometric drawings and floor plans throughout this book. There's a complete list of plumbing abbreviations at the end of this book.

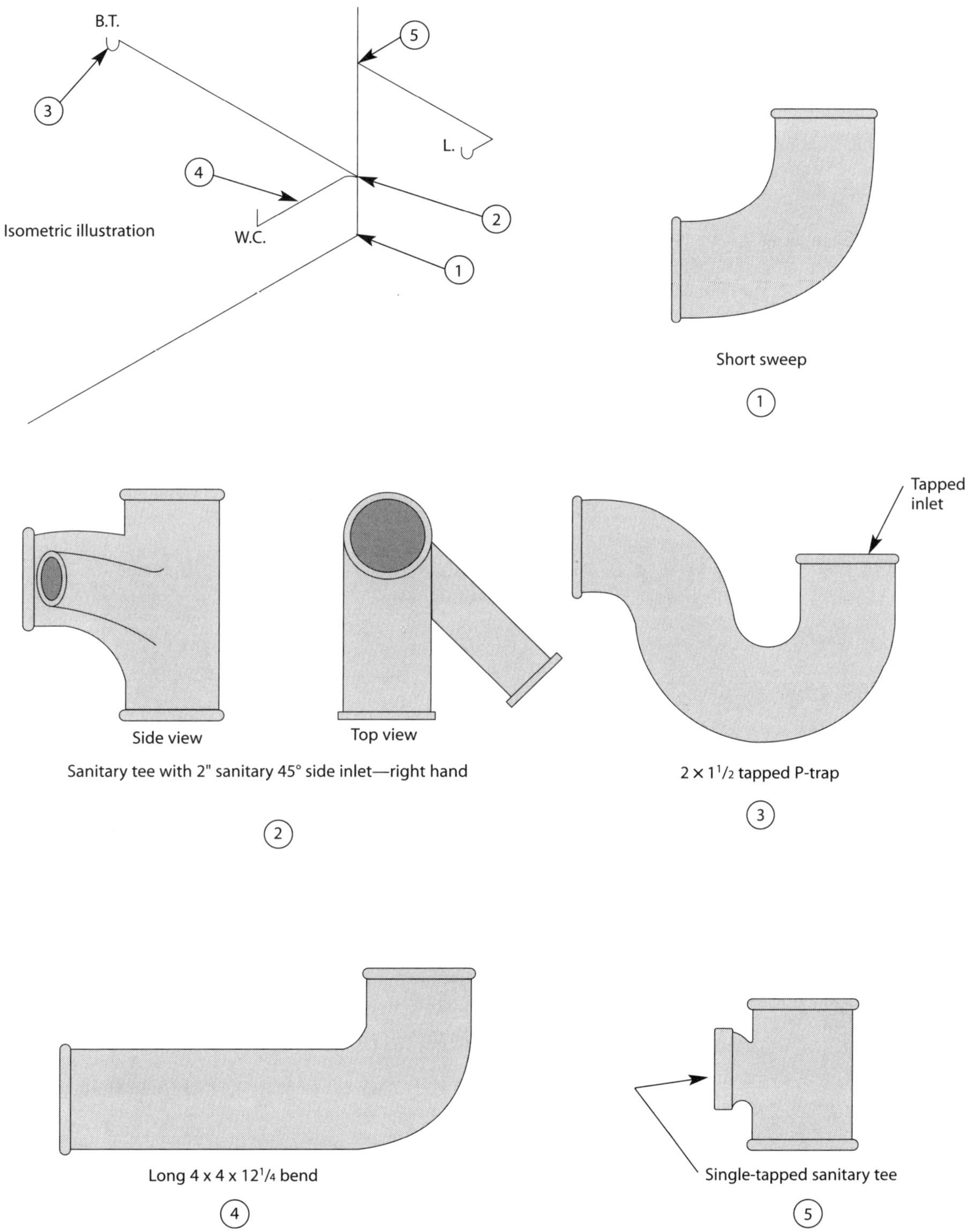


**Figure 3-5**

*Typical isometric with close-ups*

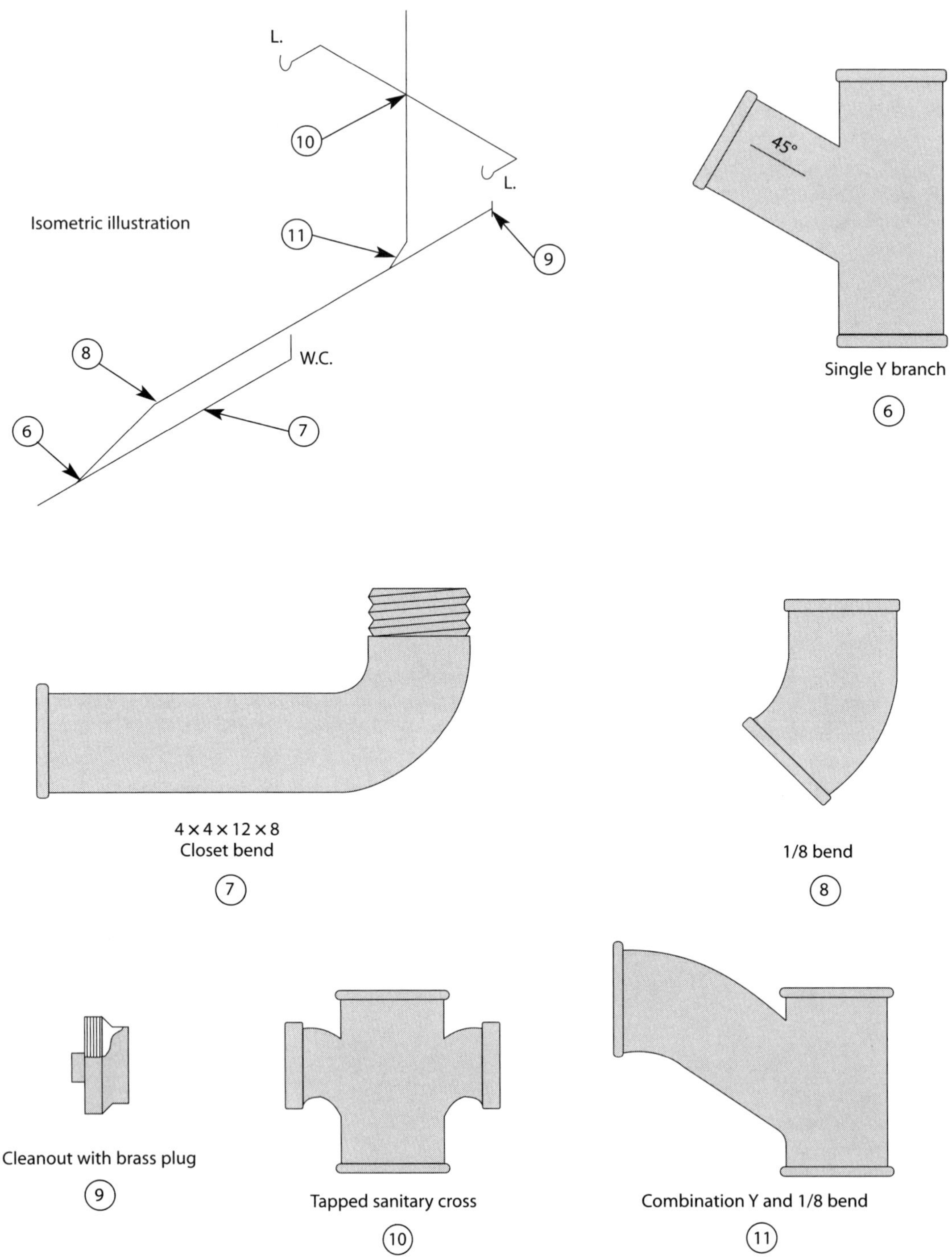


**Figure 3-6**

*Typical isometric with close-ups*

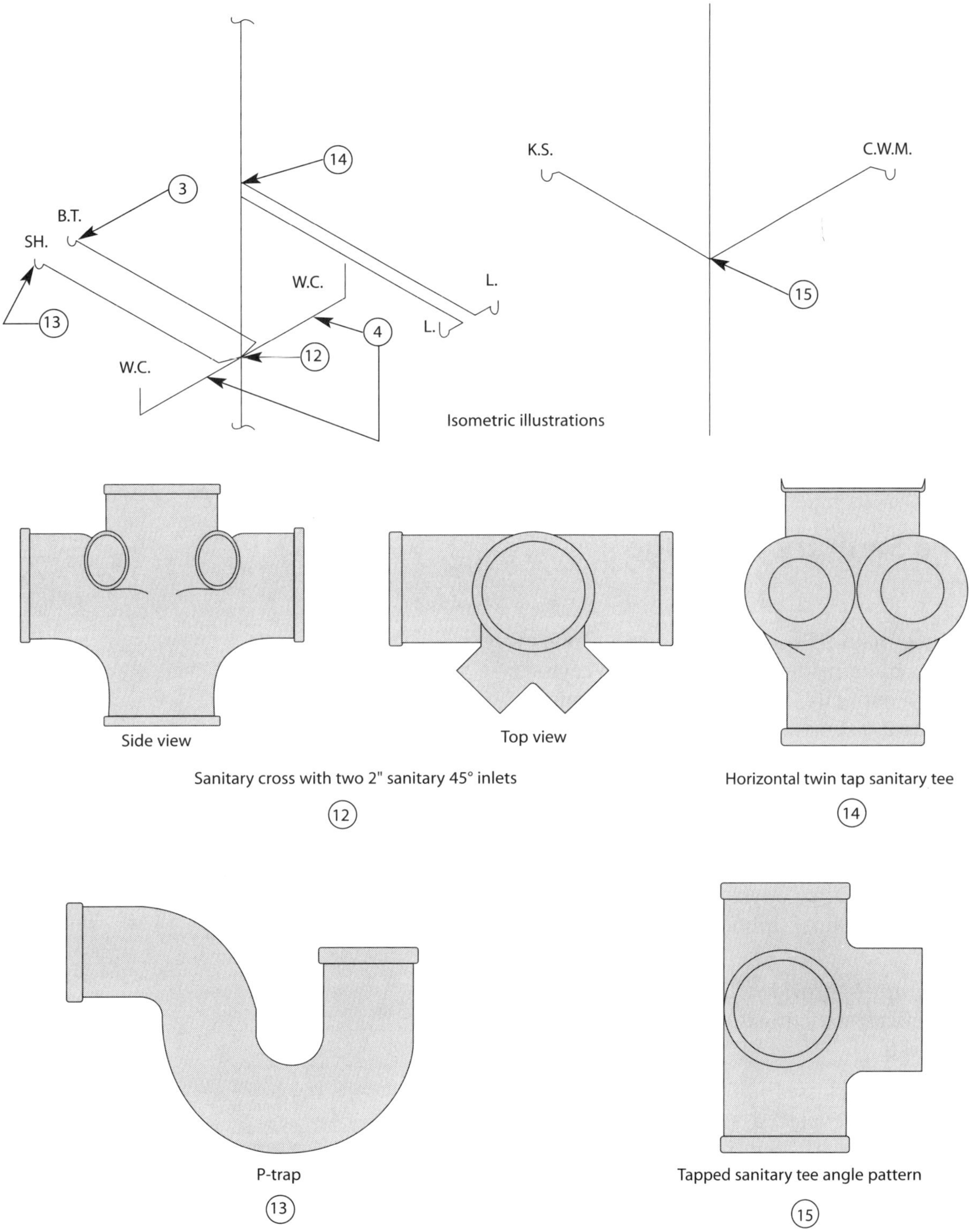


**Figure 3-7**

*Typical isometric with close-ups*

| Plumbing Fixtures | Abbreviations |
|---|---|
| Bathtub | B.T. |
| Cleanout | C.O. |
| Clothes washing machine | C.W.M. |
| Kitchen sink | K.S. |
| Lavatory | L. |
| Shower | SH. |
| Water closet | W.C. |
| Vent through roof | V.T.R. |

**Figure 3-8**

*Typical plumbing fixture abbreviations. (More complete listing at end of book.)*

## Parts of a Sanitary Drainage System

There are three major classes of piping and fittings that make up a building drainage system. Each class must do its job correctly for the system to function properly. Figure 3-9 shows the three major classes of piping and fittings in a simple drainage system.

1. *Drainage pipes* — receive and carry used water to a public or private disposal system.
2. *Vent pipes* — admit air to, and exhaust from, all parts of a drainage system. They protect fixture traps against siphoning and back pressure.
3. *Fixture traps* — provide a liquid seal protection without restricting the flow of sewage or other waste.

These three major classes are so important and complex that I've devoted a separate chapter to each one. Right now we'll focus just on the drainage system.

### Sizing Drainage Pipes

When you size individual pipes, use these factors to determine the minimum vertical and horizontal sizes for drainage pipes:

- maximum fixture unit load
- types of fixtures used
- slope (fall) of drain pipe in inches per foot
- the vertical length (height) of drain pipe

It's common sense that the more a pipe has to carry, the larger the pipe must be. That's also what the plumbing code requires. The volume is measured in drainage fixture units (dfu). Let me explain. Each fixture or device that drains waste into a building drainage system is assigned a value roughly in proportion to the liquid the fixture will discharge per

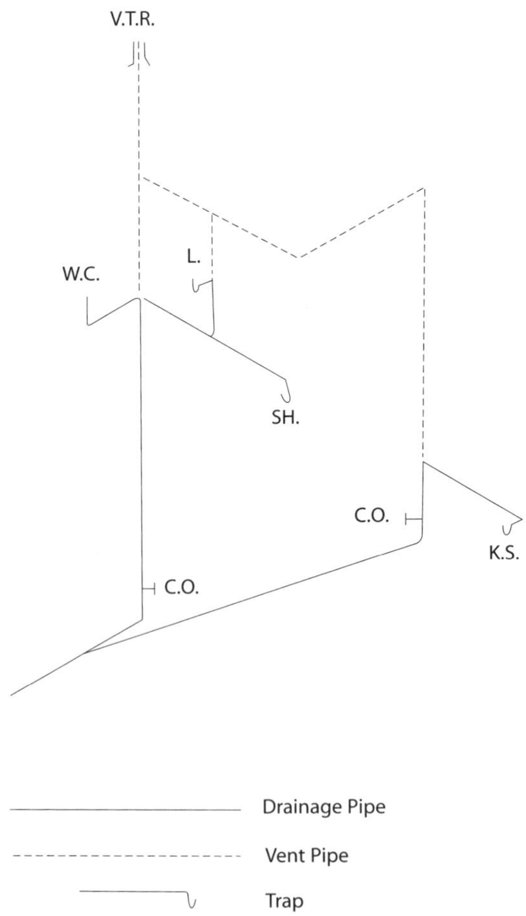


**Figure 3-9**

*Drainage and vent isometric*

minute. For example, the *IPC* assigns 3 dfu to a private water closet that uses 1.6 gallons per flush (gpf) and 4 dfu to a private water closet that uses more than 1.6 gpf. See *IPC* Table 709.1 shown in Figure 3-10. Before saving water became so important, older water closet versions were assigned 6 gpf. As the fixture load factors have downsized over recent years, the size of the pipes have become smaller as well.

You'll find several tables in the *IPC* that show how to figure the total dfu for a plumbing system. As we look at these tables, I'll explain how they work in sizing a plumbing system. Let's try a small job; the principle is the same for any size building.

Look at the isometric drawing in Figure 3-11. You can use *IPC* Table 709.1 shown in Figure 3-10 to figure the total dfu for this system. The individual pipe

**TABLE 709.1**
**DRAINAGE FIXTURE UNITS FOR FIXTURES AND GROUPS**

| FIXTURE TYPE | DRAINAGE FIXTURE UNIT VALUE AS LOAD FACTORS | MINIMUM SIZE OF TRAP (inches) |
|---|---|---|
| Automatic clothes washers, commercial[a, g] | 3 | 2 |
| Automatic clothes washers, residential[g] | 2 | 2 |
| Bathroom group as defined in Section 202 (1.6 gpf water closet)[f] | 5 | — |
| Bathroom group as defined in Section 202 (water closet flushing greater than 1.6 gpf)[f] | 6 | — |
| Bathtub[b] (with or without overhead shower or whirlpool attachments) | 2 | $1^1/_2$ |
| Bidet | 1 | $1^1/_4$ |
| Combination sink and tray | 2 | $1^1/_2$ |
| Dental lavatory | 1 | $1^1/_4$ |
| Dental unit or cuspidor | 1 | $1^1/_4$ |
| Dishwashing machine,[c] domestic | 2 | $1^1/_2$ |
| Drinking fountain | $^1/_2$ | $1^1/_4$ |
| Emergency floor drain | 0 | 2 |
| Floor drains | 2 | 2 |
| Kitchen sink, domestic | 2 | $1^1/_2$ |
| Kitchen sink, domestic with food waste grinder and/or dishwasher | 2 | $1^1/_2$ |
| Laundry tray (1 or 2 compartments) | 2 | $1^1/_2$ |
| Lavatory | 1 | $1^1/_4$ |
| Shower | 2 | $1^1/_2$ |
| Sink | 2 | $1^1/_2$ |
| Urinal | 4 | Note d |
| Urinal, 1 gallon per flush or less | 2[e] | Note d |
| Wash sink (circular or multiple) each set of faucets | 2 | $1^1/_2$ |
| Water closet, flushometer tank, public or private | 4[e] | Note d |
| Water closet, private (1.6 gpf) | 3[e] | Note d |
| Water closet, private (flushing greater than 1.6 gpf) | 4[e] | Note d |
| Water closet, public (1.6 gpf) | 4[e] | Note d |
| Water closet, public (flushing greater than 1.6 gpf) | 6[e] | Note d |

For SI: 1 inch = 25.4 mm, 1 gallon = 3.785 L.

a. For traps larger than 3 inches, use Table 709.2.
b. A showerhead over a bathtub or whirlpool bathtub attachments does not increase the drainage fixture unit value.
c. See Sections 709.2 through 709.4 for methods of computing unit value of fixtures not listed in this table or for rating of devices with intermittent flows.
d. Trap size shall be consistent with the fixture outlet size.
e. For the purpose of computing loads on building drains and sewers, water closets or urinals shall not be rated at a lower drainage fixture unit unless the lower values are confirmed by testing.
f. For fixtures added to a dwelling unit bathroom group, add the DFU value of those additional fixtures to the bathroom group fixture count.
g. See Section 406.3 for sizing requirements for fixture drain, branch drain, and drainage stack for an automatic clothes washer standpipe.



**Figure 3-10**

*Drainage fixture units for fixtures and groups*

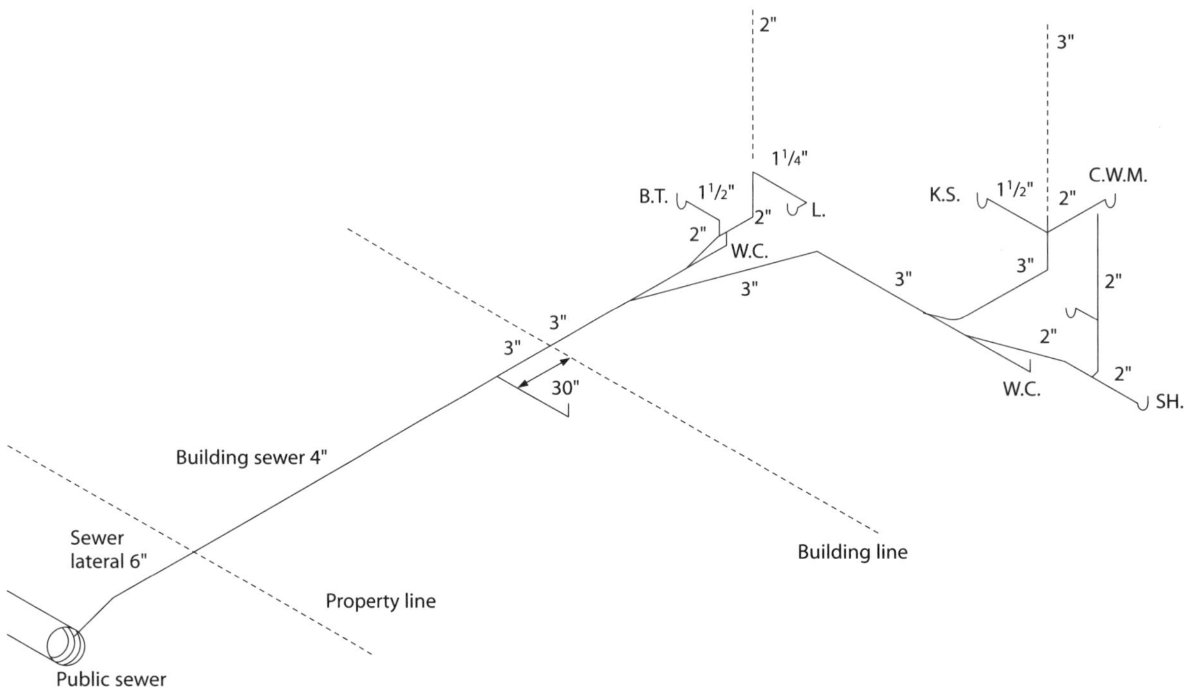


**Figure 3-11**

*Residential plumbing drainage system*

sizes in some instances will be from other tables, footnotes or sub-sections. We'll consider these as we go, but right now let's just look at the total dfu load.

First, we'll assume these are water-saving water closets using 1.6 gpf. On the third line of *IPC* Table 709.1, a bathroom group as defined in *IPC* Section 202, Chapter 2, Definitions, has a water closet, lavatory, bathtub or shower, with or without a bidet, an emergency floor drain, or both. In the example, we have two bathroom groups, one kitchen sink and one clothes washing machine. Their combined dfu is 14. Figure 3-12 shows how they're listed.

## *Special Fixtures*

Naturally, codes can't list every type of fixture, appliance, equipment or device that discharges some kind of waste into a building drainage system. Some special fixtures you use aren't listed in *IPC* Table 709.1. These fixtures are usually found in commercial buildings — they include items such as service sinks, mop sinks, bottle coolers, milk or soft drink dispensers, ice-making machines and coffee urns, to name a few. *IPC* Table 709.2 (shown in Figure 3-13) and a few sub-sections take care of this problem.

*IPC* Table 709.2 lists dfu values for drain or trap sizes up to 4 inches. Use this table to size fixtures not listed in *IPC* Table 709.1. For example, some service sinks are manufactured with a 3-inch waste outlet and a 3-inch special trap. *IPC* Table 709.2 assigns 5 dfu to a 3-inch fixture drain or trap. When you figure the dfu for individual drains in a building, add all special fixture dfu first. Then add that total to the dfu for the other fixtures in the building.

## *Continuous and Semi-Continuous Flow Devices*

When you figure total dfu, watch for devices not in *IPC* Tables 709.1 and 709.2. These devices (pumps, air conditioning equipment and equipment with indirect waste piping) can discharge directly or

| Figure type and quality | Fixture unit (dfu) as load factors |
|---|---|
| Bathroom group consisting of: 1 water closet, 1 lavatory and 1 bathtub | 5 |
| Bathroom group consisting of: 1 water closet, 1 lavatory and 1 shower | 5 |
| 1 kitchen sink, domestic | 2 |
| 1 clothes washing machine, residential | 2 |
| **Total Fixture Units:** | **14** |

**Figure 3-12**

*Calculating residential fixture loads*

TABLE 709.2
DRAINAGE FIXTURE UNITS FOR FIXTURE DRAINS OR TRAPS

| FIXTURE DRAIN OR TRAP SIZE (inches) | DRAINAGE FIXTURE UNIT VALUE |
|---|---|
| $1^1/_4$ | 1 |
| $1^1/_2$ | 2 |
| 2 | 3 |
| $2^1/_2$ | 4 |
| 3 | 5 |
| 4 | 6 |

For SI: 1 inch = 25.4 mm.



**Figure 3-13**

*Drainage fixture units for fixture drains or traps*

indirectly into a drainage system. Use the manufacturer labels on these devices to get the flow rate information required by the *IPC*. *IPC* Section 709.3 says that the continuous and semi-continuous flow into a drainage system must be computed on a basis that 1 gpm of flow equals 2 dfu. So, if you know the flow rate, you can easily calculate the dfu for any device.

The final information you need to get the total dfu for a building is the dfu for indirect waste receptors. These waste arrangements are generally found in restaurants, cafeterias, clubs, lounges and similar establishments. *IPC* Section 709.4 requires that "the drainage fixture unit load of an indirect waste receptor receiving the discharge of indirectly connected fixtures shall be the sum of the drainage fixture unit values of the fixtures that discharge to the receptor, but not less than the drainage fixture unit value given for the indirect waste receptor in *IPC* Tables 709.1 or 709.2."

I'll try to simplify this statement with an example. Let's say you're installing a 4-inch waste receptor (floor drain or floor sink) in a restaurant, and you have three pieces of equipment with indirect waste pipes draining into it. One drain pipe is 1-inch, one is $^3/_4$-inch and one is $^1/_2$-inch. Each drain is less than $1^1/_4$-inches. So according to *IPC* Table 709.2, each drain is 1 dfu, for a total of 3 dfu. Now, since 3 dfu for the three indirect waste pipes is less than 6 dfu for the 4-inch receptor, the code requires you to use the 6 dfu to figure the total dfu when you size the drainage system.

List the numbers and types of fixtures and their dfu, including special fixtures and devices, to accurately figure the maximum dfu load for any building.

## Sizing Drainage Pipes for Single-Family Residences

Now that we have some knowledge of how to use the tables and sub-sections of the code to figure total dfu, let's advance to sizing some drainage piping. *IPC* Table 710.1(1), Building Drains and Sewers, shown in Figure 3-14, lists the pipe size, slope (fall) per foot and maximum dfu for each horizontal pipe. The most acceptable slope per foot for horizontal pipe is either $^1/_4$ inch or $^1/_8$ inch. *IPC* Section 704.1 requires that horizontal drainage piping be installed at uniform slopes in accordance with Table 704.1. This table is shown in Figure 4-34 in Chapter 4.

You may run into certain construction problems where you can only use a $^1/_{16}$- or $^1/_2$-inch slope, though both are legal and listed in *IPC* Table 710.1(1). These slopes are considered special, so when you need them, be sure to check them out

**TABLE 710.1(1)**
**BUILDING DRAINS AND SEWERS**

| DIAMETER OF PIPE (inches) | MAXIMUM NUMBER OF DRAINAGE FIXTURE UNITS CONNECTED TO ANY PORTION OF THE BUILDING DRAIN OR THE BUILDING SEWER, INCLUDING BRANCHES OF THE BUILDING DRAIN[a] | | | |
|---|---|---|---|---|
| | Slope per foot | | | |
| | 1/16 inch | 1/8 inch | 1/4 inch | 1/2 inch |
| 1 1/4 | — | — | 1 | 1 |
| 1 1/2 | — | — | 3 | 3 |
| 2 | — | — | 21 | 26 |
| 2 1/2 | — | — | 24 | 31 |
| 3 | — | 36 | 42 | 50 |
| 4 | — | 180 | 216 | 250 |
| 5 | — | 390 | 480 | 575 |
| 6 | — | 700 | 840 | 1,000 |
| 8 | 1,400 | 1,600 | 1,920 | 2,300 |
| 10 | 2,500 | 2,900 | 3,500 | 4,200 |
| 12 | 3,900 | 4,600 | 5,600 | 6,700 |
| 15 | 7,000 | 8,300 | 10,000 | 12,000 |

For SI: 1 inch = 25.4 mm, 1 inch per foot = 83.3 mm/m.

a. The minimum size of any building drain serving a water closet shall be 3 inches.



**Figure 3-14**
*Building drains and sewers*

with your plumbing inspector before installing them. He'll probably agree with you and give his stamp of approval unless he's able to come up with an alternative.

Let's look back at the isometric drawing shown in Figure 3-11 to see how to size residential DWV pipes. We'll begin with the most distant bathroom, the one with the shower. We need to size the building drain before sizing the branch drains. The code says in *IPC* Table 710.1(1), Footnote (a) that the minimum size building drain for a water closet is 3 inches. Now, we have to size the building drain at 3 inches all the way to the building sewer, which begins 30 inches beyond the building wall line.

Next we'll size the building sewer. As you know, most codes require a minimum 4-inch building sewer. The *IPC* doesn't consider sizing a building sewer specifically but leaves it up to the total dfu load. We've already figured out that this residence's dfu is 14. According to *IPC* Table 710.1(1), the building sewer could be sized 2 inches, since this size pipe at 1/4-inch slope per foot can carry up to 21 dfu. However, the controlling factor here is the building drain. The code says in Section 704.2 that the size of drainage piping can't be reduced in the direction of flow. So the minimum size sewer for this residence must be 3 inches.

Having now sized the main building drain and the building sewer, let's move on to sizing the branch drain and wet vents. Our first step is to size the horizontal branch drain and vertical waste pipe up to the kitchen sink and clothes washer. It has to carry 4 dfu. *IPC* Table 710.1(1) permits up to 21 dfu on a 2-inch pipe at 1/4-inch slope per foot. So, a 2-inch pipe would meet code specifications here. *IPC* Table 916.1 (shown in Chapter 4, Figure 4-28) allows up to 12 dfu on a 2-inch waste pipe to be vented by a 1 1/4-inch vent stack — if the length isn't more than 30 feet. So the stack vent could be sized at 1 1/4 inches.

Now we need to size the wet vents serving the two bathrooms. *IPC* Section 903.1 says that every building where plumbing is installed shall have at least one stack (main stack) that's at least one-half the required size of the building drain. In this case, we could use a 1 1/2-inch stack. Now it gets a little tricky, for there are controlling factors in another section and table of the code that must be considered. Let's look at each of these.

*IPC* Section 909.1 says that within a bathroom, the wet vent is considered to be the pipe connected to the dry vent. It extends in the direction of flow in the drain pipe to the most downstream fixture drain connection, to the horizontal branch drain. Only the fixtures within the bathroom group can connect to it. Other fixtures in the area must connect downstream of the wet vent.

In *IPC* Table 909.3 (look ahead to Chapter 4, Figure 4-21), a 2-inch wet vent can carry up to 4 dfu. Here's where code interpretation comes in. Does the wet vent serve only the bathtub and lavatory (or shower)? Or does it also serve as a vent for the water closet? If it includes all three fixtures, then the wet vent would have to be 2$^1/_2$ inches, because *IPC* Table 709.1 shows 5 dfu for each bathroom.

What we end up doing in a case like this is resorting to accepted engineering practice. As shown in Figure 3-11, we size the pipes larger than the minimum requirements. A main vent in a building should be at least 3 inches. A 2-inch wet vent is considered adequate for a single bathroom. And certainly, the inspector won't turn down a 4-inch sewer.

The code isn't easy to understand, even for those of us who have spent a lifetime studying and interpreting it. In many sections the requirements aren't clearly stated, so a lot is left up to good engineering practice. When you're in doubt, call or visit your plans examiner or plumbing chief to find out how to interpret a code section so your work will be acceptable to them.

## Sizing Drainage Pipes for Multi-Story Buildings

You need *IPC* Table 710.1(2) shown in Figure 3-15 to compute the dfu, pipe size and permitted length of soil and waste stacks for multi-story buildings. Look at the five-story building isometric shown in Figure 3-16. It's just a simple design to help you understand how to use *IPC* Tables 710.1(1) and 710.1(2). The fixtures in the example include lavatories and showers. This is how I used the code to size the pipes, as illustrated.

**TABLE 710.1(2)**
**HORIZONTAL FIXTURE BRANCHES AND STACKS[a]**

| DIAMETER OF PIPE (inches) | MAXIMUM NUMBER OF DRAINAGE FIXTURE UNITS (dfu) | | | |
|---|---|---|---|---|
| | | Stacks[b] | | |
| | Total for a horizontal branch | Total discharge into one branch interval | Total for stack of three branch intervals or less | Total for stack greater than three branch intervals |
| 1$^1/_2$ | 3 | 2 | 4 | 8 |
| 2 | 6 | 6 | 10 | 24 |
| 2$^1/_2$ | 12 | 9 | 20 | 42 |
| 3 | 20 | 20 | 48 | 72 |
| 4 | 160 | 90 | 240 | 500 |
| 5 | 360 | 200 | 540 | 1,100 |
| 6 | 620 | 350 | 960 | 1,900 |
| 8 | 1,400 | 600 | 2,200 | 3,600 |
| 10 | 2,500 | 1,000 | 3,800 | 5,600 |
| 12 | 2,900 | 1,500 | 6,000 | 8,400 |
| 15 | 7,000 | Note c | Note c | Note c |

For SI: 1 inch = 25.4 mm.

a. Does not include branches of the building drain. Refer to Table 710.1(1).

b. Stacks shall be sized based on the total accumulated connected load at each story or branch interval. As the total accumulated connected load decreases, stacks are permitted to be reduced in size. Stack diameters shall not be reduced to less than one-half of the diameter of the largest stack size required.

c. Sizing load based on design criteria.



**Figure 3-15**

*Horizontal branches and stacks*

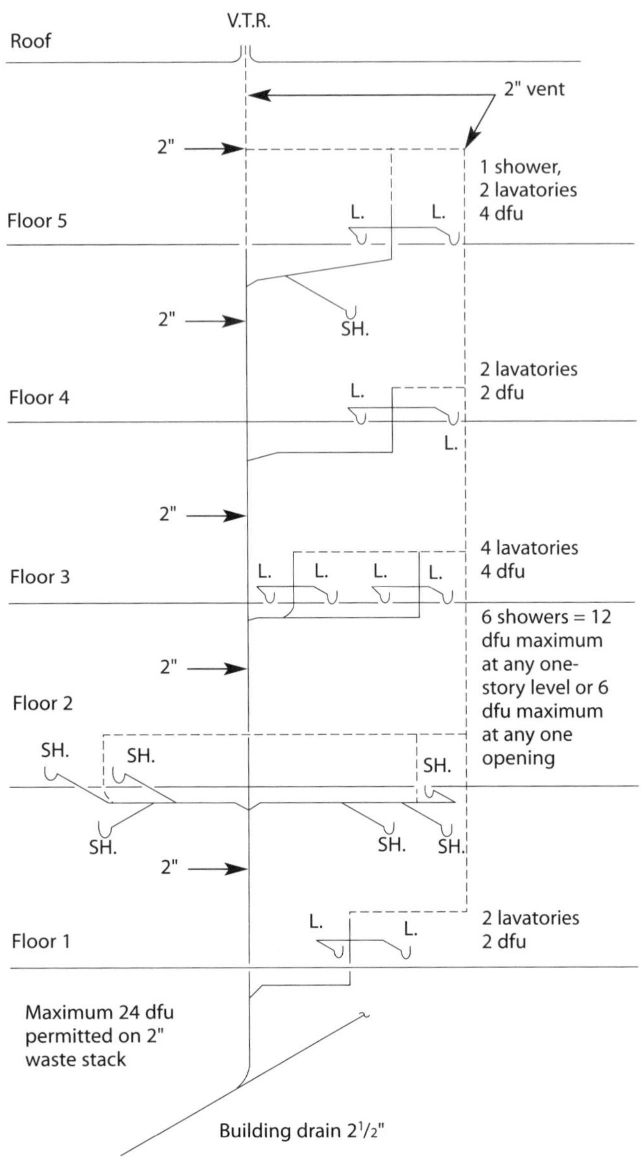


**Figure 3-16**

*Plumbing drainage system for a multi-story building*

Let's start with the vertical waste pipe. We'll use *IPC* Table 710.1(2) to size the vertical waste pipe for our example. The table tells us that the maximum dfu for a 2-inch waste pipe with more than three branch intervals is 24. Also the maximum dfu for any horizontal branch drain at any one story is 6. Remember, this is a waste stack, not a soil stack. This means, according to code definitions (*IPC* Section 202), the pipe carries liquid waste containing no fecal matter.

From *IPC* Table 709.1 we know that each shower in Figure 3-16 is 2 dfu and each lavatory is 1 dfu. From *IPC* Table 710.1(2), we see that you can install 12 showers (24 dfu) on a 2-inch stack, but you must install them on different levels if the building is more than three stories in height. If the building is three stories or less, 10 dfu (5 showers) are the maximum you may install on a 2-inch stack.

You can install three showers (6 dfu) on a horizontal branch drain at each story, or six showers at the same floor level using two openings, as shown on the second floor in Figure 3-16. You could install 24 lavatories (24 dfu) on this 2-inch waste stack at different floor levels, or six lavatories (6 dfu) on a horizontal branch drain using one opening. I mixed the two fixture types in Figure 3-16 so you can see several ways of meeting code requirements.

Now we need to size the building drain serving the stack. We'll use $^1/_4$-inch slope per foot. In *IPC* Table 710.1(1), we see that a 2-inch pipe at $^1/_4$-inch slope can carry up to 21 dfu. So the 2-inch horizontal waste pipe won't be large enough for 24 dfu. We have to move up to the $2^1/_2$-inch pipe size, which is just large enough to meet minimum code requirements. However, you can't always get $2^1/_2$-inch pipes, so you may have to upgrade to the next pipe size — 3 inch.

You can use this same procedure to size soil or waste piping vertically or horizontally, once you know the type of fixtures (which tell you the dfu) and the total dfu for each section of pipe.

### *Sizing Complicated Branch and Building Drains*

Let's move on to a more complicated type building. We'll size the building drain and branch drains for the multi-story building isometric drawing shown in Figure 3-17. We have the total dfu listed at the base for each stack. To size the building drain and branch drains, we'll use *IPC* Table 710.1(1). Again we'll assume a slope of $^1/_4$-inch per foot.

We'll begin with Stack A-1 and work down to the building line. Stack A-1 has 130 dfu, so a 4-inch building drain is okay. Stack B-1 has 350 dfu and will require a 5-inch drain. At the junction of stacks A-1 and B-1 the total dfu has climbed to 480. According to *IPC* Table 710.1(1), we can still use the 5-inch drain.

Stack B-2 has 350 dfu, so its branch drain must be sized at 5 inches. Total combined dfu is now 830, which means we need a 6-inch building drain. Stack C adds 48 dfu at its junction with the building drain, bringing the total up to 878 dfu. Now we have to move up to an 8-inch drain. Stack B-3 has 350 dfu, requiring a 5-inch branch drain. At this juncture the building drain has to accommodate 1228 dfu, but the 8-inch pipe will handle that amount. The branch drain serving the last stack, A-2, has 130 dfu and needs a 4-inch pipe.

This brings the total dfu for the building to 1358. Using *IPC* Table 710.1(1), the building drain, now sized at 8 inches, can continue at this size to the point of disposal.

### *Sizing Vertical Stacks in a Multi-Story Building*

Now let's try sizing vertical stacks in a multi-story building. Let's look at the four-story building shown in Figure 3-18. We'll begin with Stack A at the top floor

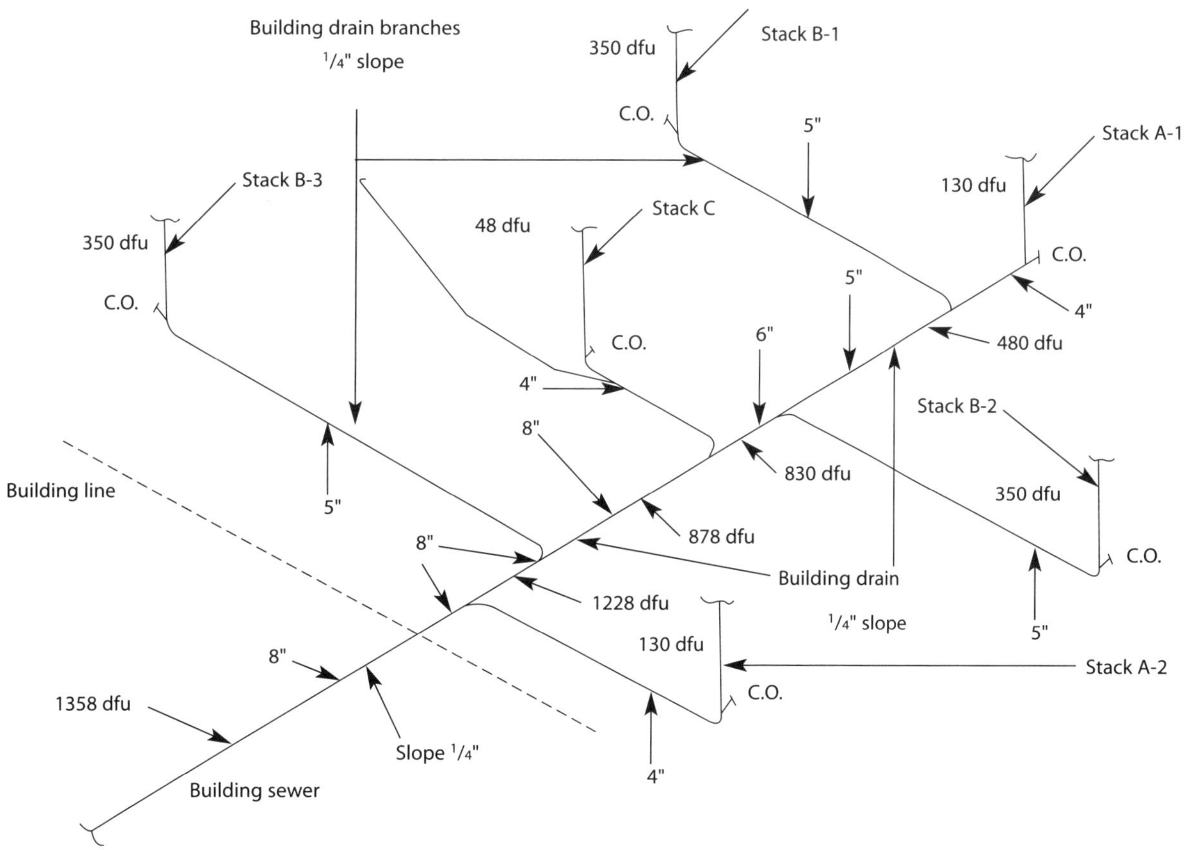


**Figure 3-17**

*Sizing a building drain and sewer*

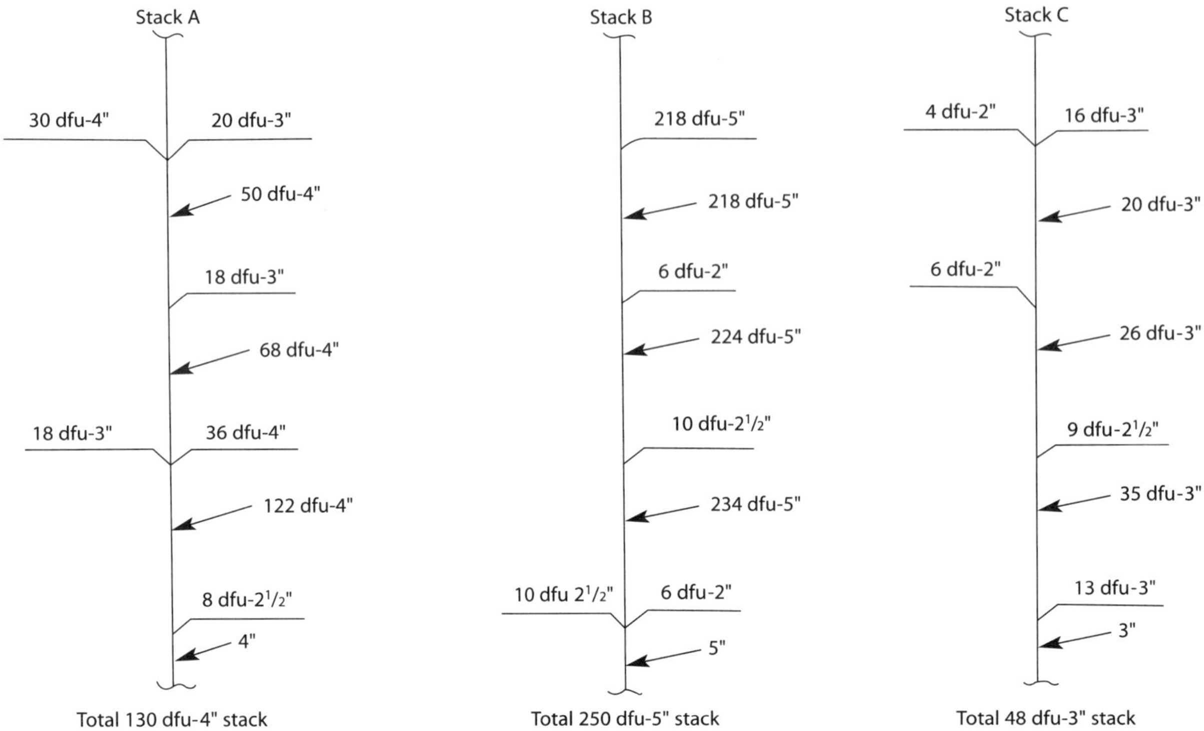


**Figure 3-18**
*Sizing vertical stacks*

and work down each floor to the stack base. Use *IPC* Table 710.1(2) — note that the horizontal branch drains control the size of the stacks.

One branch drain on the top floor has 30 dfu and the other 20 dfu. Using *IPC* Table 710.1(2), the branch drain with 20 dfu can use a 3-inch pipe. But, the branch drain with 30 dfu requires a 4-inch pipe. The third floor branch drain with 18 dfu must be 3 inches. The second floor branch drain with 18 dfu can also be 3 inches, but the other second floor branch drain, with 36 dfu, must be 4 inches. Finally, the first floor branch drain in Stack A with 8 dfu can use a $2^1/2$-inch pipe.

Apply this same procedure to size Stacks B and C. Remember that the branch drains control the size of certain stacks. Drainage pipes, either horizontal or vertical, can't be reduced in size in the direction of flow. Stack B is an example of this. The top floor branch drain, having 218 dfu, must use a 5-inch pipe. The stack has a total of 250 dfu (218 + 6 + 10 + 10 + 6), which would normally require a 4-inch stack. But in this case, the 5-inch branch drain controls the size of the stack.

The only exception to this rule is shown in Figure 3-19. According to *IPC* Section 704.2, a 4- × 3-inch reducing closet bend isn't considered a reduction in pipe size, even through a soil or waste stack can't be smaller than the largest horizontal drain pipe connecting to it. Note that in Footnote (a) of *IPC* Table 710.1(1), the minimum size pipe that can serve a water closet is 3 inches. However, Footnote (c) of *IPC* Table 710.1(2) says the sizing load is based on the design criteria. This gives considerable leeway to the designer and much to interpretation, if questions arise.

The *IPC* allows stacks to be smaller as the total accumulated dfu decreases. However, they can't be reduced to less than one-half of the diameter of the largest stack size required. See *IPC* Table 710.1(2), Footnote (b). Most local codes frown on reducing

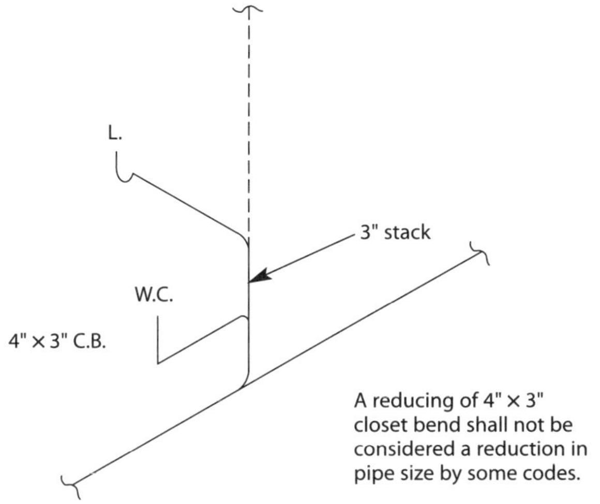


**Figure 3-19**
*Reducing closet bend*

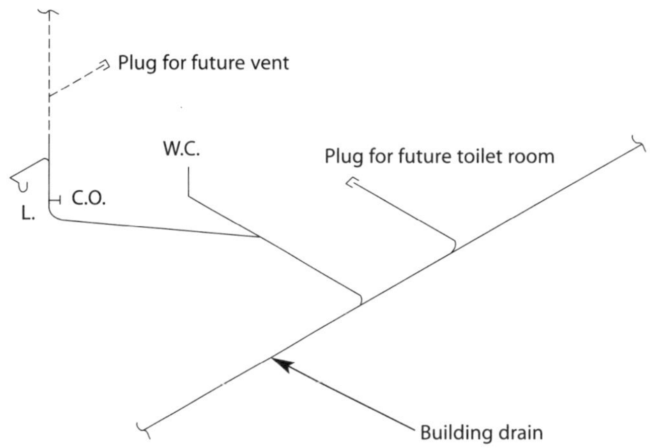


**Figure 3-20**
*Drainage and vent pipe plugged for future installation*

the size of waste and soil stacks in multi-story buildings. Before installing such a system, check the plumbing isometrics on the approved set of job plans or discuss the matter with your plumbing inspector, just to be safe.

### *Allowing for Fixtures in the Future*

You may be required to make provisions for installing fixtures in the future. If so, you need to consider the number and type of such fixtures when you size drain and vent pipes. Use an approved cap or plug to terminate pipes for future use, as shown in Figure 3-20. Refer to *IPC* Sections 704.4 and 710.2 for requirements.

### *Vertical Offsets in Drainage Pipes*

Any vertical stack with an offset of 45 degrees or less is considered straight, and you size it as straight. Refer to *IPC* Section 710.1(2) and Figure 3-21.

### *Horizontal Offsets in Vertical Drainage Stacks*

To size a vertical drainage stack having an offset of more than 45 degrees, as shown in Figure 3-22, you must comply with *IPC* Section 711. The building

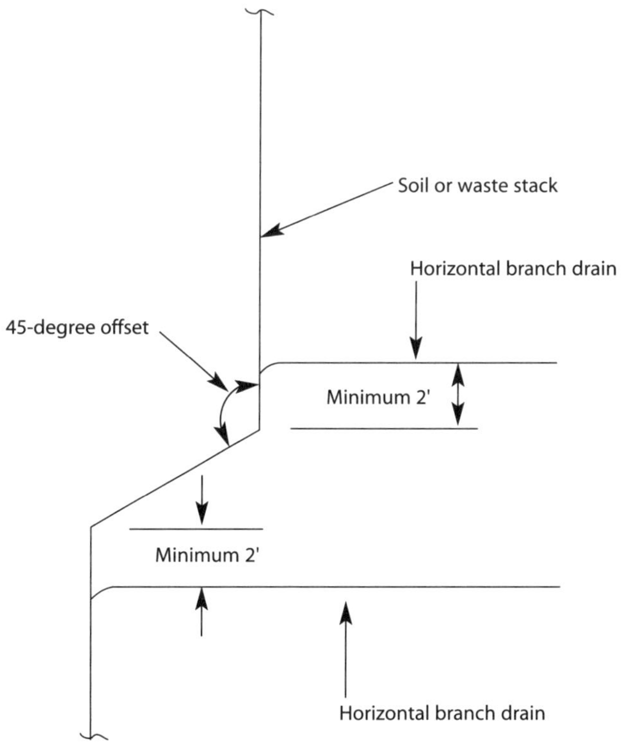


**Figure 3-21**
*45-degree offset in a vertical stack*

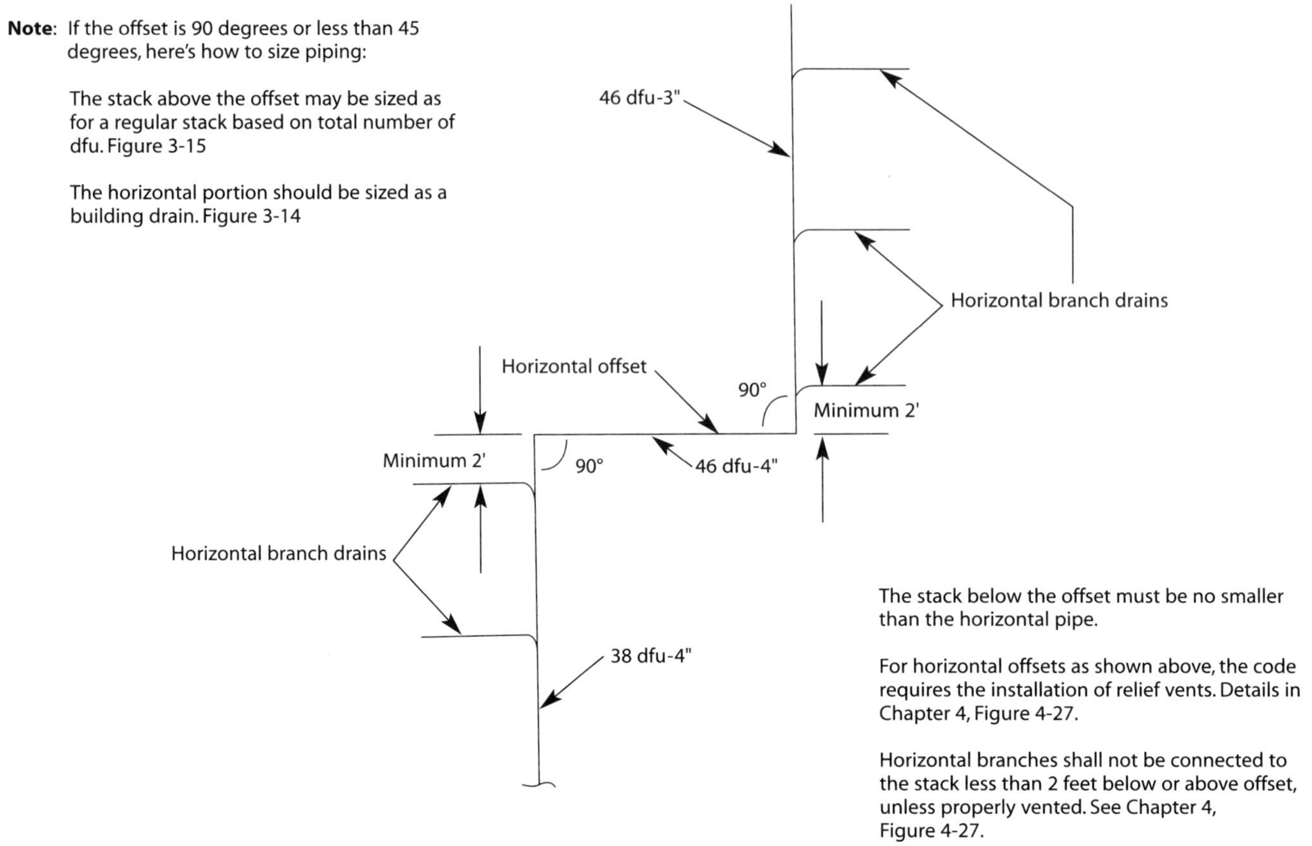


**Figure 3-22**
*Horizontal offset*

horizontal branch lines can't connect to the vertical stack within 2 feet below, or above, the offset unless they're properly vented. We'll discuss the venting method in Chapter 4.

Here are some other code requirements for horizontal and vertical offsets in vertical drainage stacks:

- You may omit vents for vertical offsets of more than 45 degrees, provided you size the stack and its offset one pipe size larger than required for a building drain with ½-inch slope per foot. See the ½-inch slope column in *IPC* Table 710.1(1).
- Don't connect a horizontal drainage branch pipe within 2 feet above or below such horizontal offset.
- Size the vertical stack above an offset using the total dfu for the pipe. See column five of *IPC* Table 710.1(2).
- Size the horizontal portion of the stack as a building drain using the total dfu assigned for the pipe. Again, refer to the ½-inch slope column in *IPC* Table 710.1(1).
- According to the *IPC*, that portion of a stack below the offset can be the same size as the horizontal offset or based on the total dfu for the entire stack.
- If a horizontal offset occurs in a soil or waste stack below the lowest horizontal drainage branch pipe, size the required diameter of the offset and the stack below it as a building drain. See *IPC* Table 710.1(1).

- A horizontal building drain may connect to a horizontal stack offset, provided it can connect 10 times the diameter of the drainage stack downstream from the upper stack.

- Horizontal drainage branches must not connect to a building drain unless they connect a minimum of 10 pipe diameters downstream from the base of the stack. See *IPC* Section 704.3. You used the same principle of separation here as for the vent stack connection shown in Figure 4-4 in the next chapter.

**TABLE 712.4.2**
**MINIMUM CAPACITY OF SEWAGE PUMP OR SEWAGE EJECTOR**

| DIAMETER OF THE DISCHARGE PIPE (inches) | CAPACITY OF PUMP OR EJECTOR (gpm) |
|---|---|
| 2 | 21 |
| $2^1/_2$ | 30 |
| 3 | 46 |

For SI: 1 inch = 25.4 mm, 1 gpm = 3.785 L/m.



**Figure 3-23**
*Minimum capacity of sewage pump or sewage ejector*

## *Building Drains Below Grade*

If you have a building sub-drain that you can't discharge to a drainage system by gravity flow, you must discharge its waste into an approved sump. You must then have automatic pumping equipment that will lift the contents and discharge them into the building gravity drainage system. Be sure the sump and pumping equipment are large enough for the volume and type of liquid they'll have to carry. On most jobs, the architect will specify the equipment. If an architect isn't involved with the job, check it out with the plumbing department.

Remember, you can only have waste that requires lifting discharge into a sump. Size, install and vent the sub-drainage system for this waste like a standard gravity system. All other drains must discharge by gravity into the regular building drainage system.

## *Sewage Ejectors*

You must locate sump pits and receiving tanks so they can be inspected, repaired and cleaned. The drainage pipe must flow into the sump by gravity. The sump pit or receiving tank must have a minimum diameter of 18 inches, with a 24-inch depth, unless otherwise approved. It has to be made of tile, concrete, steel, plastic or other approved materials. Make sure the pit bottom is solid and provides permanent support for the pump(s). The sump must be adequately vented and have a gas-tight, removable cover. A sump that's constructed of metal must be thick enough to serve the intended purpose and be protected from external corrosion.

Duplex pumping equipment is generally required for public use (but not required by the *IPC*). This permits the pumps to discharge waste alternately. Also, in case of repair, one pump can remain in service, avoiding a shut down of businesses.

Just in case this isn't an area within your expertise, let's review a couple of additional requirements for sumps or receiving tanks that receive body waste. First, you have to use at least a 2-inch discharge pipe. See *IPC* Table 712.4.2 shown in Figure 3-23. This will handle up to 21 gpm, which equals 42 dfu. Keep in mind that the code for semi-continuous flow allows 2 dfu for each gpm of flow. Second, there must be a local vent large enough to maintain atmospheric pressure in the sump under normal operating conditions. The minimum size vent for this is $1^1/_4$ inches. See *IPC* Section 916.5.2. According to *IPC* Table 916.5.1, shown in Figure 3-24, a $1^1/_4$-inch vent would be large enough for the sump shown in Figure 3-25, as long as the building is not more than 31 feet high. A $1^1/_2$-inch vent is required if the building height is less than 75 feet.

Footnote (a) of *IPC* Table 916.5.1 is very difficult to understand. Here is how I interpret it. *IPC* Table 916.5.1, Footnote (a) says that if a more precise value isn't available, an allowance of 50 percent of the developed length must be assumed for entrance losses, friction due to fittings, direction changes and diameter. I would suggest that if you know the capacity of the pump, the gpm and the height of the building, you divide the length of vent pipe in half. That's what I did in the two examples I used. With this method you should be within the intent of the code.

**TABLE 916.5.1**
**SIZE AND LENGTH OF SUMP VENTS**

| DISCHARGE CAPACITY OF PUMP (gpm) | MAXIMUM DEVELOPED LENGTH OF VENT (feet)[a] Diameter of vent (inches) | | | | | |
|---|---|---|---|---|---|---|
| | $1^1/_4$ | $1^1/_2$ | 2 | $2^1/_2$ | 3 | 4 |
| 10 | No limit[b] | No limit | No limit | No limit | No limit | No limit |
| 20 | 270 | No limit | No limit | No limit | No limit | No limit |
| 40 | 72 | 160 | No limit | No limit | No limit | No limit |
| 60 | 31 | 75 | 270 | No limit | No limit | No limit |
| 80 | 16 | 41 | 150 | 380 | No limit | No limit |
| 100 | 10[c] | 25 | 97 | 250 | No limit | No limit |
| 150 | Not permitted | 10[c] | 44 | 110 | 370 | No limit |
| 200 | Not permitted | Not permitted | 20 | 60 | 210 | No limit |
| 250 | Not permitted | Not permitted | 10 | 36 | 132 | No limit |
| 300 | Not permitted | Not permitted | 10[c] | 22 | 88 | 380 |
| 400 | Not permitted | Not permitted | Not permitted | 10[c] | 44 | 210 |
| 500 | Not permitted | Not permitted | Not permitted | Not permitted | 24 | 130 |

For SI: 1 inch = 25.4 mm, 1 foot = 304.8 mm, 1 gallon per minute = 3.785 L/m.

a. Developed length plus an appropriate allowance for entrance losses and friction due to fittings, changes in direction and diameter. Suggested allowances shall be obtained from NSB Monograph 31 or other approved sources. An allowance of 50 percent of the developed length shall be assumed if a more precise value is not available.

b. Actual values greater than 500 feet.

c. Less than 10 feet.



**Figure 3-24**
*Size and length of sump vents*

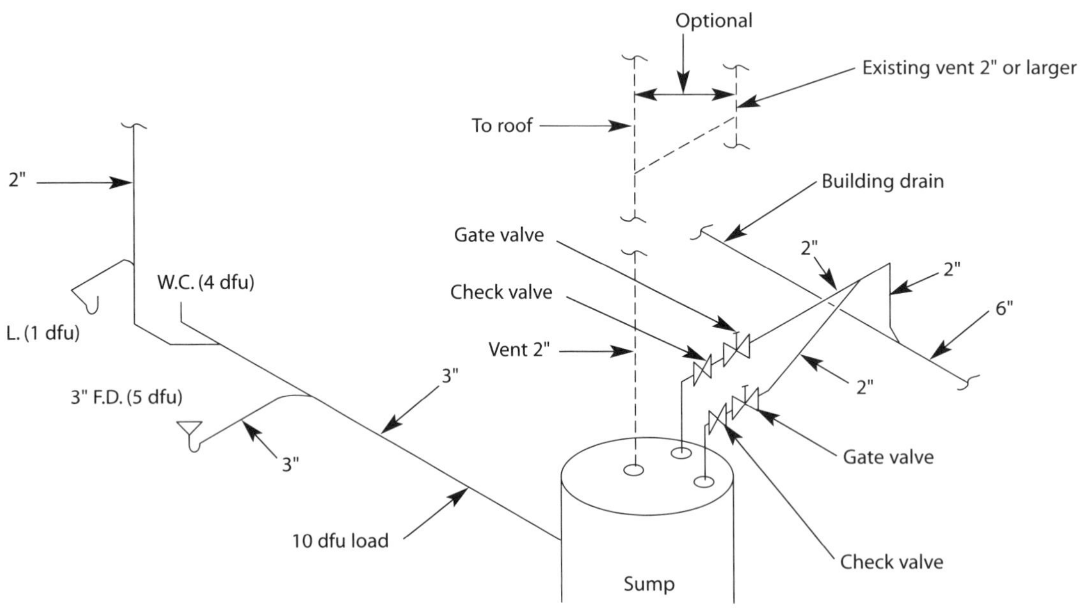


**Figure 3-25**
*Sewage ejector "public use" sump*

To pipe the sump shown in Figure 3-25, you must first install a swing check valve, and then a gate valve in each of the twin discharge lines between the pump and the building gravity system. You have to connect the sump discharge pipe into the top of the building gravity drain through a wye fitting. Be sure this connection is at least 10 feet from the base of any soil or waste stack. However, an exception to this rule says that you only need a check valve on the discharge pipe from the pump if it's installed to service a one- or two-family house.

Sumps receiving clear water waste from floor drains, air conditioning condensate drains and the like don't need to be vented or have a cover. Also, don't forget that, according to *IPC* Table 712.4.2, the minimum capacity of a sewage pump or ejector is based on the required diameter of the discharge pipe. The exception to this is if you install grinder pumps or grinder ejectors that receive water closet discharge. Then, the discharge opening must be at least $1^1/_4$ inches.

Good luck with the sumps, pumps and ejectors!

# Chapter 3 — Questions and Answers

**3-1 The determining factor in sizing all vertical and horizontal drainage piping is the**

(A) maximum drainage fixture unit (dfu) load on the piping.

(B) number of water closets installed.

(C) approved occupancy rating of the building.

(D) length of vertical and horizontal drains.

***Answer:* A**

*Code response: Section 709.1*
*This book, page 32*

**3-2 The *IPC* says you can figure out the drainage system requirements for semi-continuous flow into a drainage system on the basis that 1 gpm of flow is equal to ______ dfu(s).**

(A) 1

(B) 2

(C) 3

(D) 4

***Answer:* B**

*Code response: Section 709.3*
*This book, page 35*

**3-3 The *IPC* says a horizontal stack offset must be sized as required**

(A) by the code official.

(B) for the building's sewer.

(C) for a vertical stack offset.

(D) for the building's drain.

***Answer:* D**

*Code response: Section 710.1.1*
*This book, page 42*

**3-4 According to the *IPC*, the maximum dfu for a 2-inch waste pipe with more than three branch intervals is**

(A) 12.

(B) 16.

(C) 22.

(D) 24.

***Answer:* D**

*Code response: Table 710.1(2)*
*This book, pages 37 and 38*

**3-5 The *IPC* allows you to install _____ shower(s) on a 2-inch horizontal branch drain at any one-story opening.**

(A) 1

(B) 2

(C) 3

(D) 5

***Answer:* *C***

*Code response: Table 710.1(2)*
*This book, page 38*

**3-6 To carry 350 dfu to a building's drain, a _______ horizontal drainage pipe at a $^1/_4$-inch fall per foot is required.**

(A) 2-inch

(B) 3-inch

(C) 4-inch

(D) 5-inch

***Answer:* *D***

*Code response: Table 710.1(1)*
*This book, pages 36, 38 and 39*

**3-7 Two elements that must be considered regarding the installation of future fixtures are**

(A) the number and type of such fixtures.

(B) the size and length of the vent pipe.

(C) the size and material of the vent pipe.

(D) the size and slope of the drainage pipe.

***Answer:* *A***

*Code response: Section 710.2*
*This book, page 41*

**3-8 The closest you can connect a horizontal drainage pipe above or below a vertical stack offset, is**

(A) 12 inches.

(B) 18 inches.

(C) 24 inches.

(D) 30 inches.

***Answer:* *C***

*Code response: Sections 711.1 and 711.2*
*This book, pages 41 and 42*

**3-9 When using a sump for a building sub-drain, remember, the sump may only receive**

(A) waste that can't drain by gravity.

(B) waste from a water closet.

(C) clear water waste.

(D) waste from low dfu fixtures.

***Answer: A***

*Code response: Section 712.1*
*This book, page 43*

**3-10 You can't connect a sewage ejector discharge pipe to a building drainage system closer than _____ feet from the base of any stack.**

(A) 6

(B) 10

(C) 12

(D) 15

***Answer: B***

*Code response: Section 712.3.5*
*This book, page 45*

◆ *Chapter 4* ◆

# Understanding Vent Systems

There are two basic vent system requirements: First, every fixture trap must be protected from back pressure and siphonage. Second, every vent pipe must let air circulate freely throughout the entire drainage system. Every sanitary or waste system needs a vent system designed to let all gases, pressures and odors escape into the atmosphere. At the same time, the design must let air into the system. To do both of these, the air pressure in the system has to be within 1 inch of the water column of the atmospheric pressure. For example, consider a fixture trap with a 2-inch minimum water seal as required by code. The vent has to let enough air circulate to limit the air pressure variations in the fixture drains to no more than 1 inch of the water column above or below atmospheric pressure. Code tables for vent size, location and length are based on this rule.

## Types of Vents

The *IPC* lists the types of vents you need to design and install drain, waste and vent systems. Let's take a look at each type.

### *Vent Stacks*

According to the *IPC*, every building where plumbing is installed must have at least one vent stack, sized at least one-half of the required size of the building drain. This is considered the main vent, although *IPC* Section 903.1 doesn't call it that. A vent stack is the principal pipe that the vent branches on each floor are connected to, as shown in Figures 4-1 and 4-2. It's often confused with a stack vent, which is an extension of a soil or waste stack (dry section) up through the roof of a building, as shown in Figure 4-3. Most local jurisdictions require a stack vent to run full size as directly as possible from its connection to the soil or waste stack, through the roof, to the open air.

You must connect a vent stack at or below the lowest horizontal branch at the base of the drainage stack, as shown in Figures 4-1 and 4-2. The connection point to the building drain must be within a distance 10 times the diameter of the drainage stack, and downstream from it, per *IPC* Section 903.4. Figure 4-4 shows this. For a 4-inch vent stack, you need at least a 40-inch separation between the two stacks.

### *Vent Header*

A vent header is a horizontal single pipe that receives the connection of two or more vent pipes and then connects to the main vent stack or stack vent that extends to the open air above the roof. Figure 4-5 shows several different types of vents connecting to a single vent stack using a vent header. You must size a vent header in accordance with *IPC* Section 916.2.

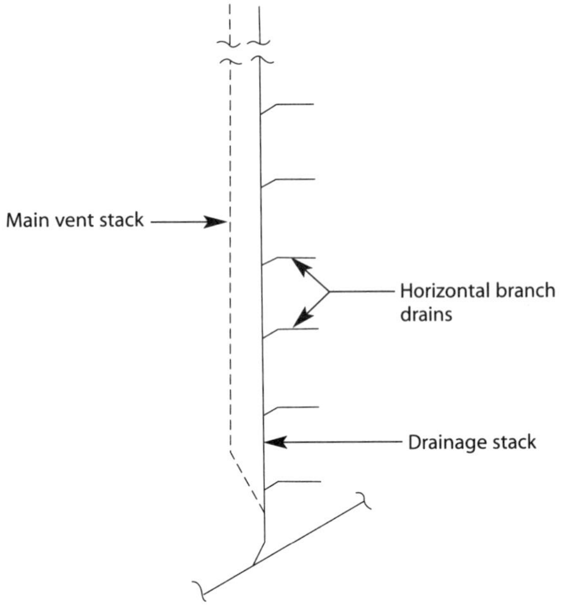


**Figure 4-1**
*Main vent stack properly connected*

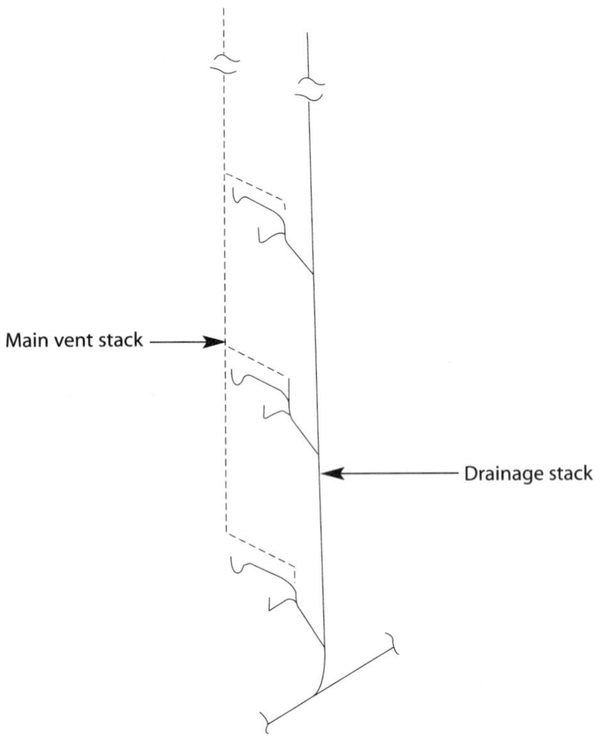


**Figure 4-2**
*Alternate design for main vent stack*

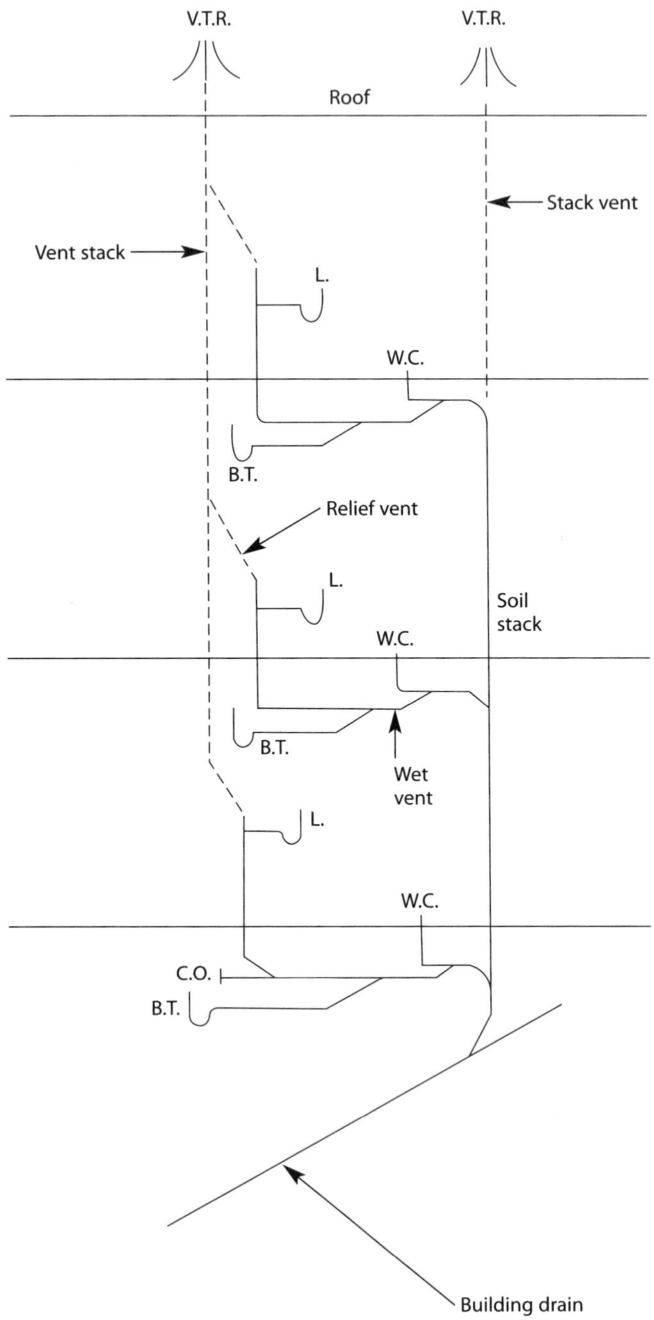


**Figure 4-3**
*Alternate method of wet venting single bathrooms in a multi-story building*

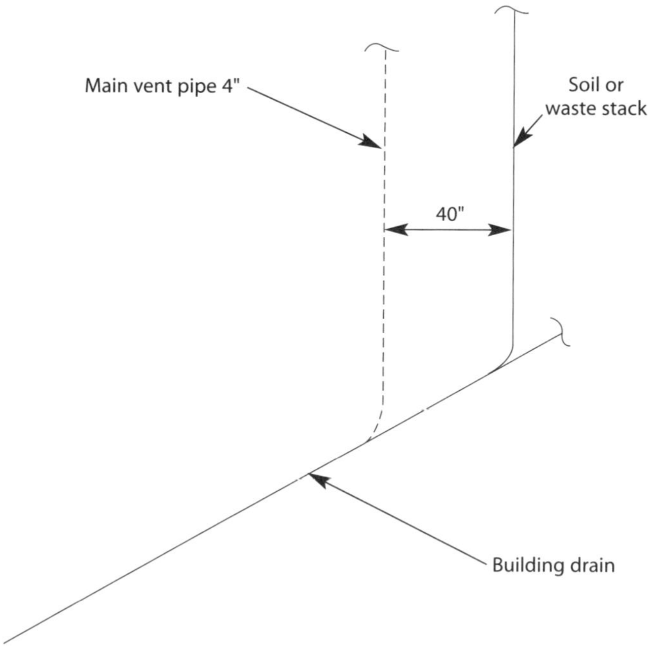


**Figure 4-4**

*Maximum 40" separation of the two pipes, according to the International Plumbing Code*

## Crown-vented Trap

*IPC* Section 1002.3 lists a crown-vented trap as a prohibited trap. However, there is some leeway for its use in special circumstances. During the years I worked in plumbing, I never used a crown-vented trap. But *IPC* Section 906.3 explains its requirement, which is basically that you can't install a crown vent within 2 pipe diameters of a trap weir (the top most water level located on the discharge side of a trap). Look ahead to Chapter 5, Figure 5-3 to see a crown-vented tubular P-trap.

## Individual and Common Vents

An individual vent, shown in Figure 4-6, is a pipe you install to vent a single fixture. *IPC* Section 907.1 says you must connect it to the fixture drain of a trap or to the trapped fixture being vented. It's also known as a back vent, as shown in Figure 4-7. The vertical portion connecting to the fixture drain of the trap and extending through the roof of the building is the stack vent, as shown in Figure 4-8.

A common vent is like an individual vent except it must vent two fixture traps located on the same floor level. The vent connection must be at the interconnection of the fixture drains, as shown in Figure 4-9, and also in B, in Figure 4-10.

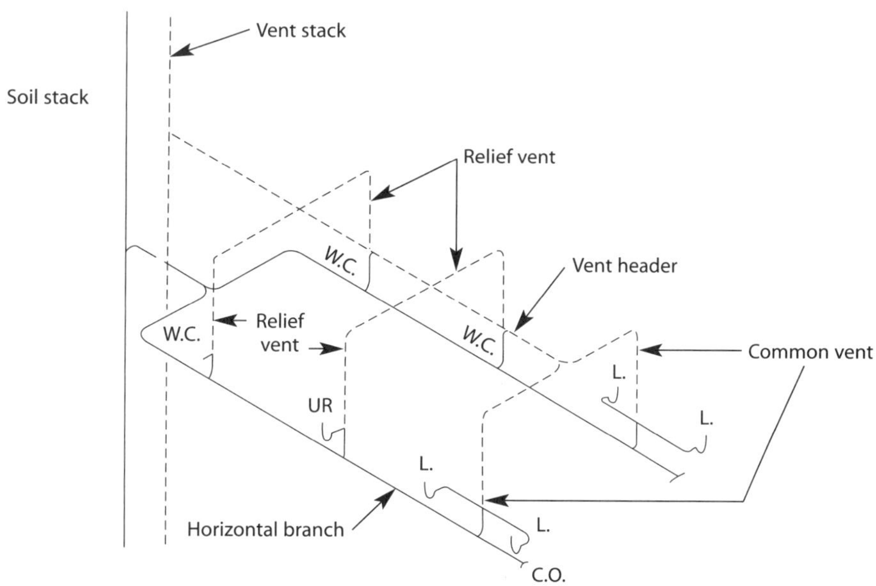


**Figure 4-5**

*Battery of fixtures properly vented using a vent header*

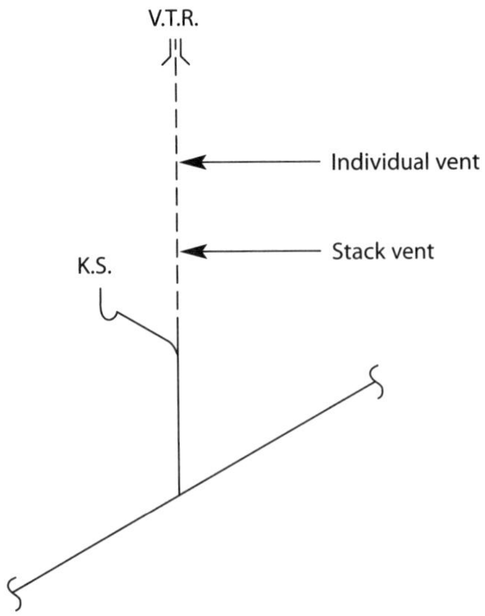


**Figure 4-6**
*Individual vent*

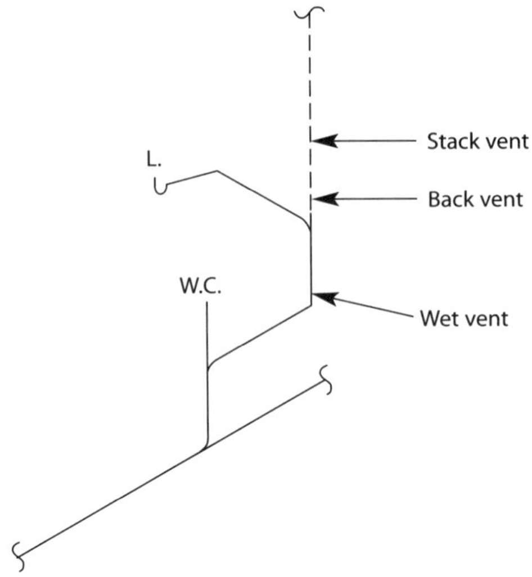


**Figure 4-7**
*Back vent*

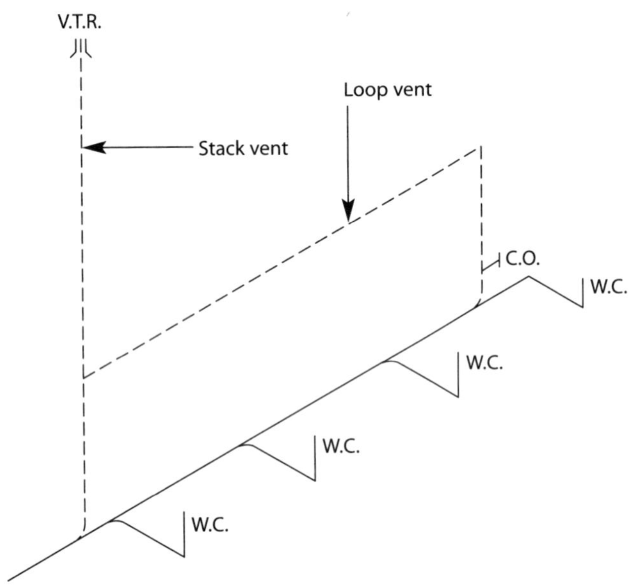


**Figure 4-8**
*Loop vent*

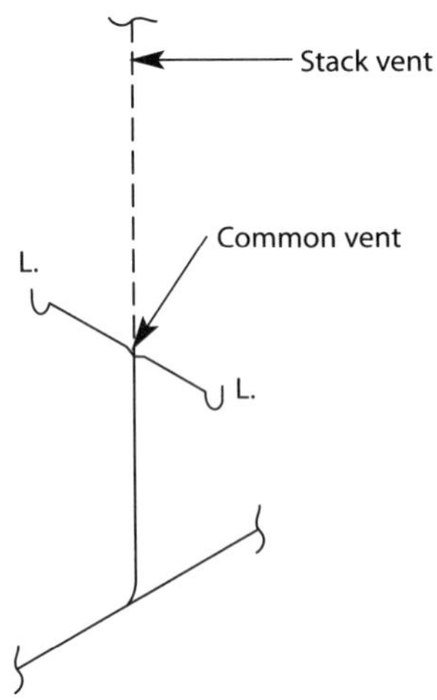


**Figure 4-9**
*Common vent*

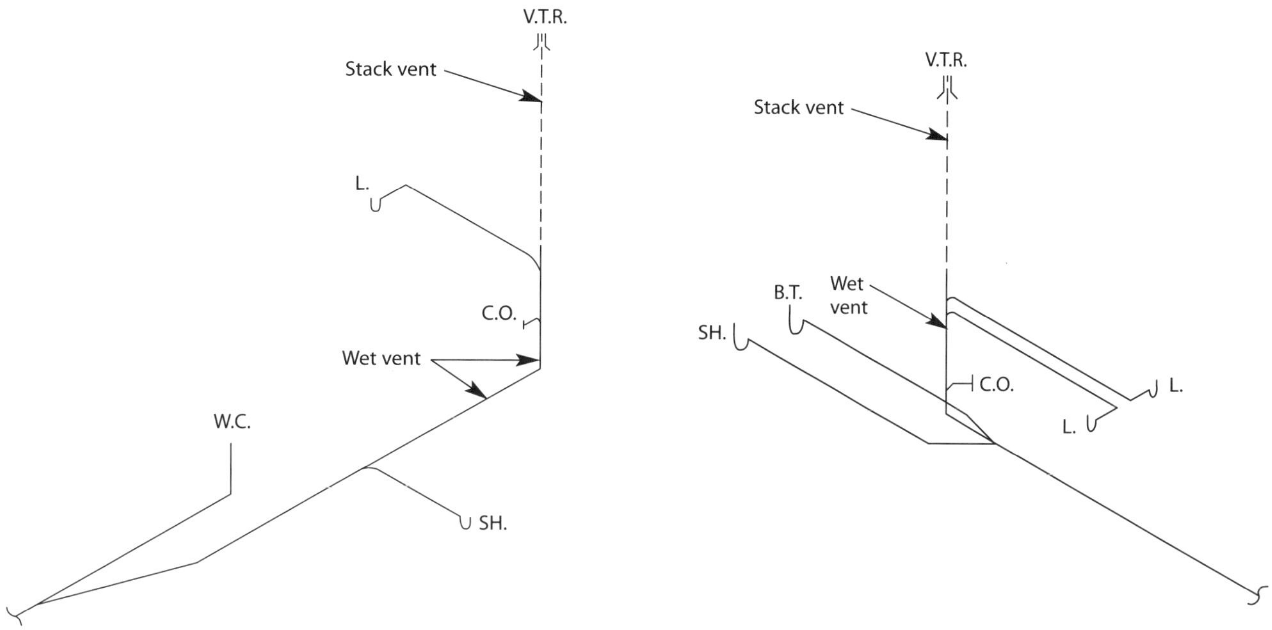


**A** Flat wet vented system

**B** Wet vented shower and bathtub

**Figure 4-10**

*Two types of wet vents*

## *Fixture Drain Connection Levels*

When you vent fixture traps located at different levels, the code requires special sizing of the pipe between the fixture traps. *IPC* Table 908.3, shown in Figure 4-11, shows the pipe sizes you need between two fixture drains based on the dfu of the upper fixture. In Figure 4-12, a kitchen sink with 2 dfu is the upper fixture and a lavatory with 1 dfu is the lower fixture. According to *IPC* Table 908.3, the sink requires a 2-inch vent pipe. That means the wet vent pipe between the two fixture traps will have to be at least 2 inches. This pipe serves as a vent for the lower fixture (lavatory) as well.

**Table 908.3**
**COMMON VENT SIZES**

| PIPE SIZE (inches) | MAXIMUM DISCHARGE FROM UPPER FIXTURE DRAIN (dfu) |
|---|---|
| $1^1/_2$ | 1 |
| 2 | 4 |
| $2^1/_2$ to 3 | 6 |

For SI: 1 inch = 25.4 mm.

From the *IPC*® ©2003, with permission of the ICC.

**Figure 4-11**

*Common vent sizes*

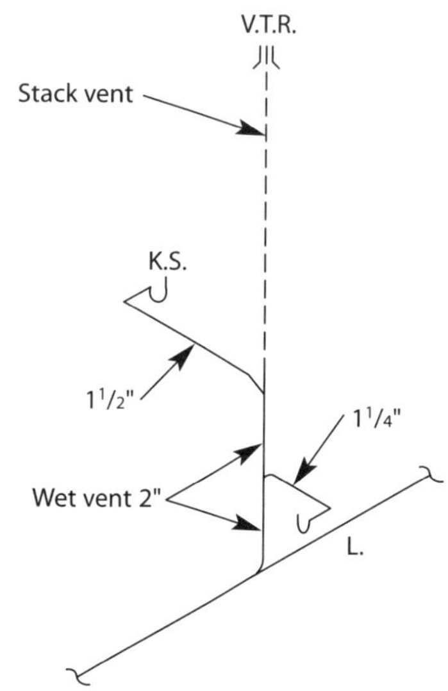


**Figure 4-12**

*Wet venting fixture*

## Sizing Vents for Future Use

Sometimes you have to rough-in plugged outlets so that additional plumbing can be installed in the future. Here are the *IPC* Section 905.6 requirements:

- Size a future vent connection at least one-half the size of the rough-in drain you install. For example, if you install a 4-inch rough-in drain, install a 2-inch vent.
- Tag any capped or plugged vent outlet to identify it as a vent to keep it from being used as a waste pipe some time in the future.

Look back to Figure 3-20 in Chapter 3 to see plugged outlets installed for future use.

## Wet and Dry Venting

A major part of plumbing is installed using the wet vent system. Wet venting requirements are in *IPC* Section 909. A wet vent is a waste pipe that vents and carries waste from any combination of fixtures within a single or double bathroom group that's located on the same floor level. It must extend from the connection of the dry vent to the most downstream fixture drain connection to the horizontal branch drain.

Only the fixtures within a bathroom group can connect to a wet vent branch drain. You must size a dry vent in a wet-vented system according to the largest required diameter pipe in the system. And if there are other fixtures nearby that aren't in the bathroom group, you must connect them downstream of the wet vent connection. You may install a wet vent on the flat as well as in stacked systems. I've included several drawings, acceptable almost anywhere, to show wet vent systems. Look at Figure 4-10 and Figures 4-12 through 4-20. You can use *IPC* Table 909.3, shown in Figure 4-21, to size wet vent pipes.

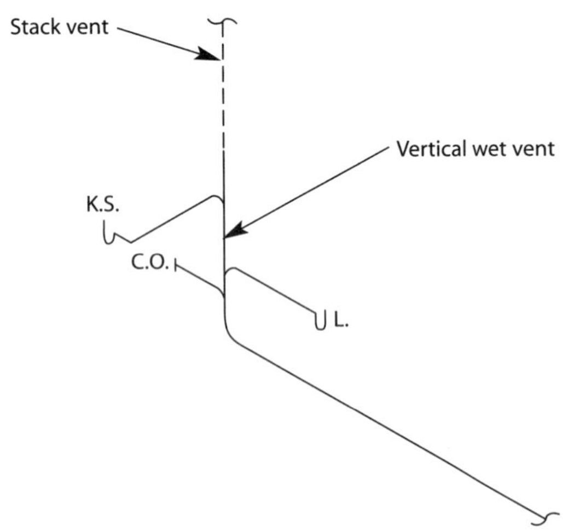


**Figure 4-13**
*Lavatory wet venting*

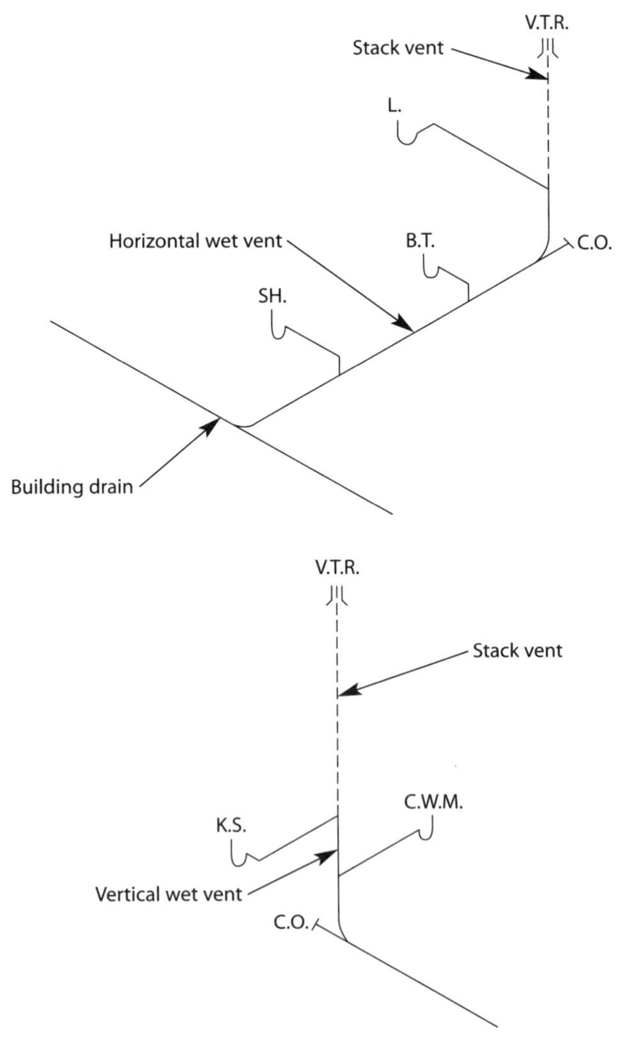


**Figure 4-14**
*Horizontal and vertical wet venting*

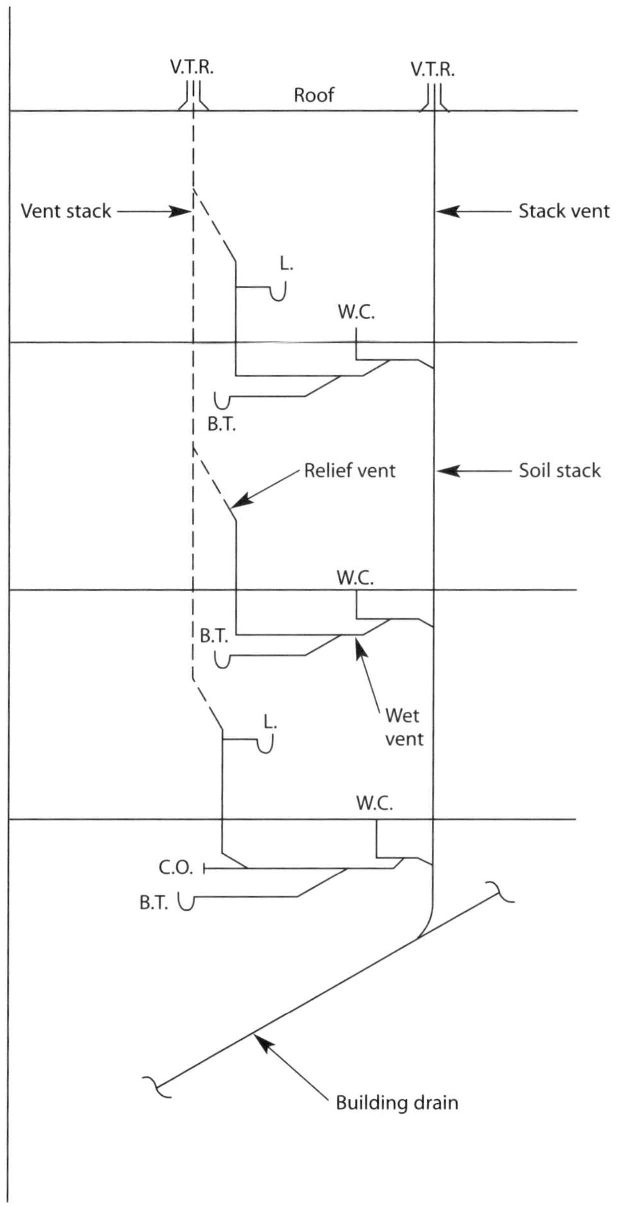


**Figure 4-15**
*Alternate method of wet venting single bathrooms in multi-story buildings*

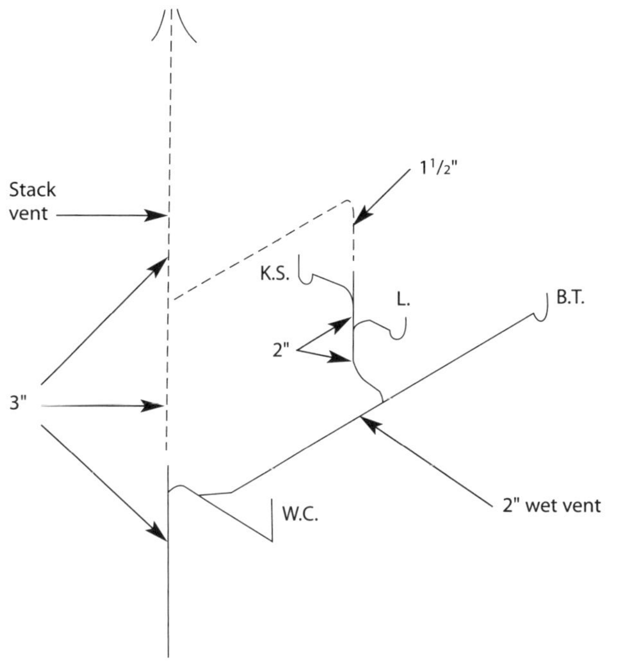


**Figure 4-16**
*Wet vented single bathroom and kitchen sink, top floor only*

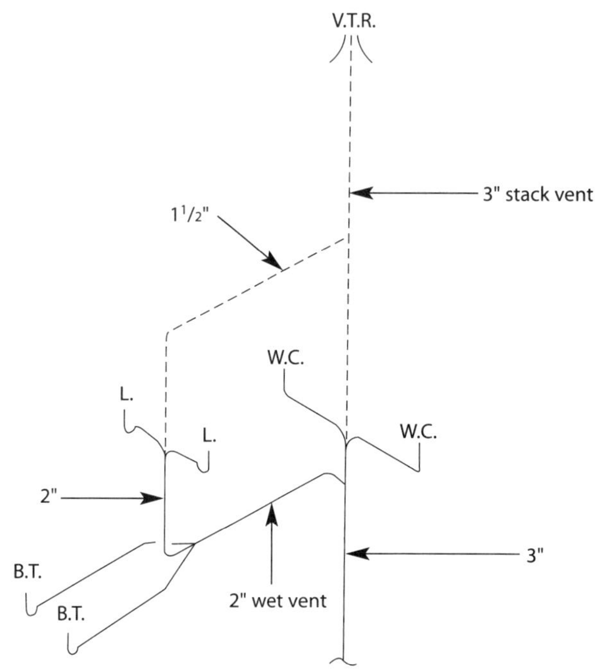


**Figure 4-17**
*Wet vented double bath, top floor only*

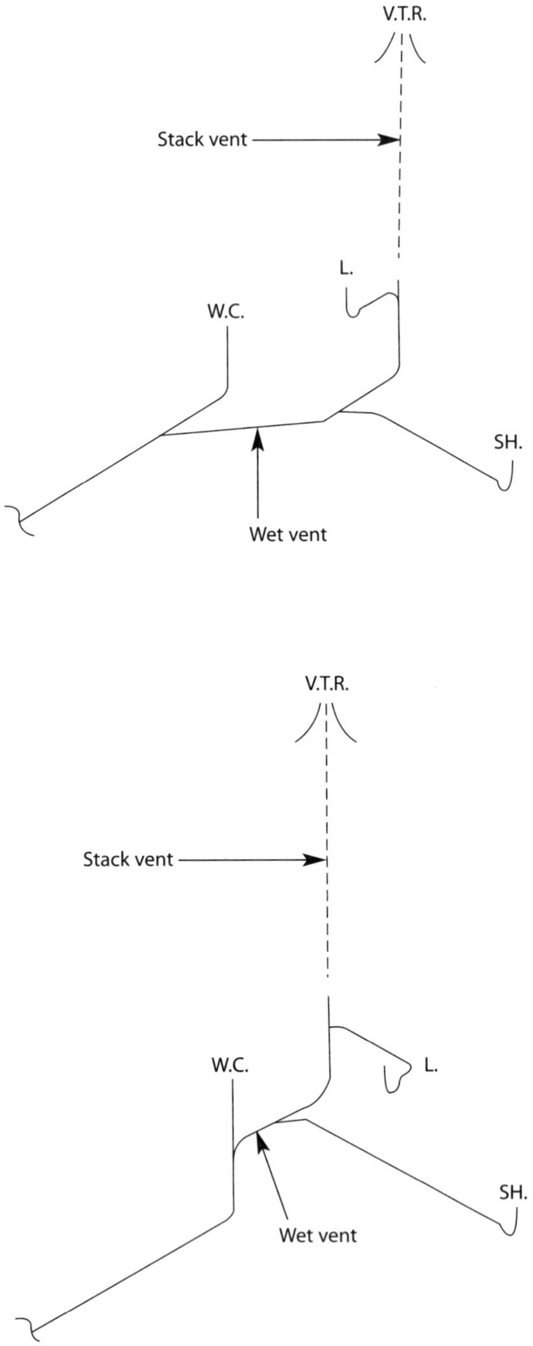


**Figure 4-18**
*Flat and stacked wet-vented bathroom groups*

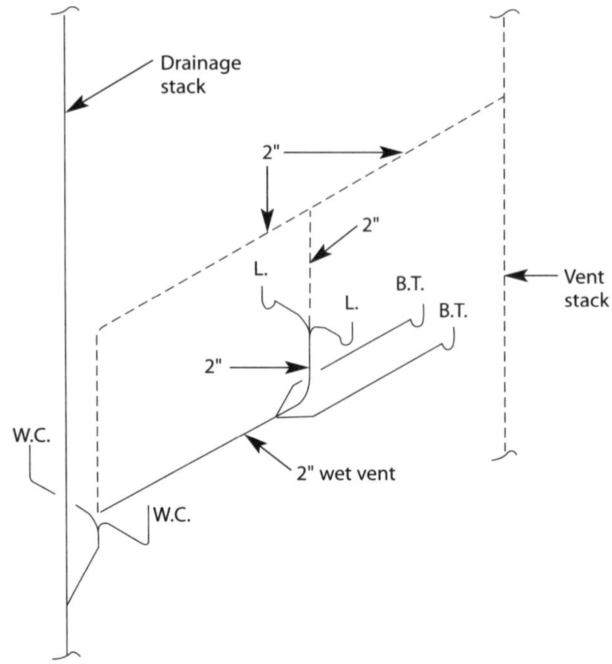


**Figure 4-19**
*Wet-vented double bath, lower floors*

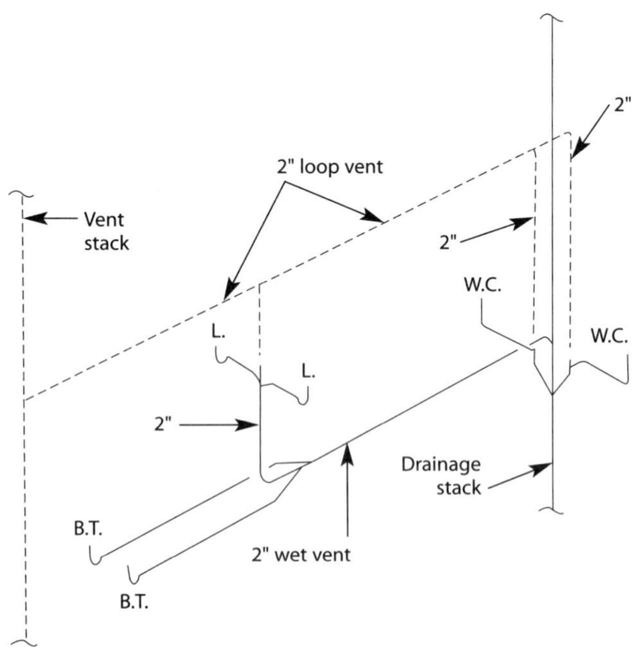


**Figure 4-20**
*Alternate method of wet venting double bath, lower floors*

**TABLE 909.3**
**WET VENT SIZE**

| WET VENT PIPE SIZE (inches) | DRAINAGE FIXTURE UNIT LOAD (dfu) |
|---|---|
| $1^1/_2$ | 1 |
| 2 | 4 |
| $2^1/_2$ | 6 |
| 3 | 12 |

For SI: 1 inch = 25.4 mm.

From the *IPC*® ©2003, with permission of the ICC.

**Figure 4-21**
*Wet vent sizes*

## Waste Stack Venting

You may be more familiar with the phrase "vertical combination waste and vent stack" than the phrase "waste stack venting" used here and in the code. Waste stack venting describes the vertical pipe that receives discharge from fixtures other than high dfu fixtures such as water closets, urinals, etc. The code prohibits offsets in the vertical portion of the drainage pipe. Each fixture drain must connect separately to the waste stack. The waste stack must be the same size throughout its length.

You can use a waste stack vent in a high-rise building to receive waste and to vent specific types of plumbing fixtures. You can actually use it to your advantage in any multi-story building where plumbing fixtures are located directly over each other on different floor levels. Since you need only one vertical pipe riser in the installation, it's both economical and practical.

In waste stack venting, the stack vent must go through the roof. Also, the stack vent can't be smaller than the waste stack. If you install a 3-inch waste stack, the stack vent must also be 3 inches. You can offset a stack vent if necessary. However, any offset must be at least 6 inches above the flood level rim of the highest fixture, as shown in Figure 4-22. You must grade the vent pipe to drain back to the drainage pipe by gravity. Figure 4-23 shows two ways to use a waste stack vent. The fixtures in the figures are all kitchen sinks.

Use *IPC* Table 910.4, shown in Figure 4-24, to size waste stack vents according to total dfu. However *IPC* Table 910.4 doesn't mention length for the stacks. Check with your inspector to find out the approved length for your area before you install this type of system.

## Yoke Vent

A yoke vent (also referred to in the code as a relief vent) is a pipe connecting to a soil or waste stack through a wye. The upper end must connect to the vent stack through a wye and not be less than 3 feet above the floor. A yoke vent prevents pressure changes in the stack.

In a high-rise building, the code requires you to have a yoke vent for soil or waste stacks every ten floors, beginning at the top floor of the building and working down. See Figure 4-25. A yoke vent must be the same size as the vent stack it connects to.

## Circuit Venting

According to *IPC* Section 911, a circuit vent will serve more than two fixtures but not more than eight. It must rise vertically from between the two most horizontal upstream fixture drains. In a multistory installation, you install a relief vent between the horizontal branch drain, the vent stack and the most downstream fixture drain of the circuit vent. A relief vent is required when the drainage stack

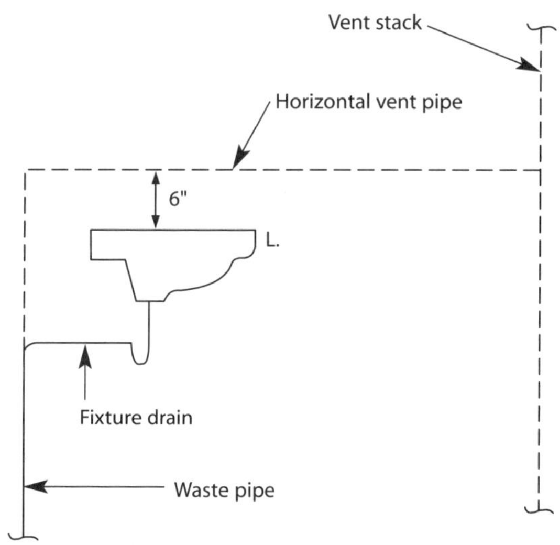


**Figure 4-22**
*Minimum height of vent pipe above floor-level rim of fixture*

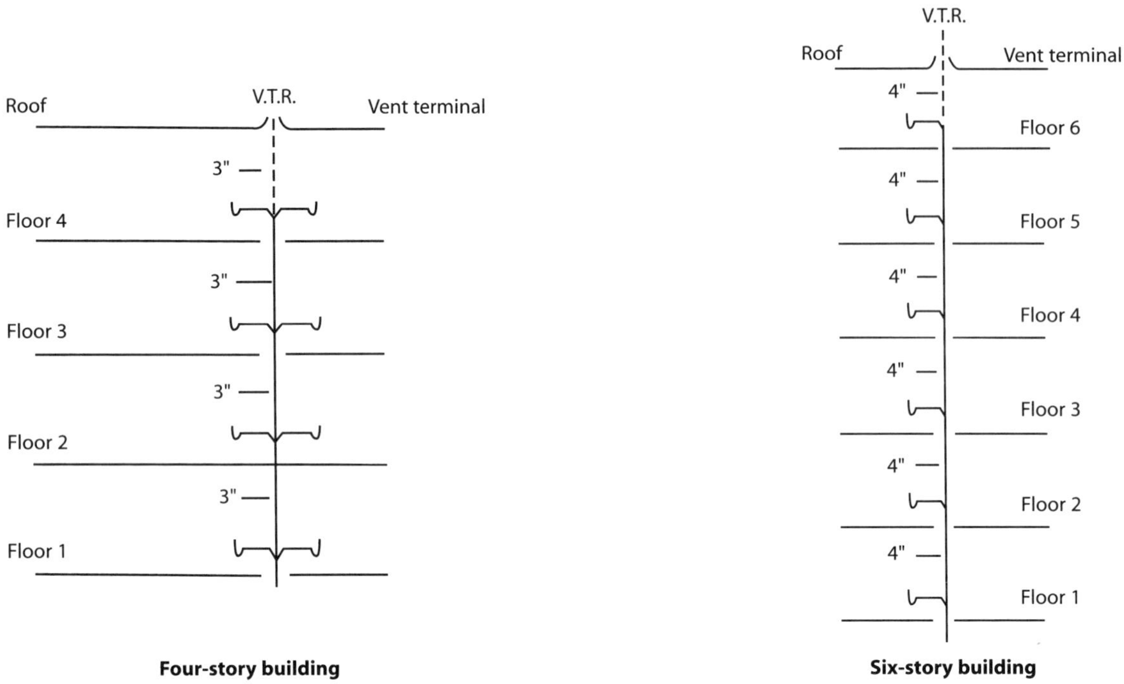


**Figure 4-23**
*Two ways to use a waste stack vent*

receives the discharge of soil or waste from upper horizontal branches. In this type of installation, you loop the circuit vent back and connect it to a vent stack, as shown in Figure 4-26. In a one-story building or on the top floor of a multi-story building, a circuit vented system would loop back and connect to a stack vent. (Don't confuse the terms "vent stack" and "stack vent." They're not interchangeable. A vent stack is a vent pipe that continues up and through two or more floors that other vents connect to. A stack vent is the vent pipe that extends through the roof from the topmost fixture. Both are shown in Figure 4-25.)

**Table 910.4**
**WASTE STACK VENT SIZE**

| STACK SIZE (inches) | MAXIMUM NUMBER OF DRAINAGE FIXTURE UNITS (dfu) | |
|---|---|---|
| | Total discharge into one branch interval | Total discharge for stack |
| $1^1/_2$ | 1 | 2 |
| 2 | 2 | 4 |
| $2^1/_2$ | No limit | 8 |
| 3 | No limit | 24 |
| 4 | No limit | 50 |
| 5 | No limit | 75 |
| 6 | No limit | 100 |

For SI: 1 inch = 25.4 mm.

From the *IPC*® ©2003, with permission of the ICC.

**Figure 4-24**
*Waste stack vent sizes*

A fixture drain or fixture branch for fixtures located within the same circuit-vented horizontal branch may discharge into the relief vent. No more than 4 dfu may discharge into a relief vent. Also note the two lavatories connected to the relief vent in Figure 4-26. Other fixtures on the same floor level as the circuit-vented fixtures may discharge into the horizontal branch drain. *IPC* Section 911.5 says that the fixtures must be either individually or common vented.

To size circuit vents:

- A circuit vent can't be less than $1^1/_4$ inches in diameter under any circumstances.

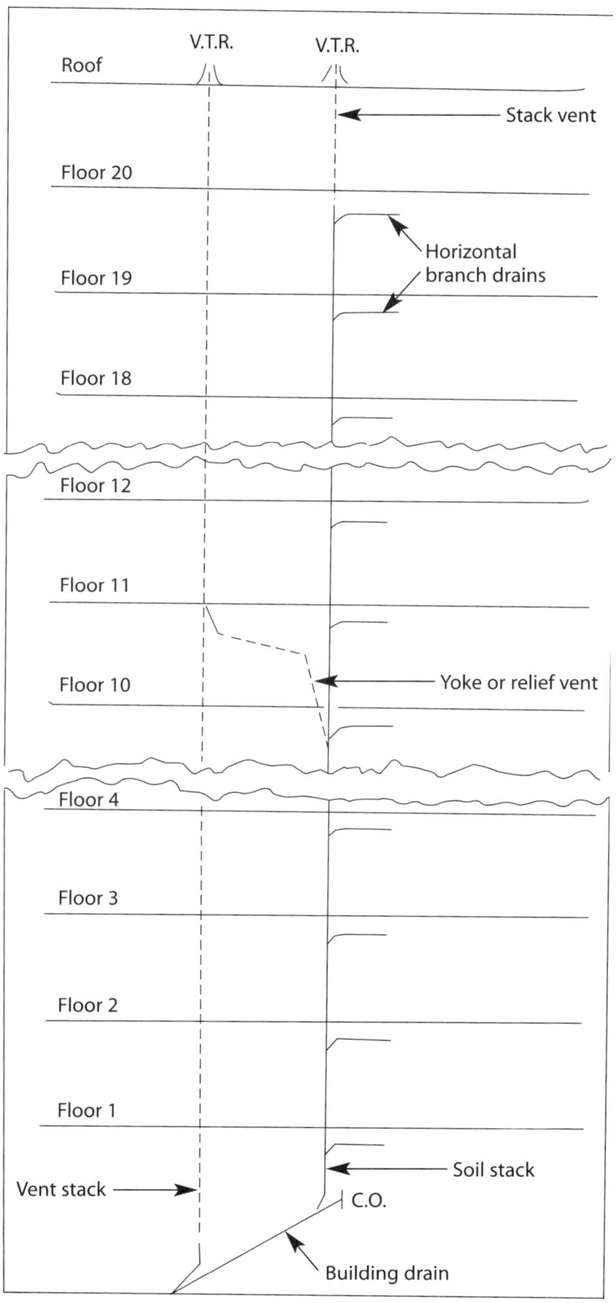


**Figure 4-25**
*Yoke vent placement*

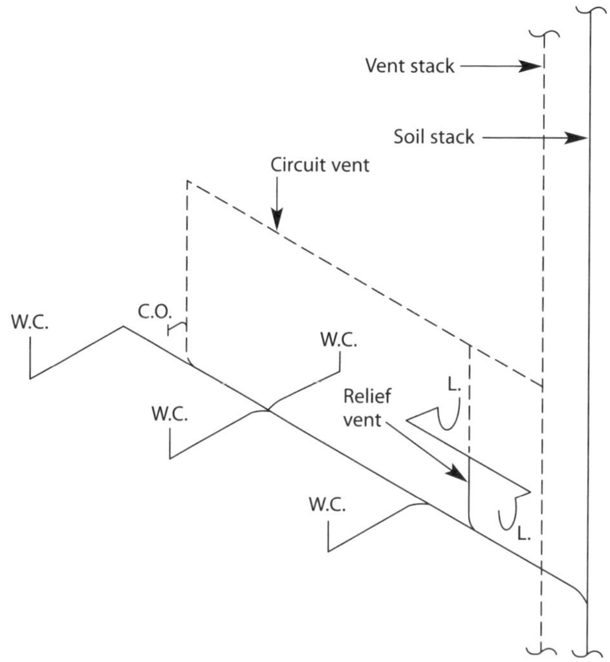


**Figure 4-26**
*Circuit vent*

- A circuit vent can't be less than one-half the size of the horizontal branch drain it connects to.
- For developed length over 40 feet, increase vent pipe size by one size for the entire developed length. For example, if a horizontal branch drain is 4 inches and the developed length is less than 40 feet, you can use a 2-inch circuit vent. If the developed length is more than 40 feet, you need a 3-inch circuit vent.

## Venting Stack Offsets

As I mentioned in Chapter 3 and Figure 3-22, some high-rise buildings may require a horizontal offset in the soil or waste stack. If there are five or more branch intervals (floors) above a stack offset, the *IPC* says you must provide a relief vent (main vent) at the base of the upper section of the drainage

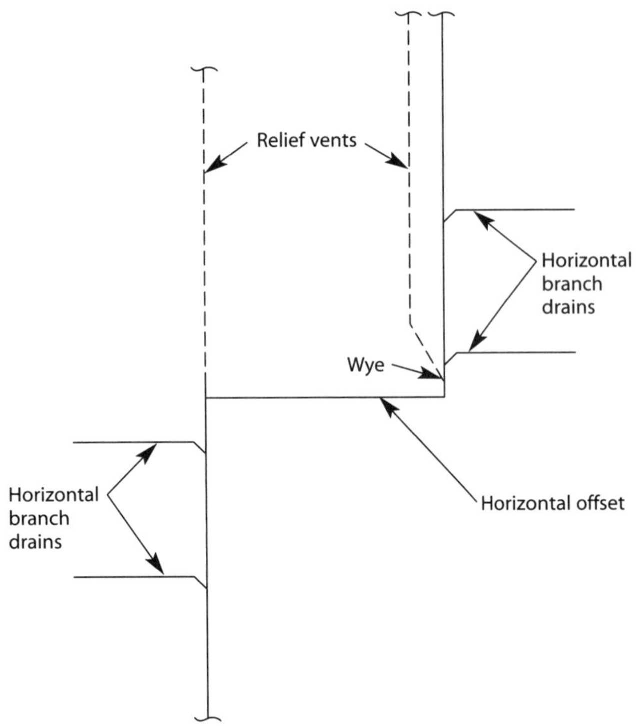


**Figure 4-27**
*Venting horizontal offset in a vertical stack*

stack. You must connect it at or below the lowest horizontal branch interval through a wye, as shown in Figure 4-27. You must also vent the lower section. To do this, extend the vertical portion of the drainage pipe. Use *IPC* Table 916.1, shown in Figure 4-28, to size the offsets.

## Venting Island Fixtures

Island fixture venting is limited to domestic sinks and lavatories located away from walls and partitions. You can also use this type of venting in a residential kitchen sink with a dishwasher and/or food waste grinder connection.

One way to use an island fixture vent is shown in Figure 4-29. The island sink is located at the end of a horizontal waste pipe. The vent is a common vent extending vertically to the underside of the drainboard and then offsetting and turning vertically downward to connect to the vertical drainpipe. This is called a return vent. Understandably, you must have full-size cleanouts that are accessible at the base of the return vent and waste pipe. The relief vent as shown may or may not be required by your local plumbing department. Your pipe sizes must be compatible with *IPC* Section 913.

Figure 4-30 shows another acceptable way you can install waste and vent pipes for an island sink. Here you install the building drain to pass directly beneath the island sink location in the kitchen. Use a common vent to connect to the sink drain connection. Extend the vent vertically to the underside of the drainboard before turning it horizontally. Then turn it vertically downward and connect it to the top of the building drain. Install two 3-inch cleanouts, one at the base of the vertical drain and one at the base of the return vent. This is so they're accessible for rodding in both directions. Size the drainage pipe and vent pipe to meet code requirements.

## Horizontal Combination Drain and Vent Systems

The plumbing fixtures approved for horizontal combination drain and vent systems are floor drains, standpipes, sinks and lavatories. These are usually found in restaurants or certain other commercial establishments, such as supermarkets or large warehouses with floor drainage. It's a one-pipe method you can use to receive waste and vent plumbing fixtures where you can't use a conventional waste and venting system. Stated another way, the trap of the fixture isn't always individually vented.

Combination drain and vent systems are covered in *IPC* Section 912 and its sub-section references. It's difficult to understand the meaning of this section, with its vague engineering jargon and very little straight talk. I'll explain the minimum code requirements that your local plumbing authority should accept, using the simple drawing in Figure 4-31. However, you should review any layout with your plumbing chief or field inspector before you install a horizontal combination waste and vent system.

**Table 916.1**
**SIZE AND DEVELOPED LENGTH OF STACK VENTS AND VENT STACKS**

| DIAMETER OF SOIL OR WASTE STACK (inches) | TOTAL FIXTURE UNITS BEING VENTED (dfu) | MAXIMUM DEVELOPED LENGTH OF VENT (feet)[a] DIAMETER OF VENT (inches) | | | | | | | | | | |
|---|---|---|---|---|---|---|---|---|---|---|---|---|
| | | 1¼ | 1½ | 2 | 2½ | 3 | 4 | 5 | 6 | 8 | 10 | 12 |
| 1¼ | 2 | 30 | | | | | | | | | | |
| 1½ | 8 | 50 | 150 | | | | | | | | | |
| 1½ | 10 | 30 | 100 | | | | | | | | | |
| 2 | 12 | 30 | 75 | 200 | | | | | | | | |
| 2 | 20 | 26 | 50 | 150 | | | | | | | | |
| 2½ | 42 | | 30 | 100 | 300 | | | | | | | |
| 3 | 10 | | 42 | 150 | 360 | 1,040 | | | | | | |
| 3 | 21 | | 32 | 110 | 270 | 810 | | | | | | |
| 3 | 53 | | 27 | 94 | 230 | 680 | | | | | | |
| 3 | 102 | | 25 | 86 | 210 | 620 | | | | | | |
| 4 | 43 | | | 35 | 85 | 250 | 980 | | | | | |
| 4 | 140 | | | 27 | 65 | 200 | 750 | | | | | |
| 4 | 320 | | | 23 | 55 | 170 | 640 | | | | | |
| 4 | 540 | | | 21 | 50 | 150 | 580 | | | | | |
| 5 | 190 | | | | 28 | 82 | 320 | 990 | | | | |
| 5 | 490 | | | | 21 | 63 | 250 | 760 | | | | |
| 5 | 940 | | | | 18 | 53 | 210 | 670 | | | | |
| 5 | 1,400 | | | | 16 | 49 | 190 | 590 | | | | |
| 6 | 500 | | | | | 33 | 130 | 400 | 1,000 | | | |
| 6 | 1,100 | | | | | 26 | 100 | 310 | 780 | | | |
| 6 | 2,000 | | | | | 22 | 84 | 260 | 660 | | | |
| 6 | 2,900 | | | | | 20 | 77 | 240 | 600 | | | |
| 8 | 1,800 | | | | | | 31 | 95 | 240 | 940 | | |
| 8 | 3,400 | | | | | | 24 | 73 | 190 | 720 | | |
| 8 | 5,600 | | | | | | 20 | 62 | 160 | 610 | | |
| 8 | 7,600 | | | | | | 18 | 56 | 140 | 560 | | |
| 10 | 4,000 | | | | | | | 31 | 78 | 310 | 960 | |
| 10 | 7,200 | | | | | | | 24 | 60 | 240 | 740 | |
| 10 | 11,000 | | | | | | | 20 | 51 | 200 | 630 | |
| 10 | 15,000 | | | | | | | 18 | 46 | 180 | 570 | |
| 12 | 7,300 | | | | | | | | 31 | 120 | 380 | 940 |
| 12 | 13,000 | | | | | | | | 24 | 94 | 300 | 720 |
| 12 | 20,000 | | | | | | | | 20 | 79 | 250 | 610 |
| 12 | 26,000 | | | | | | | | 18 | 72 | 230 | 500 |
| 15 | 15,000 | | | | | | | | | 40 | 130 | 310 |
| 15 | 25,000 | | | | | | | | | 31 | 96 | 240 |
| 15 | 38,000 | | | | | | | | | 26 | 81 | 200 |
| 15 | 50,000 | | | | | | | | | 24 | 74 | 180 |

For SI: 1 inch = 25.4 mm, 1 foot = 304.8 mm.

a. The developed length shall be measured from the vent connection to the open air.



**Figure 4-28**
*Size and developed length of stack vents and vent stacks*

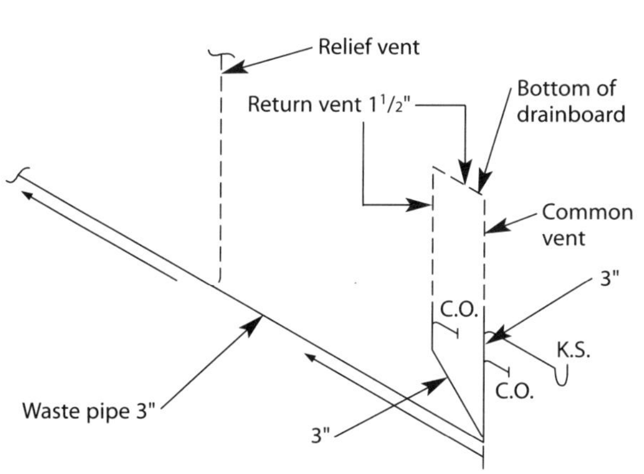


**Figure 4-29**
*Venting an island sink, example 1*

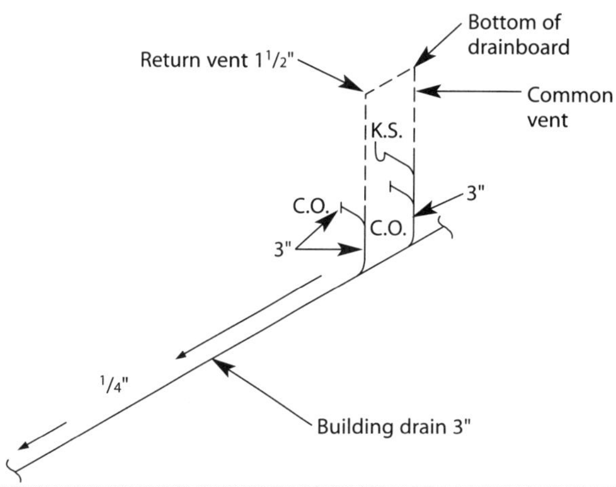


**Figure 4-30**
*Venting an island sink, example 2*

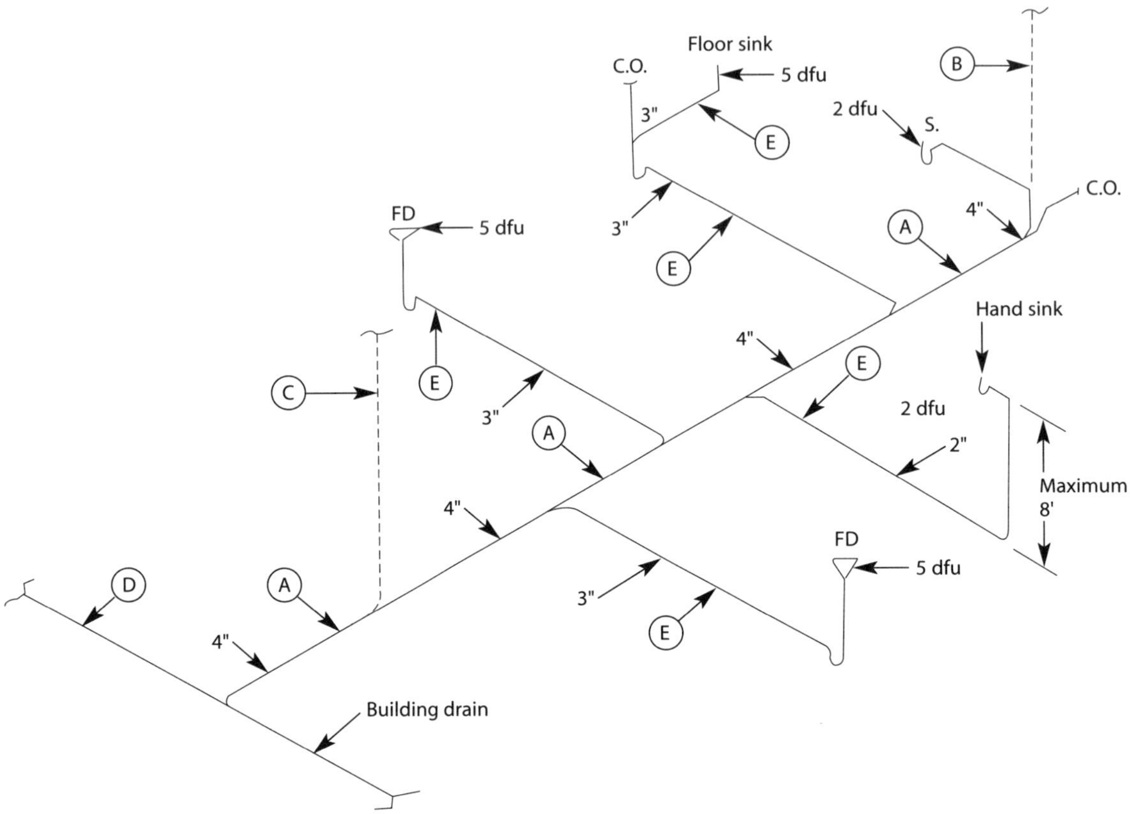


**Figure 4-31**

*Horizontal combination drain and vent system*

- Combination drain and vent pipe (**A**) is considered to be the vent for fixture branches connected to it.
- Be sure to provide a vent stack (**B**) at the upstream end of the waste pipe. A vent (**C**) at the downstream end may be an option if the waste pipe (**A**) connects into a vented building drain (**D**).
- Extend the vent pipe (**B**) vertically up and through the roof or connect it to another approved vent. If you need an offset in any vent in the system, continue it vertically at least 6 inches above the flood level rim of the highest fixture. Refer back to Figure 4-22.
- When you connect fixture branches or fixture drainpipes (**E**) to the system, don't exceed the distance specified in *IPC* Table 906.1, shown in Figure 4-32. For example, a 3-inch drainpipe with a 1/8-inch slope per foot serving a 3-inch floor drain can't be longer than 10 feet without re-venting.

**Table 906.1**
**MAXIMUM DISTANCE OF FIXTURE TRAP FROM VENT**

| SIZE OF TRAP (inches) | SIZE OF FIXTURE DRAIN (inches) | SLOPE (inch per foot) | DISTANCE FROM TRAP (feet) |
|---|---|---|---|
| $1^1/_4$ | $1^1/_4$ | $^1/_4$ | $3^1/_2$ |
| $1^1/_4$ | $1^1/_2$ | $^1/_4$ | 5 |
| $1^1/_2$ | $1^1/_2$ | $^1/_4$ | 5 |
| $1^1/_2$ | 2 | $^1/_4$ | 6 |
| 2 | 2 | $^1/_4$ | 6 |
| 3 | 3 | $^1/_8$ | 10 |
| 4 | 4 | $^1/_8$ | 12 |

For SI: 1 inch = 25.4 mm, 1 foot = 304.8 mm, 1 inch per foot = 83.3 mm/m.

From the *IPC*® ©2003, with permission of the ICC.

**Figure 4-32**

*Maximum distance of fixture trap from vent*

**Table 912.3**
**SIZE OF COMBINATION DRAIN AND VENT PIPE**

| DIAMETER PIPE (inches) | MAXIMUM NUMBER OF DRAINAGE FIXTURE UNITS (dfu) | |
|---|---|---|
| | Connecting to a horizontal branch or stack | Connecting to a building drain or building subdrain |
| 2 | 3 | 4 |
| $2^1/_2$ | 6 | 26 |
| 3 | 12 | 31 |
| 4 | 20 | 50 |
| 5 | 160 | 250 |
| 6 | 360 | 575 |

For SI: 1 inch = 25.4 mm

From the *IPC*® ©2003, with permission of the ICC.

**Figure 4-33**

*Size of combination drain and vent pipe*

**Table 704.1**
**SLOPE OF HORIZONTAL DRAINAGE PIPE**

| SIZE (inches) | MINIMUM SLOPE (inch per foot) |
|---|---|
| $2^1/_2$ or less | $^1/_4$ |
| 3 to 6 | $^1/_8$ |
| 8 or larger | $^1/_{16}$ |

For SI: 1 inch = 25.4 mm, 1 inch per foot = 0.0833 mm/m

From the *IPC*® ©2003, with permission of the ICC.

**Figure 4-34**

*Slope of horizontal drainage pipe*

- The minimum pipe size for a horizontal combination waste and vent pipe is set in *IPC* Table 912.3, shown in Figure 4-33.
- Total the dfu for the job. *IPC* Section 916.2 says the vent has to be at least one-half the diameter of the drain it serves.
- No vent can be smaller than $1^1/_4$ inches.
- Determine the minimum slope for the horizontal drainpipes according to *IPC* Table 704.1, shown in Figure 4-34. If the slope you want to use is greater than the ones in the table, discuss the matter with the plumbing chief before you install the system.
- According to *IPC* Section 912.2, the maximum vertical distance allowed between the horizontal drain and the fixture drain is 8 feet, as shown in Figure 4-31.

To learn more about sizing a horizontal combination drain and vent system, let's size the pipes in the isometric in Figure 4-31. The system has 19 dfu. First size the main drain (**A**) to receive the branch drains (**E**). *IPC* Table 912.3 says a 4-inch pipe will handle 20 dfu. So, the drain (**A**) must be 4 inches. Now we have three branch drains of 5 dfu each. According to *IPC* Table 912.3, these drains could be $2^1/_2$ inches. So why are they 3 inches? It's because of the 3-inch waste openings of the fixtures flowing into them. The drainpipes must also be at least 3 inches. For the 2 dfu hand sink, you can get by with a 2-inch drainpipe. But note that the maximum vertical distance from the horizontal drainpipe to the hand sink drain is 8 feet. To size vent pipe (**B**), note that it has to be at least one-half the diameter of the required horizontal combination drain and vent pipe, which, in this case, is 2 inches according to Figure 4-33 (*IPC* Section 916.2). That wasn't too hard, was it?

## Air Admittance Valves

Air admittance valves, covered in *IPC* Section 917, were fairly new to the plumbing profession at the time of this writing. I believe they were first mentioned in the *International Plumbing Code* published in 1997. It's possible some of you may be working with a code that includes air admittance valves, so here are highlights of the main points:

- Install air admittance valves after rough-in plumbing has passed inspection.
- Install air admittance valves according to code and manufacturer's requirements.
- Terminate and connect individual, branch and circuit vents to an air admittance valve.
- Fixtures vented by an air admittance valve must be on the same floor level.
- Install air admittance valves at least 4 inches above the horizontal branch drain or fixture drain you're venting.
- Install air admittance valves at least 6 inches above any insulation materials in an attic.
- Install air admittance valves so they're accessible in well-ventilated spaces so air may enter the valves.
- Make certain an air admittance valve is properly rated for the vent size you connect it to.
- One conventional vent stack or stack vent of any plumbing system must extend to the outside of the building.
- Don't install air admittance valves in indirect/special waste systems or in spaces used as supply or return air plenums.

# *Chapter 4 — Questions and Answers*

**4-1 According to the *IPC*, if a building requires a 6-inch building drain, one stack must be sized to at least**

(A) 2 inches.

(B) $2^1/_2$ inches.

(C) 3 inches.

(D) $3^1/_2$ inches.

***Answer:* C**

*Code response: Section 903.1*
*This book, page 49*

**4-2 According to the *IPC*, a common vent can vent ____ fixture(s) located on the same floor level.**

(A) 2

(B) 3

(C) 4

(D) 5

***Answer:* A**

*Code response: Section 908.1*
*This book, pages 51 and 52*

**4-3 When a 4-inch rough-in drain is installed for future use, the minimum size of the future vent is**

(A) $1^1/_4$ inches.

(B) $1^1/_2$ inches.

(C) 2 inches.

(D) 3 inches.

***Answer:* C**

*Code response: Section 905.6*
*This book, page 54*

**4-4 The *IPC* says that a 3-inch wet vent may handle the flow from only ______ dfus.**

(A) 6

(B) 8

(C) 10

(D) 12

***Answer:* D**

*Code response: Table 909.3*
*This book, page 57*

**4-5 A waste stack shall not receive discharge from**

(A) lavatories.

(B) showers.

(C) sinks.

(D) urinals.

***Answer: D***

*Code response: Section 910.2*
*This book, page 57*

**4-6 According to the *IPC*, no more than ______ fixtures connected to a horizontal branch drain can be circuit vented.**

(A) 4

(B) 6

(C) 8

(D) 10

***Answer: C***

*Code response: Section 911.1*
*This book, page 57*

**4-7 According to the *IPC*, a maximum load of ____ dfu may drain into a relief vent.**

(A) 2

(B) 3

(C) 4

(D) 5

***Answer: C***

*Code response: Section 911.4.2*
*This book, page 58*

**4-8 The only two fixtures the *IPC* permits to be island vented are a lavatory and a**

(A) urinal.

(B) residential kitchen sink.

(C) commercial sink.

(D) floor drain.

***Answer: B***

*Code response: Section 913.1*
*This book, page 60*

**4-9 According to the *IPC*, you can't connect a __________ to a combination drain and vent system.**

(A) 3-inch floor drain

(B) wall-hung urinal

(C) two-compartment sink

(D) cast iron lavatory

***Answer:* *B***

*Code response:* *Section 912.1*
*This book, page 60*

**4-10 If you use air admittance valves in a plumbing system, be sure to locate them**

(A) within a ventilated space.

(B) at the end of the highest pipe.

(C) in the nearest partition.

(D) at least 3 inches above the horizontal branch drain.

***Answer:* *A***

*Code response:* *Section 917.5*
*This book, page 64*

◆ *Chapter 5* ◆

# Fixture Trap Requirements

Fixture traps are important in protecting public health. So the *IPC* has strict controls on why, where, and how you install them. Let's start with the basics in *IPC* Section 1002. We'll look at requirements for fixture traps and a few requirements not mentioned in the code that are important.

The code requires that every plumbing fixture connected directly to the drainage system have a separate water seal trap. One of the few exceptions is when a fixture has a trap built into its body, like a water closet. No other trap is needed. The *IPC* says that fixture traps must be designed and constructed to provide a liquid seal. And the traps have to provide this seal without materially affecting the flow of sewage or other waste liquids. The *IPC* gives the following requirements for fixture traps:

- Trap each plumbing fixture separately with a water seal trap except for water closets, urinals or similar fixtures that have built-in traps.
- Don't double-trap any fixture. See Figure 5-1.
- Place fixture traps as close to the fixture outlet as possible.
- Fixture traps must be self-scouring, with the exception of interceptor traps.
- Don't make any trap outlet larger than the fixture drain it's connected to. For example, don't use a reducer to connect a $1^1/_2$-inch trap to a $1^1/_4$-inch drainpipe.
- Don't use a trap that has a partition in it unless the trap is built into the fixture and made of approved material that's resistant to corrosion and degradation.
- Don't use a trap that depends on movable parts.
- The *IPC* prohibits the bell traps, full S-traps, $^3/_4$ S-traps and running traps shown in Figure 5-2.

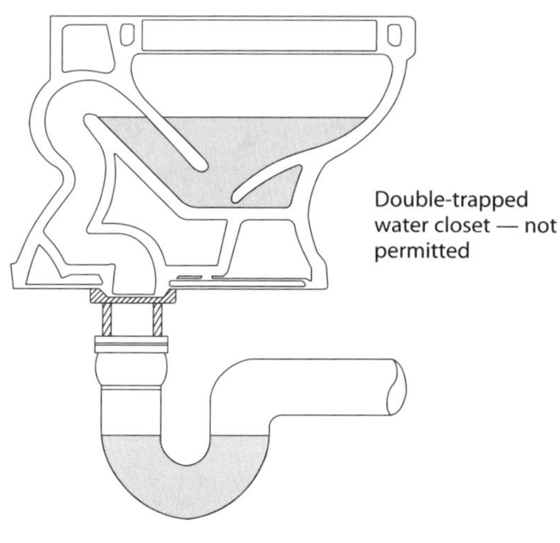


**Figure 5-1**
*No fixture can be double-trapped*

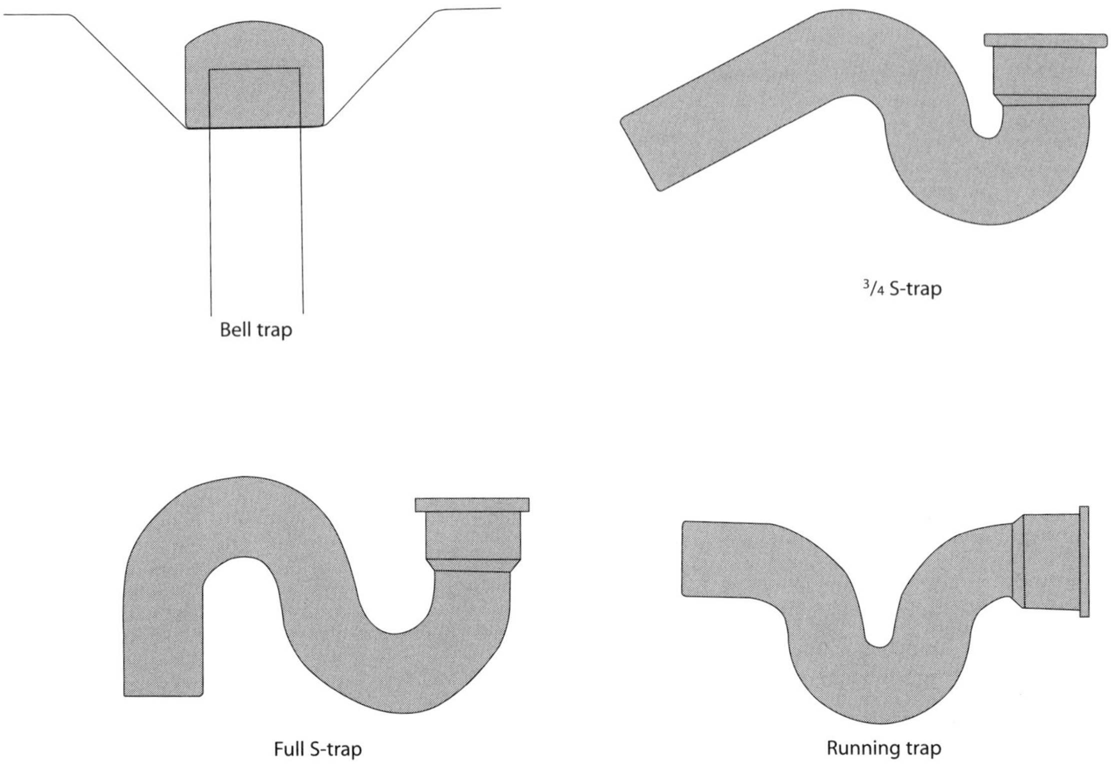


**Figure 5-2**
*Traps prohibited by most codes*

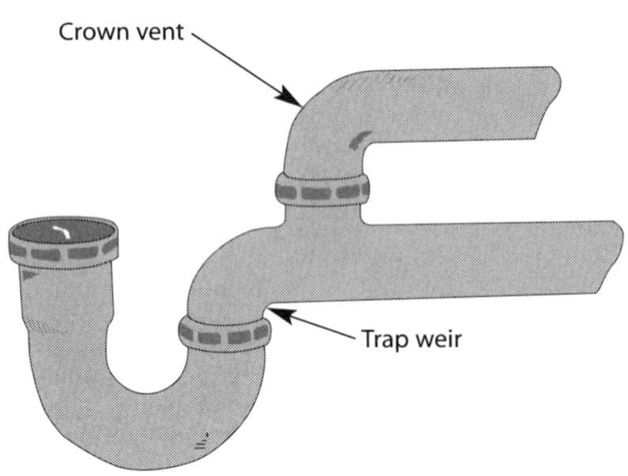


**Figure 5-3**
*Crown-vented tubular P-trap*

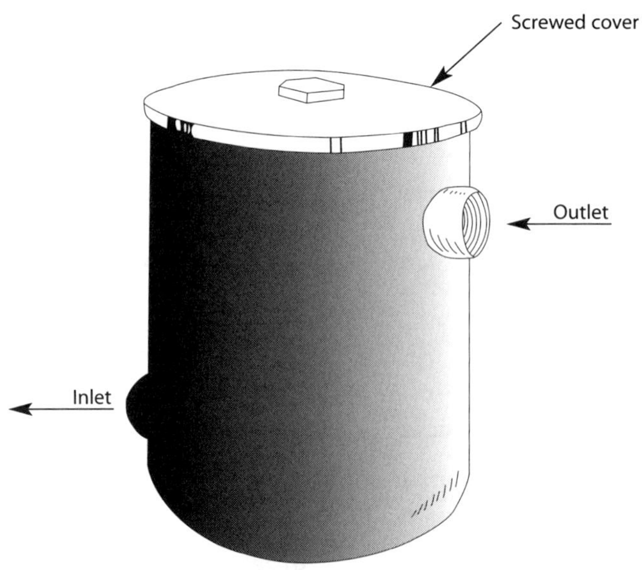


**Figure 5-4**
*Drum trap*

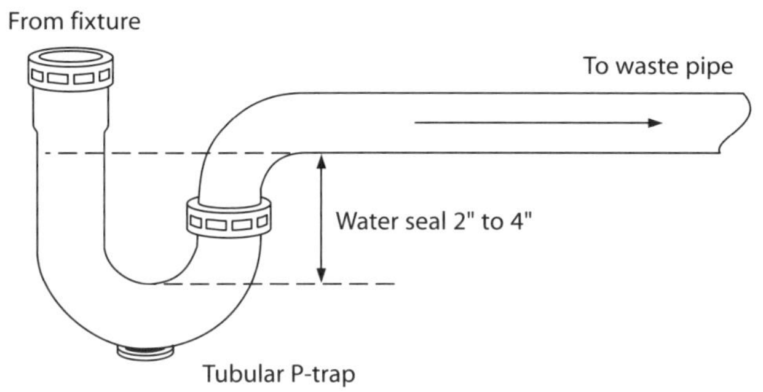


**Figure 5-5**
*2" to 4" water seal*

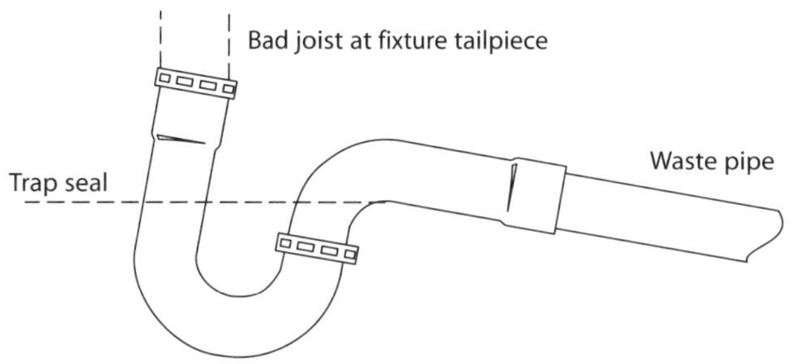


**Figure 5-6**
*Improperly installed trap*

- If special conditions warrant using a crown-vented trap, *IPC* Section 906.3 says you can't install it within two pipe diameters of a trap weir. Figure 5-3 shows you a crown-vented trap.
- Use drum traps only where approved by the local authority. Today, drum traps are only used with special fixtures for intercepting solids and serving chemical waste systems. See Figure 5-4.
- Each fixture trap must have a water seal that's 2 to 4 inches deep. See Figure 5-5. The only exception is for interceptor traps, which require deeper seals.
- Fixture traps installed below concrete floors on fill (or otherwise concealed) can't have cleanouts.
- Install all traps on the same level as their water seals. This prevents negative action and self-siphonage. Careless work creates a bad joint for the fixture tailpiece and invites a rejection from the inspector. See Figure 5-6.
- In vacation homes or other vacant buildings, protect fixture traps where water may freeze in them. The code leaves the method you use up to the local authority. If you can't wrap a trap with some kind of insulation that's acceptable, replace the water in the trap seal with mineral oil. This will protect the trap from freezing.
- Floor drain traps in public toilet rooms can lose their water seals. This lets the water evaporate and sewer gases enter the building. To protect a trap from losing its seal, use at least a 4-inch-deep seal or equip the seal with a trap primer similar to the one shown in Figure 5-7.
- The code sets the maximum vertical drop from a fixture waste outlet to a trap water seal. For kitchen sinks, lavatories, showers, bathtubs, laundry trays and similar fixtures, the maximum vertical drop is 24 inches. This helps prevent self-siphonage of the fixture trap water seal. The shorter the distance between these two points, the more efficient the fixture trap will be. See Figure 5-8.
- The maximum vertical drop for floor-connected fixtures (water closets, floor-mounted urinals and similar fixtures), is 24 inches, as shown in Figure 5-9.
- Floor drains are considered to be fixtures, so they're governed by the 24-inch tailpiece (pipe) limit shown in Figures 5-8 and 5-9. It's often difficult to meet that limit and the *IPC* doesn't regulate this type of installation. Figure 5-10 shows an installation most local authorities will accept.

  The floor drain is installed in a room where the horizontal building drain is 4 feet below the finished floor. To meet the 24-inch code requirement, the vertical rise must be 2 feet minimum.

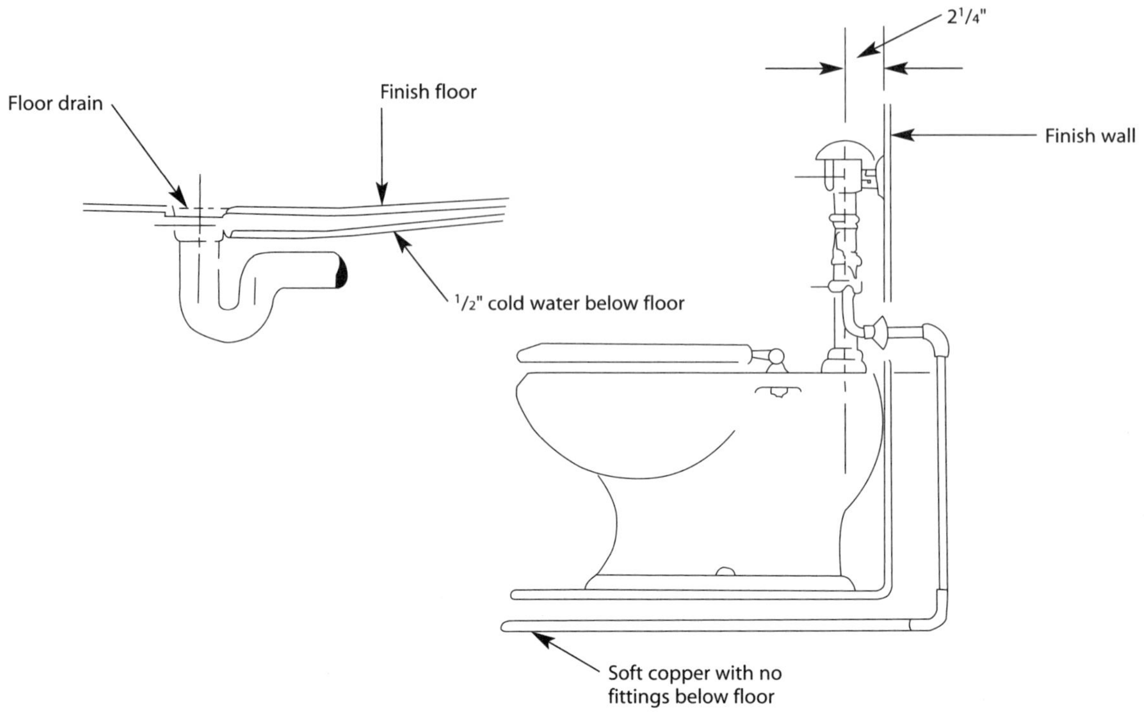


**Figure 5-7**
*Trap primer detail*

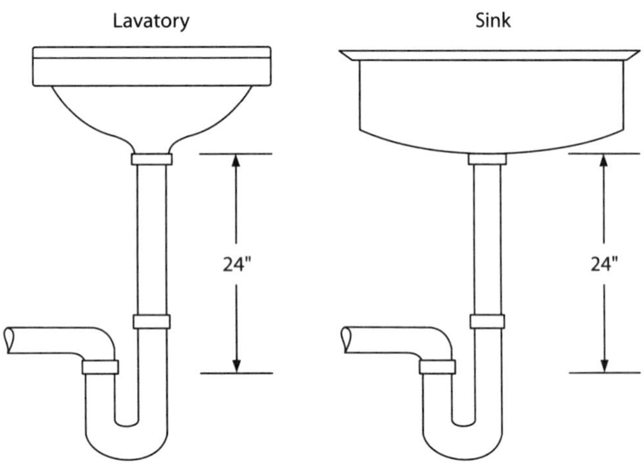


**Figure 5-8**
*Maximum vertical drop from fixture waste outlet to trap water seal*

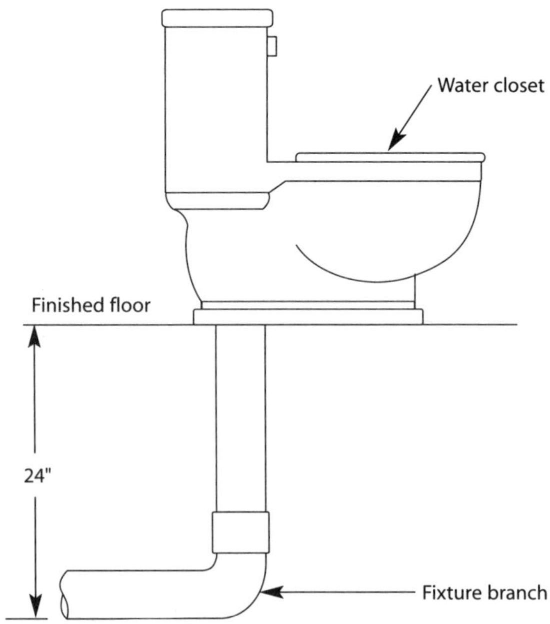


**Figure 5-9**
*Maximum vertical drop from a floor-mounted fixture with a built-in trap*

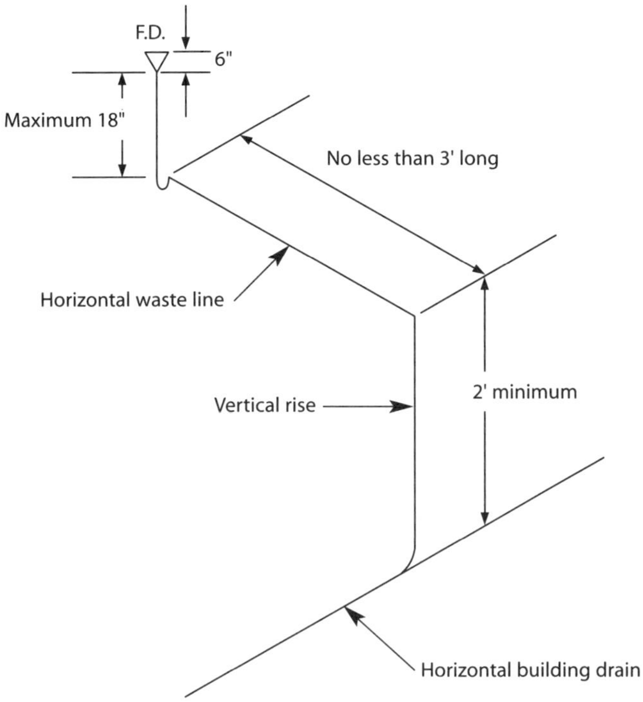


**Figure 5-10**
*Special floor drain installation showing maximum vertical drop*

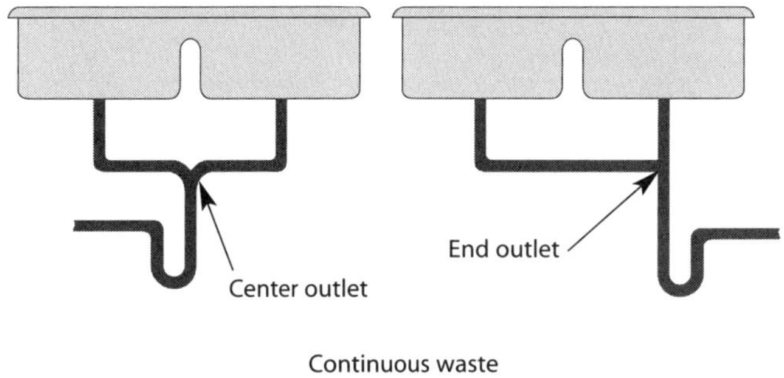


**Figure 5-11**
*Two-compartment sink with one trap, approved*

The floor drain tailpiece and floor drain body must not be more than the maximum drop of 24 inches.

To avoid making an S-trap out of the installation, the horizontal waste line should be at least 3 feet long. If you run into this type of job, check with your plumbing inspector before you begin work.

- *IPC* Section 802.4 says that a clothes washing machine standpipe must extend at least 18 inches and not more than 42 inches above a trap. (*IPC* Section 406.3 covers this in more detail.) The trap must also be accessible for rodding purposes.
- You can install two- or three-compartment sinks or laundry tubs using one trap, provided no compartment is more than 6 inches deeper than the other compartment(s) and the waste outlets are not more than 30 inches apart. See Figure 5-11.
- The code prohibits using a building trap except when local conditions warrant. The plumbing code has strict guidelines for you to follow when a building trap is required. You'll need a detailed drawing like the one in Figure 5-12 for this type of installation.
- Always provide accessible cleanouts for a building trap.
- Be sure there's a relief vent or fresh air intake on the inlet side of a building trap. The relief vent or fresh air intake can't be smaller than one-half the size of the drainpipe it connects to. It must extend above grade and terminate in a screen outlet located outside the building.

## Materials for Traps

The code is vague on what material traps can be made of and exactly where you can use certain materials. The following information should be of value to you and I believe it will meet most local code requirements.

- Materials for concealed fixture traps (bathtubs, showers, floor drains and similar fixtures) must be cast brass, cast iron, lead, ABS plastic, PVC plastics or other approved materials. Cleanouts aren't allowed on concealed fixture traps.

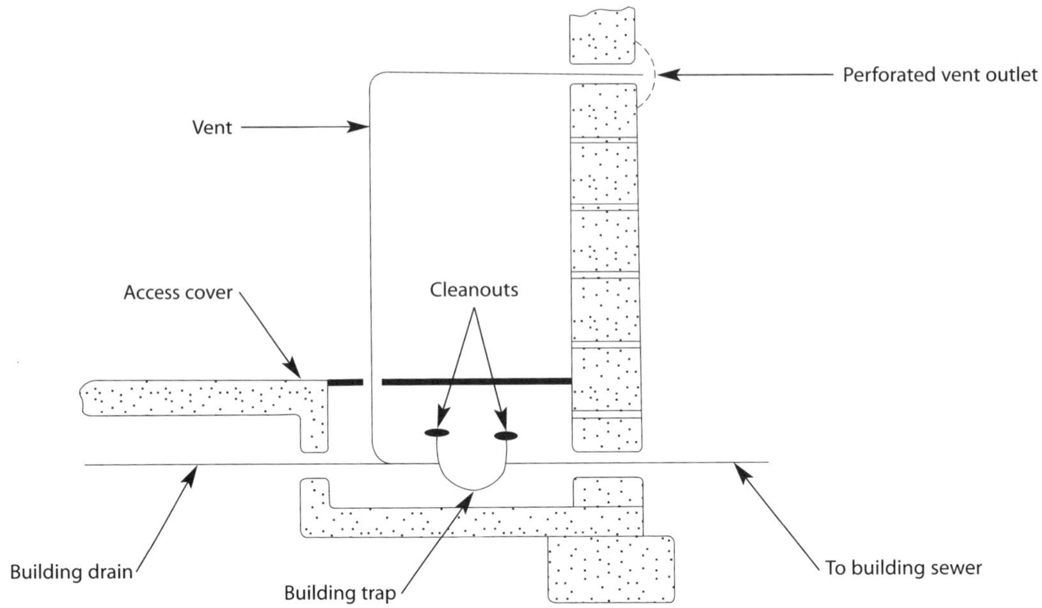


**Figure 5-12**
*Building trap detail*

- Except for fixtures with built-in traps, all exposed or accessible traps must be of cast iron, brass, lead, 17-gauge tubular brass, or copper. Some local authorities will accept 20-gauge tubular traps. In a plastic system, you must use ABS or PVC traps. It's permissible to equip exposed traps with cleanouts.
- As to materials for a chemical waste system, *IPC* Section 702.5 says only that the materials "be of an approved material that is resistant to corrosion and degradation for the concentrations of chemicals involved." However, standard materials usually approved by local authorities include borosilicate glass, high silicon cast iron, vitrified clay, and lead pipe with walls at least $^1/_8$-inch thick.
- In some below-ground installations where acid-resistant traps are made of vitrified clay or other brittleware materials, there are special code requirements for protecting the traps. Embed this type of trap in concrete that goes 6 inches beyond the bottom and sides of the trap.
- Slip joints on traps must be made with approved elastomeric gaskets. Install them on the trap inlet, trap outlet and within the trap seal. See Figure 5-13.
- Insulate exposed pipe and traps installed in a mental health facility.

## Fixture Trap Sizes and Distances

The *IPC* regulates the size of fixture traps to make sure the traps drain fixtures rapidly. *IPC* Section 1002.5 says a trap can't be larger than the drainpipe it discharges into. You can rely on the trap sizes shown in Figure 5-14.

Let's consider an example to see how to figure fixture trap distance. We'll use *IPC* Table 906.1 shown in Figure 4-32 in Chapter 4. For a $1^1/_4$-inch trap and a $1^1/_4$-inch fixture drain, with a slope of $^1/_4$ inch per foot, *IPC* Table 906.1 sets a limit from trap to vent of $3^1/_2$ feet. In some cases, because of the fixture location in a bathroom, kitchen or utility room, you may have to exceed the code-allowed distance of a

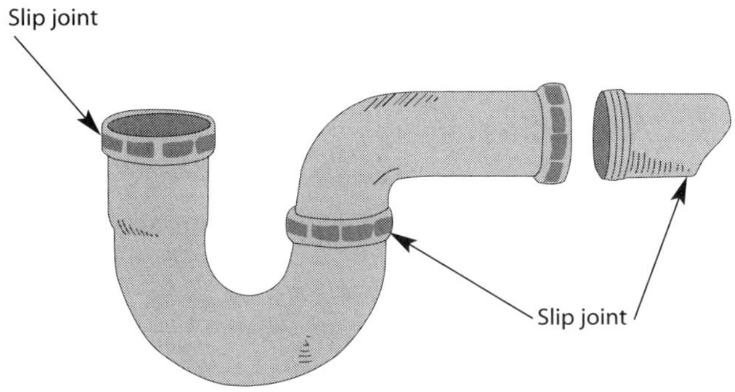


**Figure 5-13**

*Slip joints*

| Plumbing fixture type | Trap size (in) |
|---|---|
| Bathtub (with or without overhead shower) | $1^1/_2$ |
| Bidet | $1^1/_4$ |
| Clothes washing machine standpipe | 2 |
| Dental lavatory | $1^1/_4$ |
| Dishwasher, domestic | $1^1/_2$ |
| Drinking fountain | $1^1/_4$ |
| Food waste disposer, commercial | 2 |
| Food waste disposer, domestic | $1^1/_2$ |
| Kitchen sink, domestic | $1^1/_2$ |
| Lavatory, domestic | $1^1/_4$ |
| Laundry tray (1 or 2 compartment) | $1^1/_2$ |
| Shower stall or drain | $1^1/_2$ |
| Sink, commercial | $1^1/_2$ |

**Figure 5-14**

*Minimum size of traps that aren't built into a fixture*

fixture drain. If this should happen, it's likely that local authorities will accept a re-vent installed in the drain line. See Figure 5-15. Figure 5-15 A shows a fixture drain that's $1^1/_2$ feet over the limit. Figure 5-15 B shows a fixture drain that's $3^1/_2$ feet over the limit.

To measure the developed length of a fixture trap distance, you must measure from the crown weir of the fixture trap to the vent pipe. See Figure 5-16. The measurement, including offsets and turns, must be within the limits set in *IPC* Table 906.1. The size of a floor drain trap limits the distance between the trap and an individual vent or a vented drainage

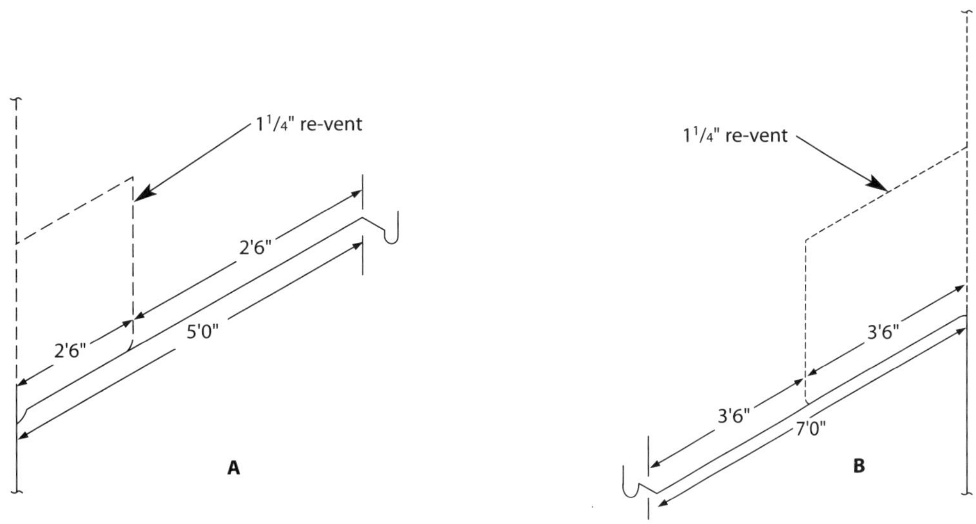


**Figure 5-15**

*Fixture drain re-vent, $1^1/_4$-inch fixture drain (fixture arm)*

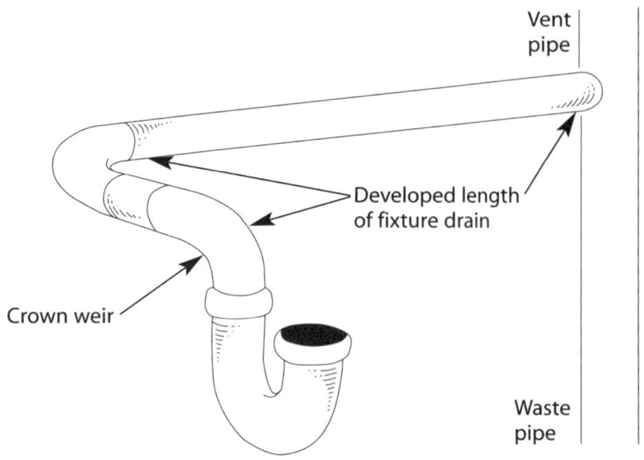


**Figure 5-16**

*Measuring the developed length of a fixture drain*

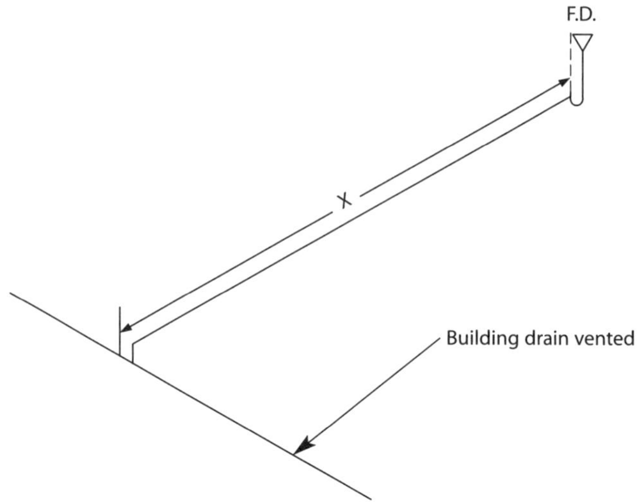


**Note:** *International Plumbing Code*
3" pipe: X is 10'0"
4" pipe: X is 12'0"

**Figure 5-17**

*Maximum distance of floor drain from building drain*

line. Refer to *IPC* Table 906.1. Figure 5-17 shows that for a 3-inch trap, the maximum distance is 10 feet. For a 4-inch trap, the maximum distance is 12 feet.

## Broken Trap Seals

The five ways a trap seal is most likely to break are through trap siphonage, back pressure, wind effect, evaporation and capillary attraction. If you install a plumbing system improperly, you can make a trap seal fail because of siphonage and/or back pressure. Figure 5-18 shows two examples of how improperly-installed fixtures may lose their trap seals. Figure 5-18A and 5-18B show a major fixture (water closet) installed above a minor fixture (lavatory). In Figure 5-18A, this creates negative pressure, which makes a vacuum by pulling water out of the lower trap. In Figure 5-18B, this creates positive pressure that's exerted at the base of the stack, forcing water out of the lower trap. Figure 5-19 shows a broken trap seal caused by installing a fixture drain that exceeds the limit set in *IPC* Table 906.1.

## Parts of a Trap

Figure 5-20 shows the parts of a one-piece trap, and Figure 5-21 shows the parts of a two-piece tubular trap. This information may be important for a plumbing examination. For example, an examination-type question might ask you to identify the dip of a trap. Would you know the correct answer?

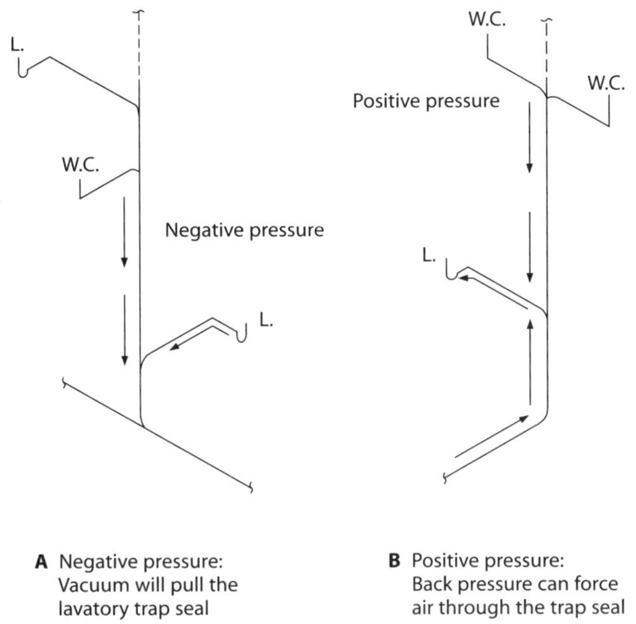


**Figure 5-18**

*Major fixtures (water closets) can't be installed above minor fixtures (lavatories)*

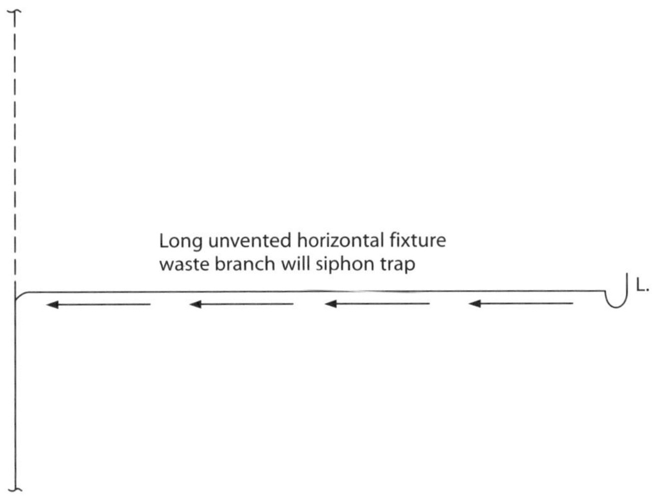


**Figure 5-19**

*This type of installation is not permitted by code*

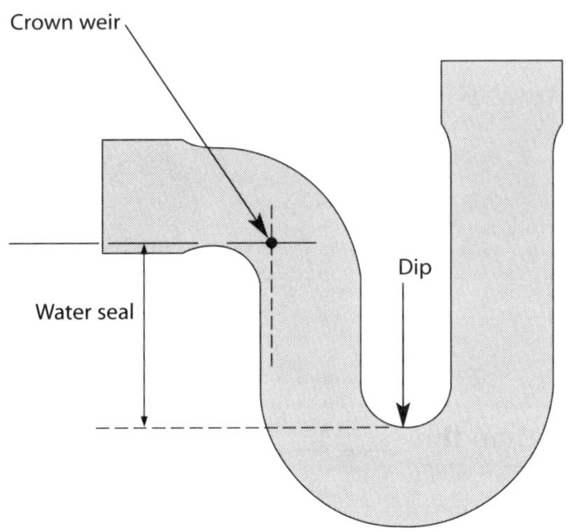


**Figure 5-20**

*One-piece trap*

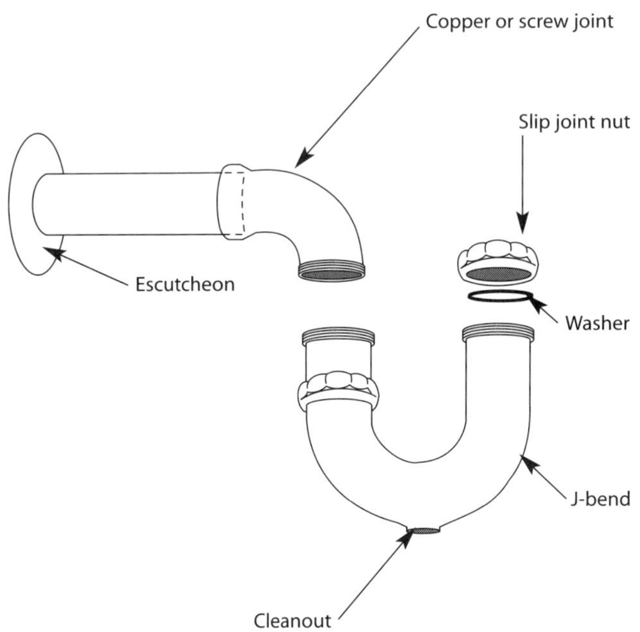


**Figure 5-21**

*Two-piece tubular trap*

# *Chapter 5 — Questions and Answers*

**5-1 The maximum vertical distance permitted by the *IPC* from a fixture outlet to its trap weir is**

(A) 12 inches.

(B) 16 inches.

(C) 20 inches.

(D) 24 inches.

***Answer: D***

*Code response: Section 1002.1*
*This book, pages 69, 71 and Figure 5-8*

**5-2 Each plumbing fixture must be separately trapped, with the exception of a**

(A) lavatory.

(B) water closet.

(C) bathtub.

(D) floor drain.

***Answer: B***

*Code response: Section 1002.1 (Exception 1)*
*This book, page 69 and Figure 5-1*

**5-3 All fixture traps must be self-scouring, except for**

(A) grease traps.

(B) built-in traps.

(C) tubular traps.

(D) traps with slip joints.

***Answer: A***

*Code response: Section 1002.2*
*This book, page 69*

**5-4 According to the *IPC*, a fixture trap shall not be larger than the __________ it discharges into.**

(A) fixture tailpiece

(B) wet vent

(C) drainage pipe

(D) fixture overflow

***Answer: C***

*Code response: Section: 1002.5*
*This book, page 69*

**5-5 When you install a trap that may lose its seal through evaporation, the *IPC* requires the seal be protected by a/an**

(A) trap primer valve.

(B) 4-inch-deep seal.

(C) swing check valve.

(D) air-conditioning condensate drain.

***Answer:*** ***A***

*Code response: Section 1002.4*
*This book, page 71*

**5-6 According to the *IPC*, a/an _______ can never be used in a plumbing system.**

(A) 17-gauge tubular trap

(B) S-trap

(C) crown-vented trap

(D) drum trap

***Answer:*** ***B***

*Code response: Sections 906.3 and 1002.3 (Subhead 5)*
*This book, page 69*

**5-7 You may install a two- or three-compartment sink on a single trap provided no compartment is more than ______ inches deeper than the other compartments.**

(A) 4

(B) 6

(C) 8

(D) 10

***Answer:*** ***B***

*Code response: Section 1002.1 (Exception 2)*
*This book, page 73*

**5-8 The *IPC* prohibits extending a clothes washing machine standpipe more than _____ inches above its trap.**

(A) 38

(B) 40

(C) 42

(D) 48

***Answer:*** ***C***

*Code response: Section 802.4*
*This book, page 73*

**5-9 Where the *IPC* permits use of a 6-inch building trap, you'll have to use at least a _______ relief vent.**

(A) 3-inch

(B) 4-inch

(C) 5-inch

(D) 6-inch

***Answer: A***

*Code response: Section 1002.6*
*This book, page 73*

**5-10 According to the *IPC*, a drum trap may be used on a**

(A) laboratory sink.

(B) bathtub.

(C) commercial lavatory.

(D) kitchen sink.

***Answer: A***

*Code response: Section 1002.3 (Exception 6)*
*This book, page 71*

◆ *Chapter 6* ◆

# Cleanouts

Years ago cleanouts weren't required on drainage pipes! As a young plumber, when my boss sent me on service calls to remove an obstruction in a drainage pipe, I used a sharp chisel and hammer to cut an opening in the pipe so I could insert a cleaning cable. After removing the obstruction, I'd close up the opening with tarpaper and a mixture of cement. And yes, sooner or later the patch job would deteriorate and let raw sewage seep out into the ground. Big health hazard! Fortunately, cleanouts are now an absolute requirement and an integral part of any drainage system. Even well-designed drainage systems are subject to stoppages. Accessible cleanouts will save you valuable time and an owner unnecessary expense.

The *IPC* specifies location, distance between, size, and other requirements for cleanouts. The two basic types of code-approved cleanouts are conventional cleanouts and manholes. Conventional cleanouts with removable plugs are required on drainage pipes in buildings or on private sewers that are up to 6 inches in diameter. Manholes are required on private sewers that are 8 inches or more in diameter. They serve as junctions to join one or more sewer lines and to allow access for clearing the lines of any accumulated solids. Manholes are common in private sewage collection systems that serve several large buildings. Figure 6-1 shows how cleanouts and manholes work together.

## Cleanout Locations

Location is everything when you design or install cleanouts for a building drain. They must be accessible so that all parts of the drainage system can be rodded. Let's look at the requirements in *IPC* Section 708.

- You must have accessible cleanouts on all horizontal drainage piping. Don't put cleanouts more than 100 feet apart. Be sure you put in a cleanout for each change of direction greater than 45 degrees. See Figure 6-2. When you have more than one 45-degree change of direction in the same building drain, use one cleanout for each 40 feet of developed length of piping.

- Cleanouts are required at the base of each waste and soil stack. If you can't extend a cleanout to an accessible location flush with the finished floor or to the outside of a building, you can use a cleanout tee in a vertical stack. Here's how to do this. If the cleanout plug is concealed, install a removable cover plate or access door to permit rodding. The raised head on the cleanout plug may be tapped to accept a long screw that holds the cover plate in place, as shown in Figure 6-3.

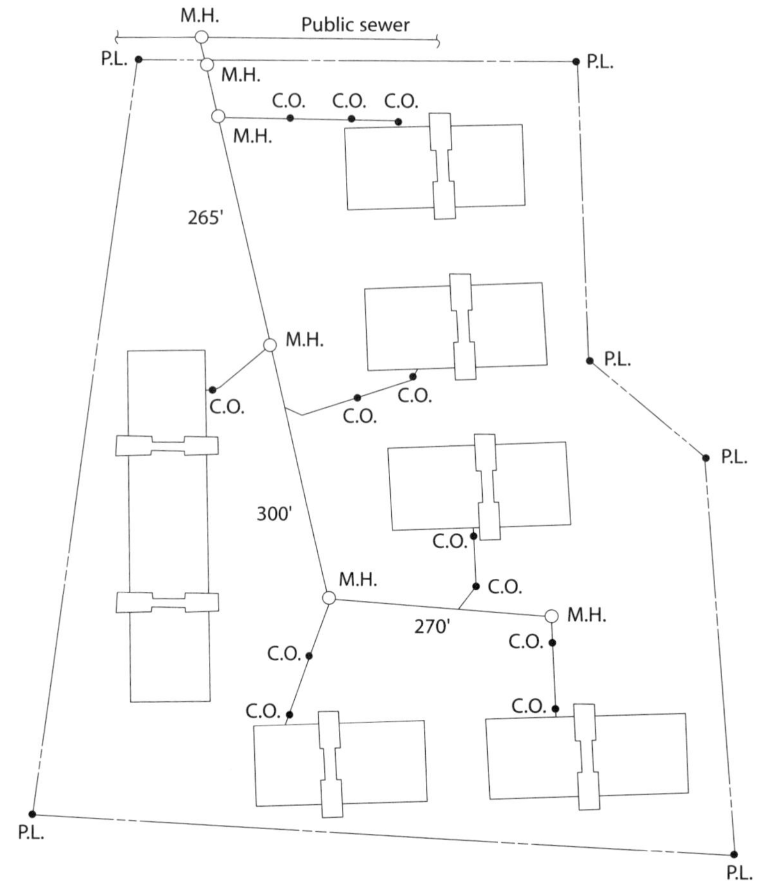


**Figure 6-1**

*Private sewage collection system*

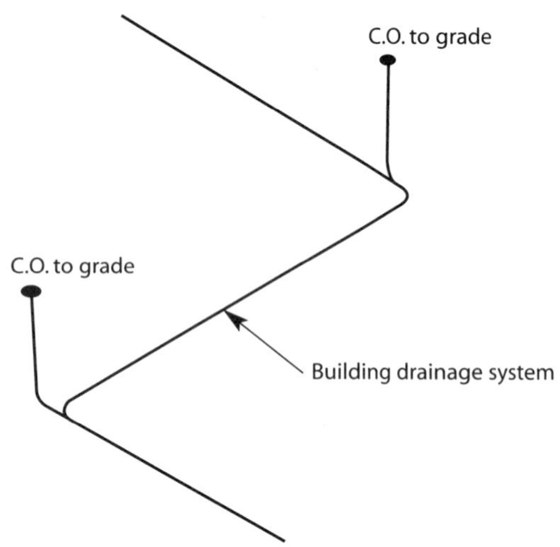


**Figure 6-2**

*Provide cleanouts at any change of direction greater than 45 degrees*

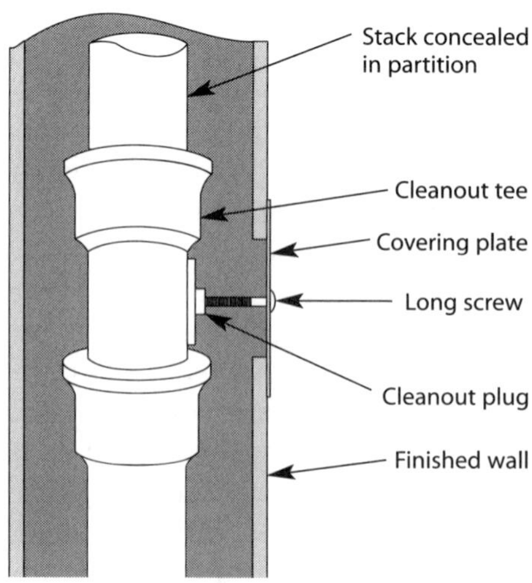


**Figure 6-3**

*Cleanout tee in a vertical stack*

- It's OK to install cleanouts in horizontal drainage lines beneath concrete flooring if they're accessible and flush with the finished floor. Cleanouts in walkways, hallways, and rooms must have countersunk plugs so people won't trip on them. See Figure 6-4. If a cleanout is subject to vehicular traffic, *IPC* Section 708.4 requires you to use covering plates, access covers or cleanouts that are approved and designed for this purpose. It's smart to check with your local authority for the type approved in your area. Figure 6-5 shows a typical installation.

- You need to locate a full-size cleanout outside a building, close to the junction of the building drain and building sewer, usually within 30 inches of the building line. At this location, you must have a two-way fitting that permits upstream as well as downstream rodding. See Figure 6-6. Your local code will determine whether or not this fitting must be brought to grade. However, you'll have to use a countersunk cleanout plug if there's any foot traffic.

*IPC* Section 708.3.5 also makes an exception that you don't have to install a cleanout at this location if there's an accessible 3-inch (or larger) cleanout in the soil stack within 10 feet of the drain and sewer connection.

- Any cleanout you install must allow a cleaning cable to be inserted in the direction of flow to clear blockages.

- *Never* use a cleanout opening to install another fixture. And don't use a cleanout as a floor drain without prior written permission from your inspector. If you change the use of a cleanout, you'll have to provide another cleanout of equal accessibility and size.

- *IPC* Section 708.7 says the minimum nominal cleanout size must be the same nominal size as the pipe it's installed in, up to 4 inches. See Figure 6-7. The code allows interior drainage pipes larger than 4 inches to use a 4-inch cleanout. Remember that building sewers 8 inches and larger must have manholes.

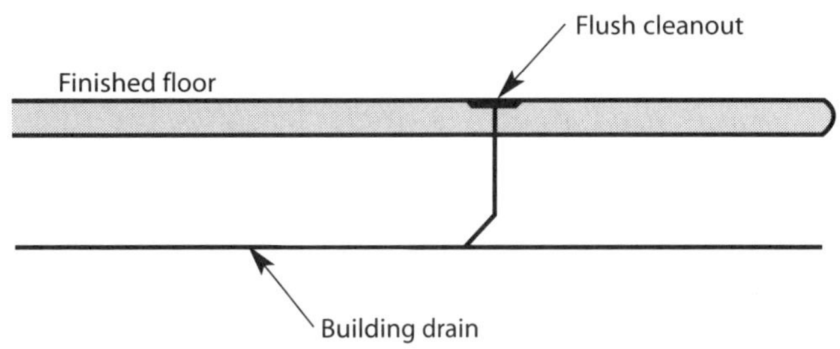


**Figure 6-4**
*Countersunk cleanout*

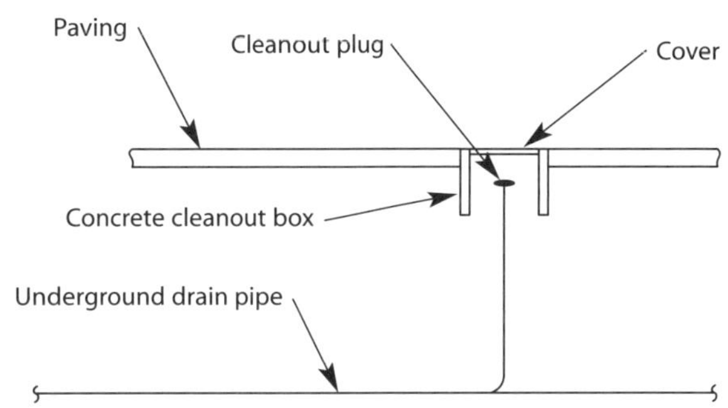


**Figure 6-5**
*Cleanout and piping installed under a vehicular traffic area*

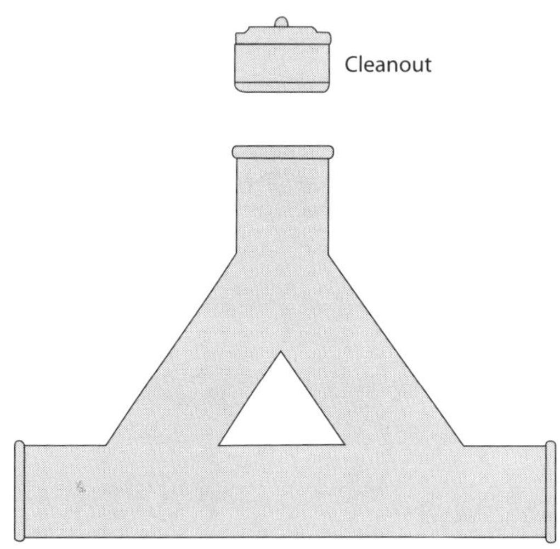


**Figure 6-6**
*"No-hub" two-way cleanout fitting*

- Allow an 18-inch clearance for rodding on each cleanout installed in a 6-inch or smaller pipe. You'll need a 36-inch clearance for an 8-inch or larger interior cleanout. See Figure 6-8.
- The required thickness for cleanout ferrule bodies is the same as that required for pipe and fittings of like material. The cleanout plug must extend at least 1/4 inch above the end of the pipe or hub. Make new installation cleanouts of heavy brass for metallic systems and plastic for plastic systems. They must have either a raised hex plug or a countersunk plug, depending on the location. A ferrule and cleanout plug must have ANSI standard tapered pipe threads. If you have to replace a damaged cleanout plug, most authorities will accept a heavy replacement lead or nylon plug (see Figure 6-9).
- *IPC* requires a manhole for any building sewer 8 inches and larger.
- A manhole must be installed at every change of direction, pipe size, alignment, grade or elevation. Straight runs must have manholes no more than 400 feet apart.
- Use only approved gas-tight manhole covers. Figure 6-10 shows a typical brick-constructed manhole made at a jobsite. Figure 6-11 shows a typical precast manhole.

## Cleanout Exceptions

You can use an exposed P-trap connected to a drainage system with a ground joint or a slip joint connection, if necessary. A trap with slip joint nuts can be taken apart to insert a cable. Some authorities will even accept a water closet as a regular cleanout, as shown in Figure 6-12. Be sure to check your own code or local authority before you install this type of system.

| Nominal pipe size (in) | Nominal cleanout size (in) |
|---|---|
| 1 1/4 | 1 1/4 |
| 1 1/2 | 1 1/2 |
| 2 | 2 |
| 3 | 3 |
| 4 and larger | 4 |

**Figure 6-7**
*Cleanout sizes*

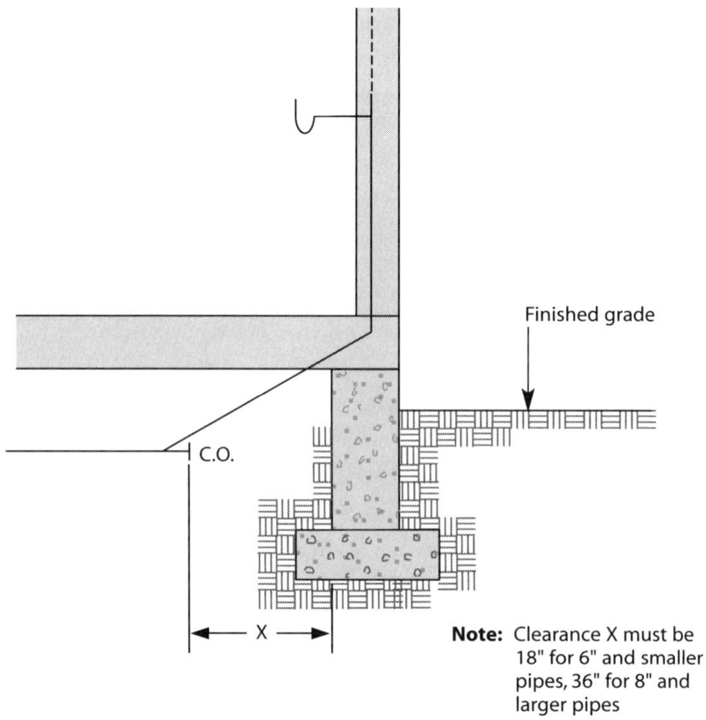


**Figure 6-8**
*Cleanout clearances*

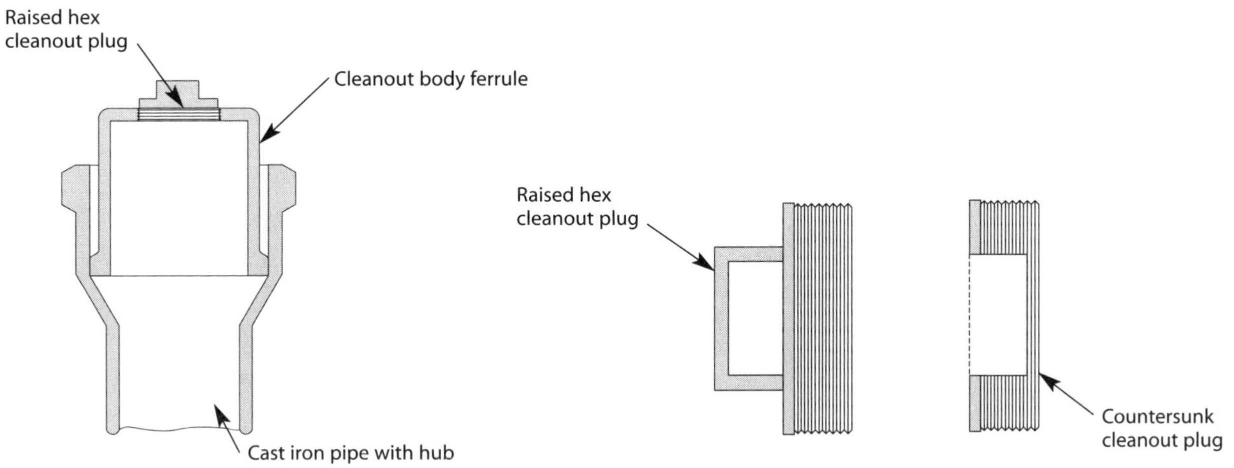


**Figure 6-9**

*Cleanout design*

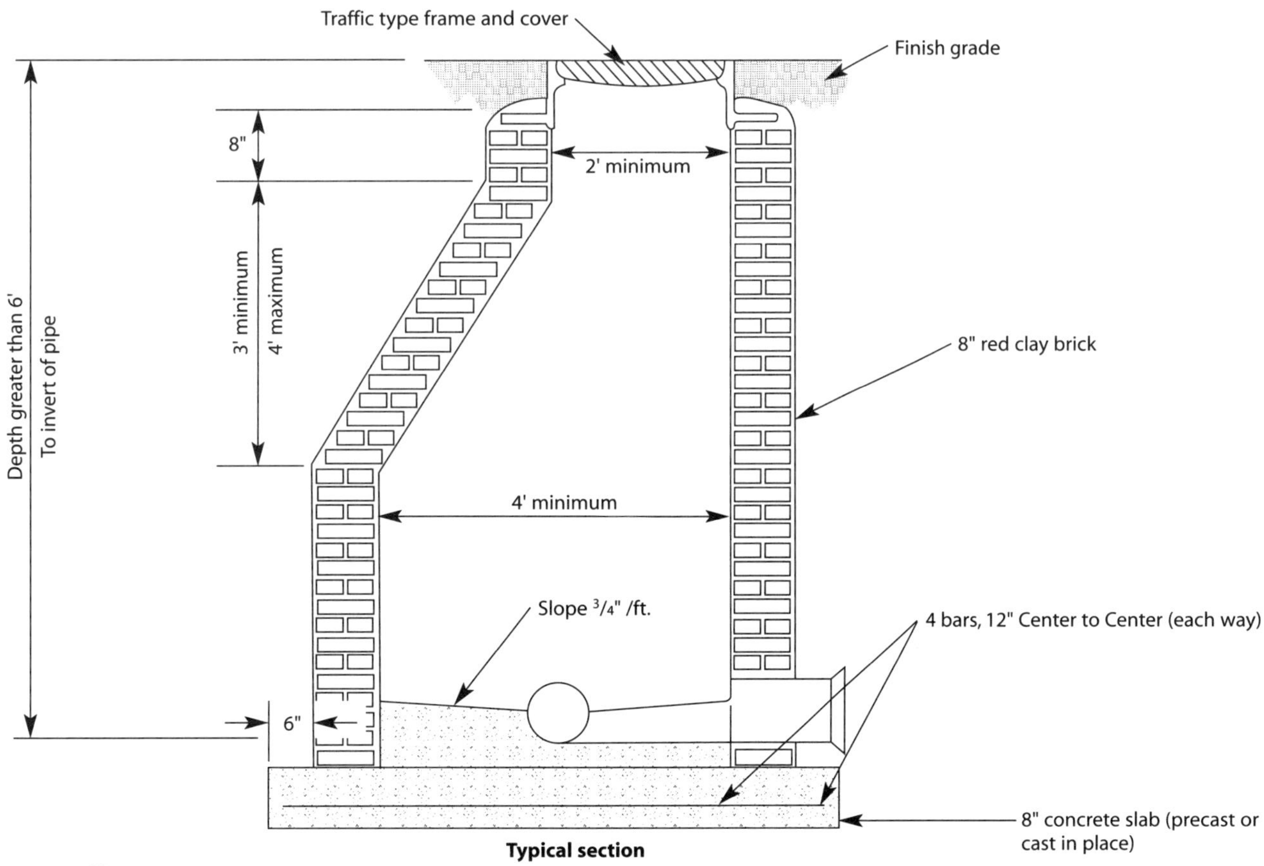


**Notes:**

1. Brick masonry construction stuccoed with 3/4" mortar inside and outside.
2. The first three sections of pipe at both influent and effluent of each manhole are a maximum 2' in length.
3. A flow channel inside the manhole will direct influent into flow stream.

**Figure 6-10**

*Standard brick manhole*

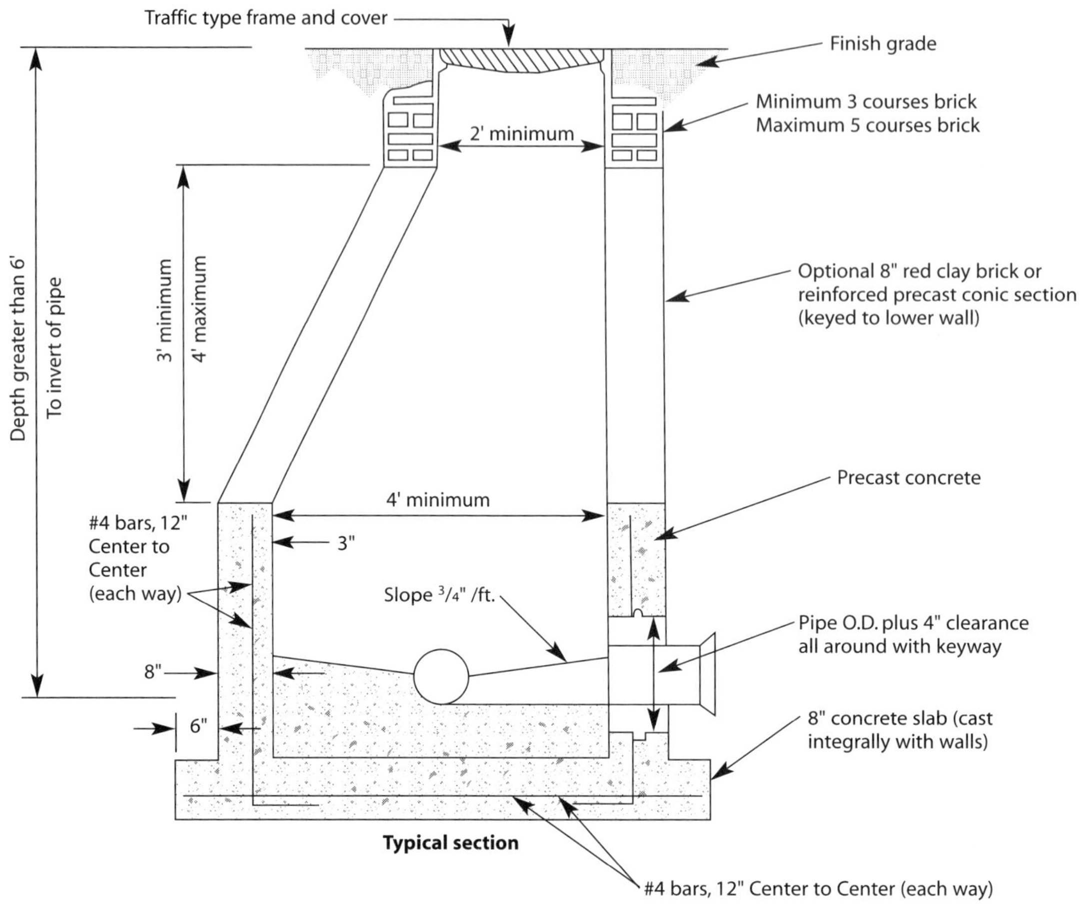


**Notes:**

1. Brick masonry construction stuccoed with 3/4" mortar inside and outside.
2. The first three sections of pipe at both influent and effluent of each manhole will be a maximum 2' in length.
3. Lift holes through precast structure are not permitted.
4. See technical specifications for placement of construction joints.
5. All openings shall be sealed with a waterproof, expanding grout.
6. A flow channel inside the manhole will direct influent into flow stream.

**Figure 6-11**
*Standard precast manhole*

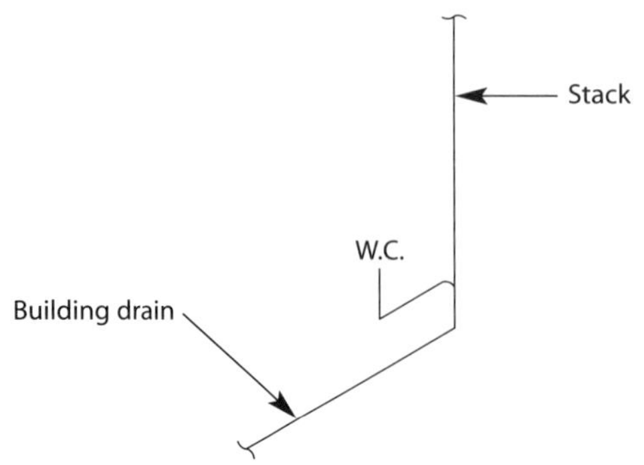


**Figure 6-12**
*Water closet outlet is considered an equivalent cleanout by some codes*

# Chapter 6 — Questions and Answers

**6-1 According to the *IPC*, a brass cleanout plug can only be used in**

(A) pipes serving acid waste systems.

(B) a cast iron drainage system.

(C) an all-plastic system.

(D) a non-metallic system.

***Answer:* B**

*Code response: Section 708.2*
*This book, page 84*

**6-2 If a cleanout plug is located where it may be tripped on, the *IPC* requires that it**

(A) be countersunk.

(B) be painted with fluorescent paint.

(C) have a square head.

(D) have a protective casing.

***Answer:* A**

*Code response: Section 708.2*
*This book, page 83*

**6-3 When you install horizontal drainage lines in a building, you must provide cleanouts every _____ feet.**

(A) 50

(B) 75

(C) 100

(D) 125

***Answer:* C**

*Code response: Section 708.3.1*
*This book, page 81*

**6-4 The smallest building sewer the *IPC* requires a manhole for is**

(A) 6 inches.

(B) 8 inches.

(C) 10 inches.

(D) 12 inches.

***Answer:* B**

*Code response: Section 708.3.2*
*This book, pages 81 and 83*

**6-5 If you install a cleanout where it may be damaged by vehicular traffic, the *IPC* requires you to install one that is**

(A) extra heavy-duty cast iron.

(B) countersunk.

(C) clearly marked.

(D) protected by a removable plate.

***Answer: D***

*Code response: Section 708.4*
*This book, page 83*

**6-6 According to the *IPC*, every cleanout you install must**

(A) be no larger than the pipe it serves.

(B) be of the same material as the pipe it serves.

(C) extend to the outside of the building.

(D) allow cleaning in the direction of flow.

***Answer: D***

*Code response: Section 708.5*
*This book, page 83*

**6-7 The *IPC* prohibits using a cleanout**

(A) as a floor drain.

(B) at the base of a vent stack.

(C) to extend a building drain.

(D) in a greasy waste system.

***Answer: A***

*Code response: Section 708.6*
*This book, page 83*

**6-8 According to the *IPC*, the minimum nominal cleanout size for an interior 6-inch drainage pipe is**

(A) 3 inches.

(B) 4 inches.

(C) 5 inches.

(D) 6 inches.

***Answer: B***

*Code response: Section 708.7*
*This book, page 83*

**6-9 The minimum clearance for cleanouts for rodding piping that's 6 inches or smaller is**

(A) 12 inches.

(B) 14 inches.

(C) 16 inches.

(D) 18 inches.

***Answer: D***

*Code response: Section 708.8*
*This book, page 84*

**6-10 When manholes are required, the *IPC* requires that they**

(A) be at least 30 feet apart.

(B) be no more than 6 feet deep.

(C) be constructed of a rust-resistant material.

(D) have secured gas-tight covers.

***Answer: D***

*Code response: Section 708.3.6*
*This book, page 84*

# ◆ Chapter 7 ◆

# Interceptors, Separators and Neutralizing Devices

In daily operation, some businesses discharge waste containing unacceptable levels of grease, oil, flammable waste, sand, plaster, lint, hair, glass, acids, and more. The *IPC* requires these materials be intercepted, separated or neutralized before they enter private or public drainage systems. The appropriate method for handling this type of waste varies with each situation. Here is my interpretation of *IPC* Section 1001, which covers this subject.

Interceptors and separators accumulate and recover objectionable substances from liquid waste. Dilution or neutralizing tanks treat corrosive liquids to make them safe before discharge into the drainage system. Code requirements for the size and type of interceptors, separators or neutralizing tanks depend primarily on the design and method of installation approved by the manufacturer. Usually, complying with this section of code is the architect's obligation, so his information should be on the approved job and office set of plans.

Even though the architect is responsible here, there are several reasons to pay close attention to this section of the code. The main reason is that plumbing exams will inevitably include questions on interceptors. Another reason is that the architect and the plumbing plans examiner don't always know the job conditions the work will be done under. Knowing the code gives you the advantage of checking the approved plans against job conditions, which may save you considerable time and money. For example, with a location change, you can see the local authorities before you begin installation if you have a question about the plans. This may help you avoid an expensive tearout and re-do.

## Grease Interceptors

Let's begin with the grease interceptor, which is probably the interceptor you'll encounter most often. In a commercial building where food is processed, prepared or served, you must install grease interceptors in the waste lines serving the facility. If you don't, the accumulated grease will clog the drainage pipes.

Let me tell you a little story about something that happened in my jurisdiction of Miami, Florida. An owner bootlegged a meat packing plant in a warehouse district without obtaining a permit. You guessed it — he didn't install a grease interceptor! The faulty work went undiscovered until a stoppage developed. And then, Wow! The stoppage created a "not to be ignored" odor over the entire warehouse complex. The owner hired a plumbing contractor to unclog the offensive pipes. He, after numerous attempts, informed the plumbing department that he couldn't find a grease interceptor. It fell to me to investigate. I searched and couldn't find a permit or approved plan for any meat packing plant at the location. The health department, fire department and the building department wrote up their violations, and it turned out to be an extremely expensive economy measure for the owner. He's not likely to do that again. Remember, if you work without a permit, sooner or later it's going to come back at you, and it isn't going to be nice.

*IPC* Section 1003.3.1 describes which businesses produce unacceptable amounts of grease. They include restaurants, hotel kitchens, bars, factory cafeterias, hospitals, school kitchens and clubs. Other common establishments not mentioned in the *IPC* are supermarkets, meat processing plants, soap factories, slaughterhouses and tallow rendering plants. Local authorities will probably insist on an interceptor for these businesses as well. Note that *IPC* Section 1003.2 says that waste that doesn't require treatment or separation *can't* pass through grease interceptors. The *IPC* doesn't require grease interceptors in single-family homes or apartment buildings.

There are two types of grease interceptors, inside and outside installations. Let's look at the inside type first.

## Inside Installation Interceptors

*IPC* Table 1003.3.4.1 gives information for determining the capacity of grease traps (interceptors). I've included a partial table in Figure 7-1. The job architect usually works closely with local authorities to determine the size, type and location of interceptors. For small restaurants or other businesses generating small amounts of grease, the architect will probably select an inside installation grease interceptor. The manufacturer supplies information regarding size. Figure 7-2 shows a factory-built cast iron unit. These come in sizes with a grease capacity from 8 pounds to the maximum 100 pounds allowed by code. They may be floor-mounted or installed below the floor, as shown in Figure 7-3.

An inside installation grease interceptor requires an approved flow control device in an accessible, visible location. See Figure 7-4. The flow-through rating can't exceed the rated capacity of the interceptor. As a rule, local authorities don't accept flow control fittings with adjustable or removable parts.

*IPC* Section 1003.3.2 says you can connect food waste grinders to a grease interceptor, but you should use an interceptor that's sized and rated for the discharge of the waste grinder. However, some local authorities don't accept this. Be sure to check with your plumbing inspector before installing this type of system.

You can omit a fixture trap for discharge into an approved grease interceptor if the interceptor is less than 5 feet from the single fixture it serves. Vent the flow control fitting and extend it at least 6 inches above the flood rim of the fixture. The typical piping and installation methods shown in Figure 7-3 will very likely meet the requirements of the *IPC* and/or your local authorities. The code doesn't give any specific location for a grease interceptor, but these guidelines usually apply:

- A grease interceptor must be in an area that's accessible for service and maintenance.
- A grease interceptor can't be located where workers have to use ladders or move bulky objects to service it.
- A grease interceptor can't be installed in a part of the building where food is handled.
- A grease interceptor should be as close as possible to the fixture it serves.

You may live where water-cooled interceptors are permitted. Such interceptors make grease solidify more quickly, but using them is questionable

| Capacity of Grease Traps | |
|---|---|
| **Total flow-through rating (gpm)** | **Grease retention capacity (pounds)** |
| 20 | 40 |
| 25 | 50 |
| 35 | 70 |
| 50 | 100 |

**Figure 7-1**
*Sizing criteria for an inside grease interceptor*

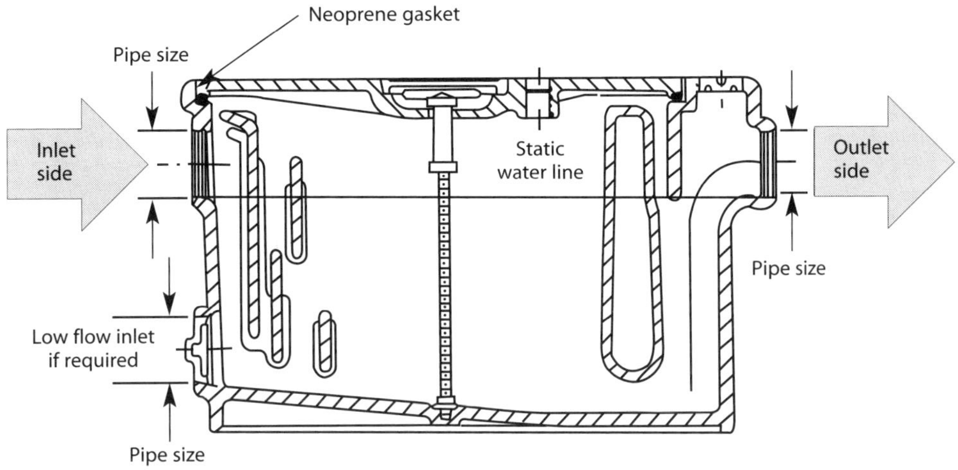


**Figure 7-2**
*Typical inside grease receptor*

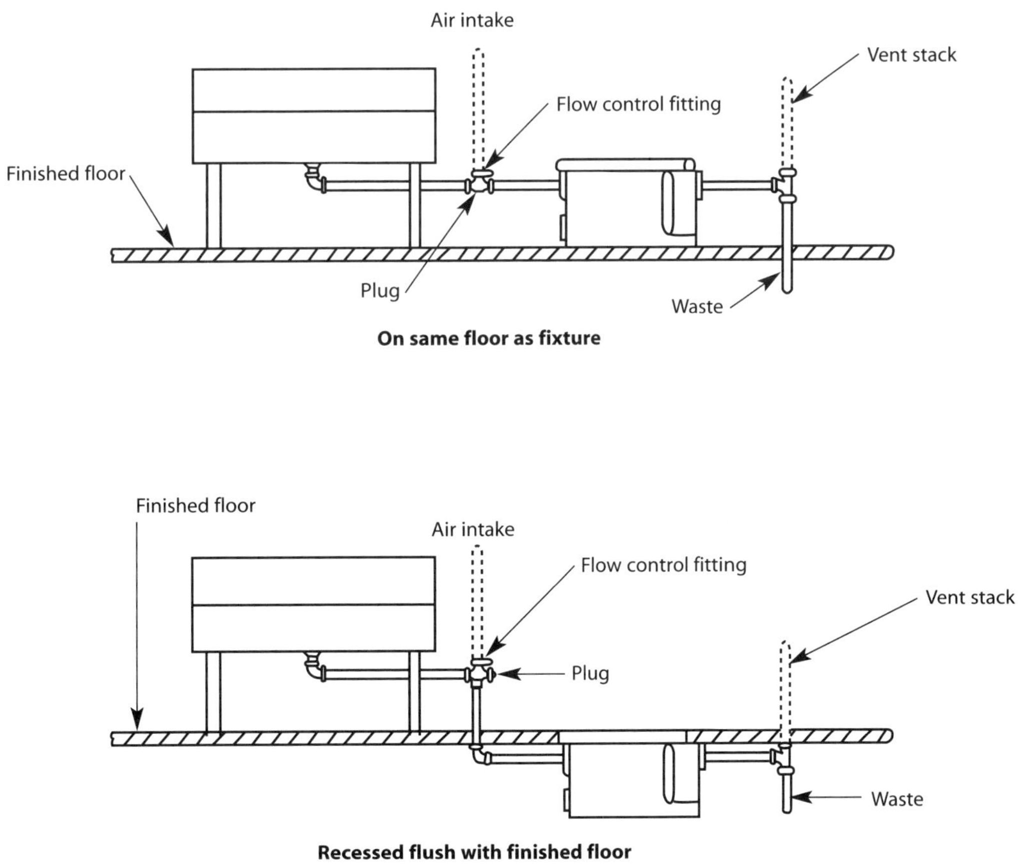


**Figure 7-3**
*Typical installations of grease interceptors*

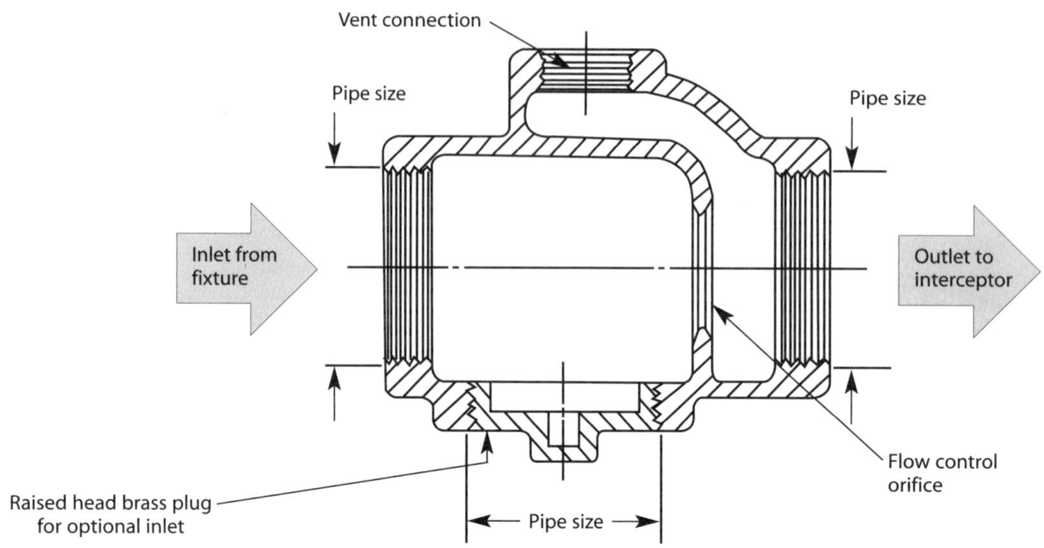


**Figure 7-4**
*Flow control fitting detail*

because there's a potential for cross-connection with the potable water supply. Most local authorities prohibit the use of this type of interceptor. Installing one in a jurisdiction where they're not permitted would be a serious offense against the plumbing code.

When you install an inside grease interceptor, make certain it's an approved size and type for the area you're working in. *IPC* Section 1003.3.4 says that grease traps (interceptors) must conform to G101 of the *Plumbing and Drainage Institute* (*PDI*). This is also found in Chapter 13 of the *IPC*, under "Referenced Standards." *IPC* Section 102.8 also refers to Chapter 13, making it clear that these "Referenced Standards" include only those testing agencies that the *IPC* recognizes. *IPC* Section 102.8 also says that if the manufacturer's installation instructions or other reference standards don't conform to the *IPC* minimum provisions, you must follow *IPC* standards. Architects, engineers and local authorities will usually be the ones using the testing agencies listed. Your only involvement with these agencies as a plumber may be a requirement to identify one or more of them on a plumbing exam.

## Outside Installation Interceptors

The *IPC* doesn't talk about outside installation grease interceptors. Many large restaurants and other grease-generating businesses may require outside grease interceptors larger than the 100-pound grease retention capacity listed in *IPC* Table 1003.3.4.1. There are many points to consider when sizing larger interceptors, so most codes leave sizing and design to the discretion of the health department or local plumbing official.

As a plumber, you probably won't be asked to install an outside grease interceptor. You'll just connect the piping to the inlet and outlet openings provided. But your master license qualifies you to do this type of work, so you should learn all you can about sizing, design and location. If sizing and design information is lacking in your own code, refer to the old *Uniform Plumbing Code (UPC)* for help.

### *Sizing Outside Installation Grease Interceptors*

In the *UPC*, a grease interceptor for a fully-equipped commercial restaurant is sized based on seating capacity (meals served) and hours of operation (8, 16 or 24 hours). The sizing formula also provides a retention period for grease to cool and separate. For example, a fully-equipped restaurant open for business 8 to 24 hours a day requires a retention time of 2.5 hours. A single-service kitchen open for the same period requires a retention time of only 1.5 hours. The *UPC* formula shown in Figure 7-5 covers both single-service and fully-equipped kitchens. Let's try using this formula to figure out the capacity for a restaurant that serves 200 meals at its peak hours each day based on an 8-hour, a 16-hour and a 24-hour period. The restaurant is equipped with a dishwasher. We'll assume each 8-hour period includes one peak hour. Using the *UPC* formula:

One 8-hour period

*200 meals × 6 gallons flow rate × 2.5 hours retention time × 1 peak hour period = 3,000 gallons capacity.*

One 16-hour period

*200 meals × 6 gallons flow rate × 2.5 hours retention time × 2 peak hour periods = 6,000 gallons capacity.*

One 24-hour period

*200 meals × 6 gallons flow rate × 2.5 hours retention time × 3 peak hour periods = 9,000 gallons capacity.*

To size the grease interceptors for the three examples above, the code breaks down the time the restaurant is open to the public. One is open for 8 hours, which equals one peak hour period. The second one is open for 16 hours, which equals two peak hour periods. The third is open for 24 hours, which equals three peak hour periods. Obviously, the more time a restaurant is open and the kitchen operating, the more grease it will generate.

Naturally, the gallon capacity of a grease interceptor determines the size of the interceptor. For example, the restaurant that's open eight hours will require an interceptor capacity of 3,000 gallons. Here's how you determine the size: 16 feet long by 5 feet wide by 5 feet deep provides the cubic feet for the adequate interceptor needed — which is 400 cubic feet. How do we know this is correct? Well, each cubic foot represents 7.5 gallons. So, we multiply 400 cubic feet by 7.5, and we get 3,000 gallons — the same answer we have in the first example.

| **Number of meals per peak hour[1]** | × | **Waste flow rate[2]** | × | **Retention time[3]** | × | **Storage factor[4]** | = | **Interceptor size (liquid capacity)** |
|---|---|---|---|---|---|---|---|---|

| | |
|---|---|
| 1. Meals served at peak hour | |
| 2. Waste flow rate | |
| A. With dishwashing machine | 6 gallon flow |
| B. Without dishwashing machine | 5 gallon flow |
| C. Single-service kitchen | 2 gallon flow |
| D. Food waste disposer | 1 gallon flow |
| 3. Retention time | |
| A. Commercial kitchen waste, dishwasher | 2.5 hours |
| B. Single-service kitchen, single serving | 1.5 hours |
| 4. Storage factors | |
| Fully-equipped commercial kitchen | 8 hours (1 peak hour) |
| | 16 hours (2 peak hours) |
| | 24 hours (3 peak hours) |
| Single-service kitchen | 1.5 peak hours |
| Self-service laundries | 1.5 hours |

**Figure 7-5**

*Sizing criteria for outside grease interceptors from the Uniform Plumbing Code*

If the required grease interceptor calculates out too large, the code will often allow more than one to be used.

### *Outside Installation Grease Interceptor Design*

The acceptable material for outside grease interceptors is governed by your local code. The most common material is precast or poured-in-place concrete. The code in your area may approve steel, fiberglass-reinforced polyester, or polyethylene. The criteria for the structural design of outside grease interceptors are generally left to the local authorities. You can also find the requirements in Appendix H of the *UPC*.

### *Outside Installation Grease Interceptor Location*

Here are code restrictions for locating outside installation grease interceptors at a building site. These are uniform across the country and meet the principal requirements of the average local code.

- Locate a grease interceptor where there's easy access for inspection, cleaning and removing intercepted grease.
- Provide a cover and reinforcement for a grease interceptor that's in a vehicular traffic area so it can withstand all anticipated loads. See *UPC* Section H 105.3.2.
- An interceptor may serve only one establishment. One exception to this rule is that your local code may accept one or more centrally-located grease interceptor(s) in a large shopping center. In this case, the grease interceptor(s) may serve several grease-generating businesses, but only if the shopping center management assumes, in writing, full maintenance responsibility.
- Locate outside installation grease interceptors more than 5 feet from a building foundation or private property line.
- You can connect waste discharge piping from an interceptor to an existing building sewer as shown in Figure 7-6.

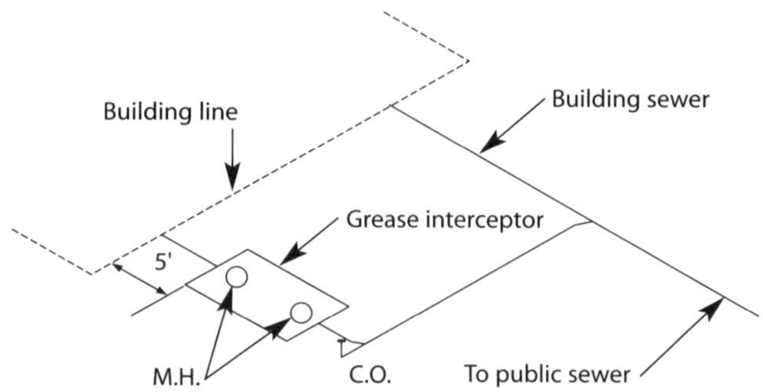


**Figure 7-6**
*Connecting an outside grease interceptor to an existing sewer line*

## Laundry Interceptors

Commercial and self-service laundries discharge solids, such as lint, string, buttons, rags and other materials harmful to the public sewage system. They must have lint interceptors to prevent these solids from entering the drainage system. Interceptors aren't required in single-family houses or in apartment buildings if there's a clothes washer in each unit. However, an apartment complex with a central laundry room is considered commercial and does require an interceptor. Laundry interceptors are covered under *IPC* Section 1003.6. We'll look at those requirements and I'll provide additional information that's acceptable to most local authorities.

In general, install an interceptor that has a removable screen or a $1/2$-inch mesh screen metal basket to collect solids. Use a screen or basket that's easy to clean like the one shown in Figure 7-7.

Some local authorities label the horizontal drainage pipes that serve self-service clothes washing machines as indirect waste pipes, as shown in Figure 7-8. This unique method of piping is economical and practical for this particular application. The indirect waste system doesn't have to be trapped or vented, like most other plumbing fixtures. The washing machine standpipes are open-ended, 3- or 4-inch diameter pipes extending to

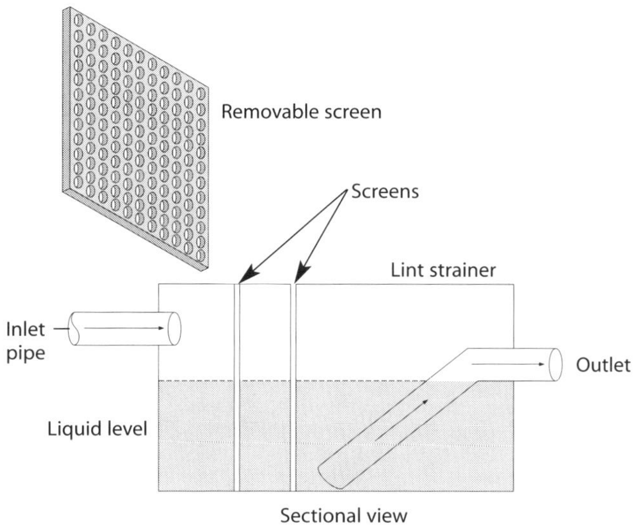


**Figure 7-7**
*Lint interceptor detail*

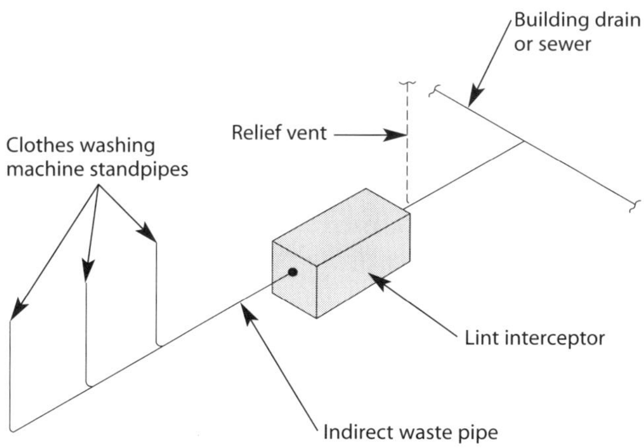


**Figure 7-8**
*Lint interceptor installed in drainage pipe from laundry*

about 26 inches above the finished floor. These drainpipes receive the discharge from washers through flexible hoses. A 3-inch standpipe can accommodate two machines; a 4-inch standpipe will serve four machines. Horizontal drainpipes collect waste from the standpipes and carry that waste to a lint interceptor. You don't have to vent this system because the standpipes permit free circulation of air.

Some local authorities require that each washing machine discharge into an individual trap, be vented, and connect separately to a horizontal drainpipe, as shown in Figure 7-9. If the code isn't clear, check with your local plumbing authority.

You can connect the outlet pipe from a lint interceptor to the regular sanitary drainage system serving other fixtures within a building. Local authorities require that you install a vent on the horizontal discharge (outlet) pipe as close as possible to a lint interceptor. The vent pipe serves the drainage pipe between the lint interceptor and the building drainage system or building sewer. It supplies and removes air as needed. This keeps wastewater flowing so the lint interceptor won't get air locked. See Figures 7-8 and 7-9.

## Sizing Lint Interceptors

It's usually the responsibility of the architect or his engineer to size lint interceptors. But you may have to calculate lint interceptor size when an engineer's seal isn't shown — for example, when a self-service laundry locates in an existing building. Unfortunately, most local codes don't provide established sizing methods. They recommend design criteria, or leave the matter up to the discretion of local plumbing officials. Commercial laundries rely on the manufacturer's specifications — and those are usually acceptable to the local authorities.

Lint interceptors for self-service laundries are sized according to the number of washing machines installed, the number of cycles, waste flow rate, retention time and storage factor. Figure 7-10 shows a formula for sizing a self-service laundry. Note: Be sure to check with your local plumbing

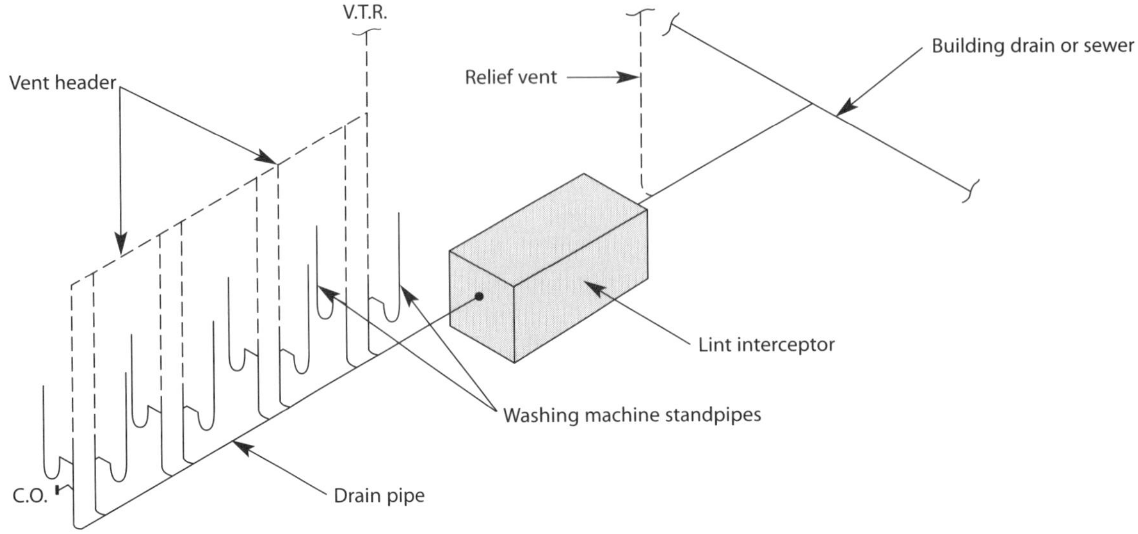


**Figure 7-9**
*Clothes washing machines individually trapped and vented (required by some codes)*

authority before you do any work. Let's plug in some numbers for an example and see how the formula works:

> *10 washers × 2 cycles per hour × 50 gallons per cycle flow rate × 2.0 hours retention time × 1.5 storage factor = 3,000 gallon liquid capacity*

Using this formula isn't that difficult once you know the number of washing machines. *Don't use this formula to size a commercial laundry.* Local authorities will size a commercial laundry to the manufacturer's specifications. This is very different from sizing a self-service laundry.

## Lint Interceptor Construction

Lint interceptors may be made of precast or poured-in-place concrete, steel or other approved materials. No matter what the material, a lint interceptor must have a removable basket or screen to trap materials. Larger units require a special basket to trap solids from wastewater that enters the lint interceptor. Smaller units, such as the one shown in

| No. clothes washing machines | × | 2 cycles per hour | × | Waste flow rate | × | Retention time | × | Storage factor | = | Interceptor size (liquid capacity) |
|---|---|---|---|---|---|---|---|---|---|---|
| 10 | × | 2 | × | 50 | × | 2.0 | × | 1.5 | = | 3,000 |

**Figure 7-10**
*Sizing criteria for lint interceptor for a self-service laundry*

Figure 7-7, may use removable screens. Remember that screens or baskets must be accessible and easy to clean.

If the inlet and outlet pipes aren't already installed with a unit, install an inlet pipe one pipe size above the liquid level. For example, if the inlet pipe is 3 inches, the inlet pipe must be installed 3 inches above liquid level. The outlet or discharge pipe should extend downward at a 45-degree angle to within about 2 inches of the bottom of the lint interceptor. This keeps lighter objects that pass through the screen and float on the water from sinking into the outlet pipe and getting into the sanitary drainage system. It also creates a liquid seal that serves as a trap, preventing sewer gases from entering the building through the washer's indirect waste pipes.

## Gasoline, Oil and Sand Interceptors

*IPC* Sections 1003.4 and 1003.5 require that you provide interceptors (or separators) to intercept and keep gasoline, grease, oil and sand from entering a drainage system. Here's a list of where they're required:

- Where motor vehicles are repaired and floor drainage is provided.
- At service stations that do oil changes using grease racks, grease pits or work racks.
- Where motor vehicles are washed commercially.
- At any public vehicle storage garage where floor drainage is provided.
- Where oil, gasoline or other volatile liquids can be discharged into a sanitary drainage system.
- At any factory which has oily or flammable waste from storage, maintenance, repair or testing processes.

Although the *IPC* says you need interceptors or separators, it doesn't provide much how-to direction. So, local code enforcement is left to provide that information. Most local authorities agree that there should be two sizing and design methods for handling volatile liquids; one for systems with small concentrations of volatile liquids and another for handling larger concentrations. *IPC* Section 1003.4.1 describes them as "light and heavy liquids with various specific gravities" and says you should separate them into "an approved receptacle." We'll look at the guidelines acceptable to most local authorities for both types.

### *Handling Small Concentrations of Volatile Liquid*

Figure 7-11 shows a typical precast or poured-in-place concrete interceptor for handling small amounts of volatile liquid. The oil interceptor with floor drainage passing through a bucket-type floor drain shown in Figure 7-12 may be approved for small installations, even though there's no separate sand interceptor. Floor drains used to drain garage areas are generally the bucket type. They usually come with at least a 4-inch outlet. The bucket, made of the same material as the floor drain, is removable for cleaning. The bottom portion of the bucket is solid to retain sand or other debris. Drainage holes near the top of the bucket let liquid waste pass out of the bucket and into the pipe(s) leading to the interceptor. See Figure 7-12.

Here are some typical design criteria that meet most local code requirements for handling small concentrations of volatile liquid. Check the design and sizing criteria with your local plumbing chief before you do any installation.

- Use an interceptor with a capacity of at least 6 cubic feet for the first 100 square feet to be drained, plus 1 cubic foot for each additional 100 square feet to be drained. See *IPC* Section 1003.4.2.2.
- An inlet pipe should enter an interceptor above the liquid level line. The evaporation space should provide enough venting for the drainage pipe(s).

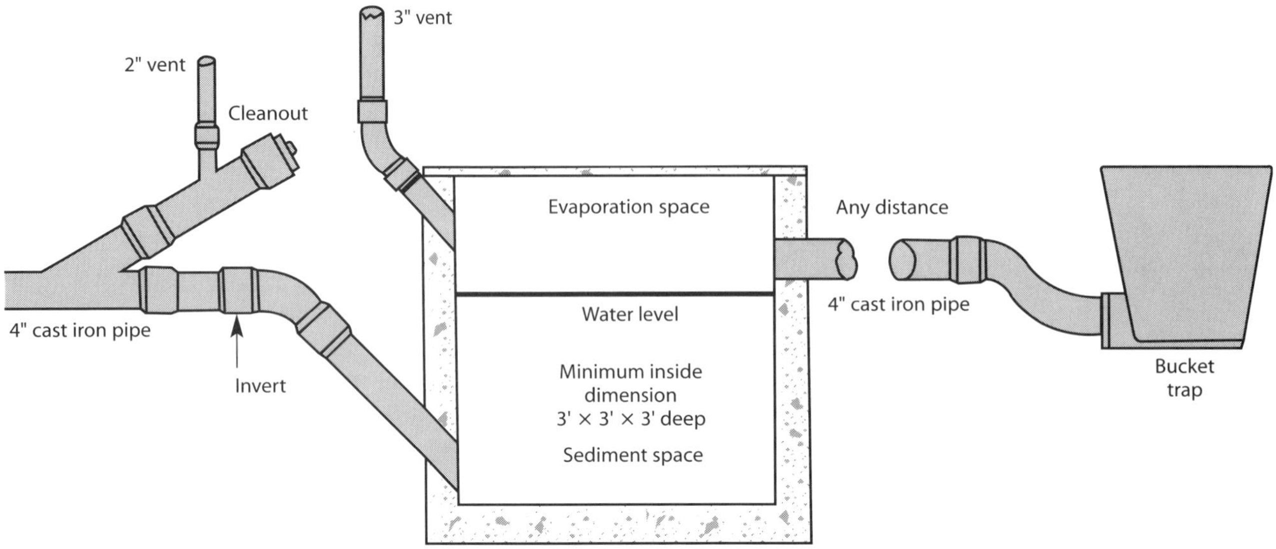


**Figure 7-11**
*Oil interceptor for small amounts of volatile liquids*

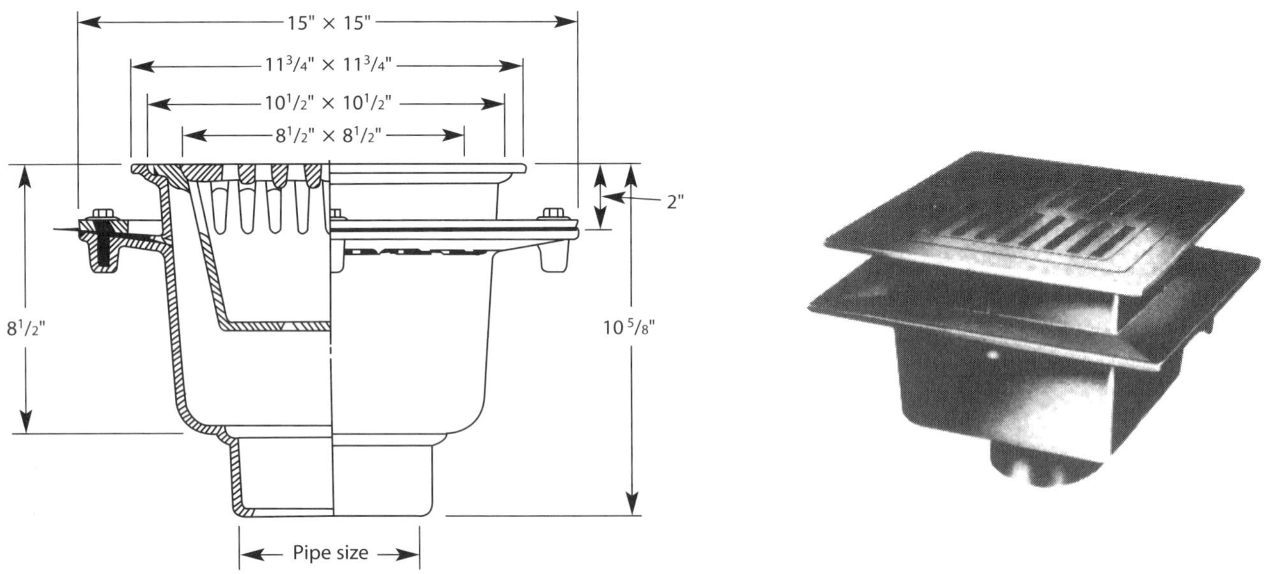


**Figure 7-12**
*Typical bucket floor drain*

- Since volatile substances contain vapors that may be harmful if inhaled, an evaporation space must be included in interceptors to vent away those vapors. Most local authorities will require at least a 3-inch vapor vent to vent the interceptor evaporation space. Terminate the vent in an approved location (most often the outside of a building) at least 12 feet above grade.
- A discharge pipe should enter an interceptor at a 45-degree angle near the bottom of the interceptor. The liquid depth should be at least 2 feet below the invert of the discharge pipe. The outlet pipe of the interceptor should provide at least an 18-inch water seal. See *IPC* Section 1003.4.2.1.
- Use a discharge pipe that's at least as big as the inlet pipe. The discharge pipe should have an accessible, full-size cleanout, brought to grade. Use at least a 2-inch-diameter vent on a discharge pipe.
- Local officials may not require you to vent a discharge pipe if it discharges into a catch basin or into a vented building sewer and isn't more than 15 feet in developed length.
- Locate an interceptor outside a building.
- Install each interceptor so it's readily accessible for removing the cover, servicing and maintenance.

Commercial garages that service or store fewer than ten vehicles and have floor drains generate small amounts of volatile liquids and sand. Service stations and small repair shops that service, but don't store, vehicles are also included in this category. However, the *IPC* doesn't require an interceptor in a parking garage that doesn't dispense gas or service, repair or wash vehicles. Parts of commercial garages used only to store vehicles also don't require an interceptor.

## *Handling Large Concentrations of Volatile Liquid*

Any business that generates large amounts of volatile liquid is required to have a waste oil storage tank. The tank receives and stores excess oil from the separator through a gravity flow line, as shown in Figure 7-13. You have to locate a waste oil storage tank outside a building. With building department

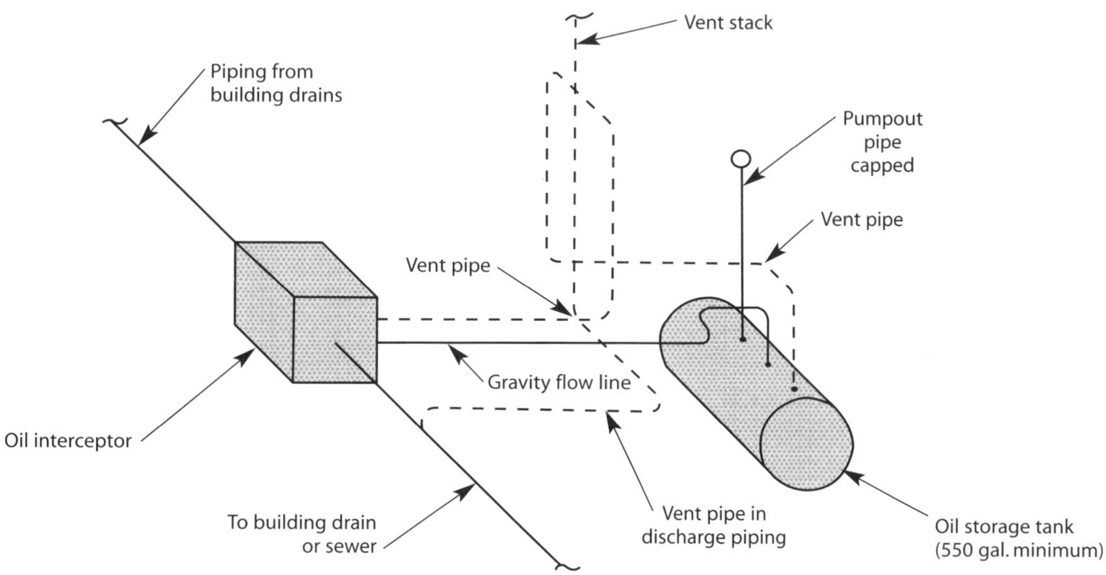


**Figure 7-13**
*Oil interceptor and oil storage*

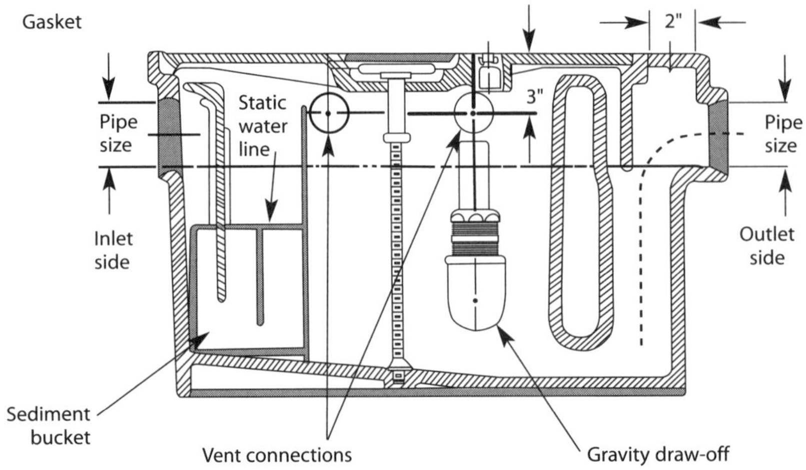


**Figure 7-14**
*Sectional view of a typical oil interceptor*

approval, the architect will determine its size and capacity. The tank must be UL-approved with a liquid capacity of at least 550 gallons. It must have at least a $1^1/_2$-inch vent terminating in the open air 12 feet above grade, or extend through the building roof. Of course, if you do the latter, you'll have to grade the horizontal vent pipe back to the interceptor so it won't collect liquid. To remove accumulated oil, you must include a 2-inch pumpout pipe, capped and brought to grade. See Figure 7-13.

Connect a draw-off line from the oil interceptor to the oil storage tank. Grade the pipe toward the oil storage tank so the contents drain into the tank by gravity. Provide a 2-inch vent to remove any accumulation of explosive vapors. Figure 7-14 shows a typical manufactured cast iron unit, accepted by most codes. It has two vent openings in the interceptor, which some local authorities require to increase the air circulation. If you use two vents, connect one vent to the vent stack at a higher elevation than the other. The unbalanced air pressure will generate the needed circulation.

## *Sand Interceptors*

*IPC* Section 1003.5 says that you must install a sand interceptor in a commercial business to prevent sand or other heavy objects from entering the building drainage system. It must be accessible for cleaning and have a water seal of at least 6 inches. Large parking garages, with automobile traffic, produce a considerable amount of sand, grit, grease and oil drippings. Without proper interceptors, the drainage system would be loaded with unacceptable amounts of sand and volatile liquids. Most local codes insist that parking garage drains first pass though an approved sand interceptor before discharging into the sanitary or storm drainage system.

Figure 7-15 shows a sand interceptor that holds sand and grit in a sediment section. The wastewater then passes on to the oil interceptor, where the

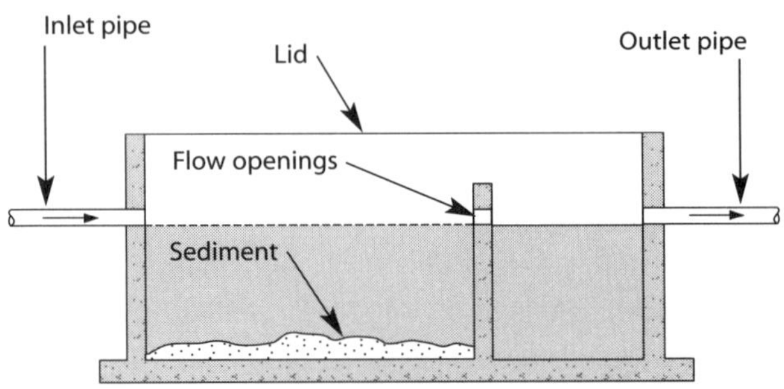


**Figure 7-15**
*Sectional view of a sand interceptor (trap)*

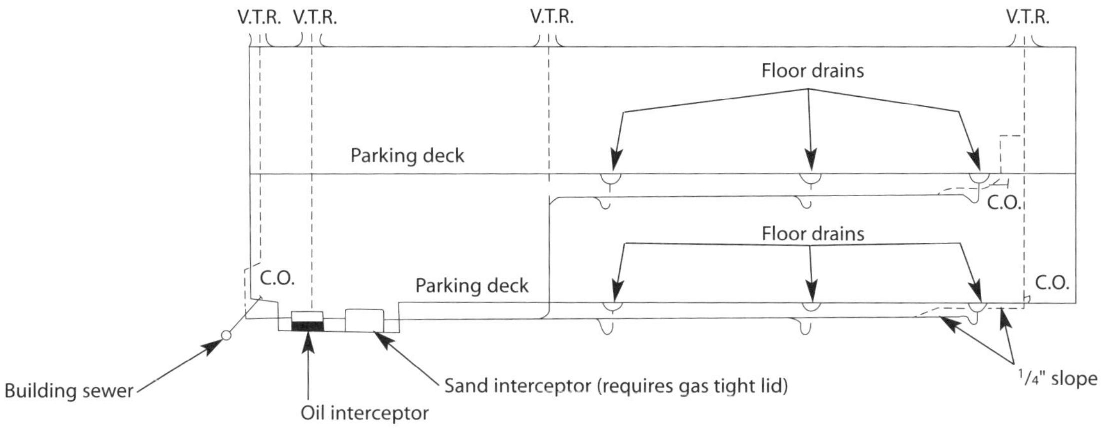


**Figure 7-16**
*Typical multi-story parking garage installation*

volatile liquids are separated from the wastewater and retained in the oil interceptor or the waste oil storage tank. The wastewater, now free of sand, grit and volatile liquids, can then discharge safely into the building's drainage system. See Figure 7-16.

When floor drainage is required for a single-story parking garage, usually you'll use a system like the one in Figure 7-11. The interceptor is small enough to catch and retain small amounts of volatile liquids, yet large enough to trap sand and grit before it reaches the building's drainage system. Figure 7-16 shows a more complex floor drainage design for multi-story parking garages using a separate sand interceptor.

Other areas that might produce more sand than the law allows are stairwells, planter drains, beach cabana showers, public pool showers, public pool deck drains, play areas and footing drains. Since these areas drain into the total drainage system, you also need to install sand interceptors.

## *Footing Drainage Systems*

Occasionally, a building official approves a site where surface water accumulation during heavy rains could weaken a foundation wall and its footing supports. If the structural engineer requests a sub-soil drainage system, the installation is the plumbing contractor's responsibility.

Figure 7-17 shows a footing drain detail that's designed to intercept surface water before it reaches a building's foundation wall or footings. Notice that the drainage pipe is located slightly above the bottom of the footing. The pipe is laid on a bed of crushed rock for support. Crushed rock around and above the pipe also facilitates easy drainage.

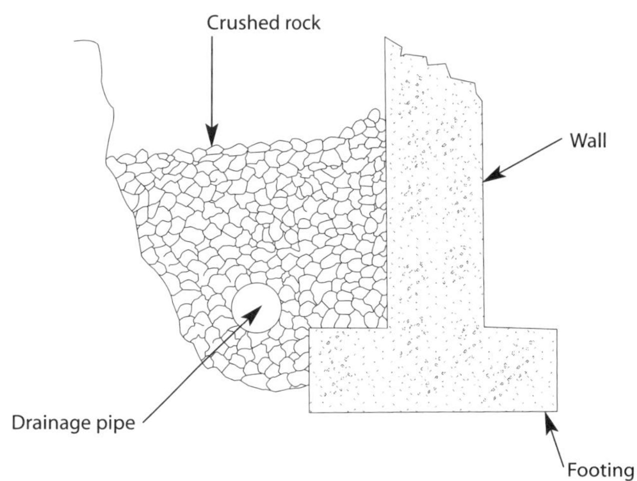


**Figure 7-17**
*Footing drain detail*

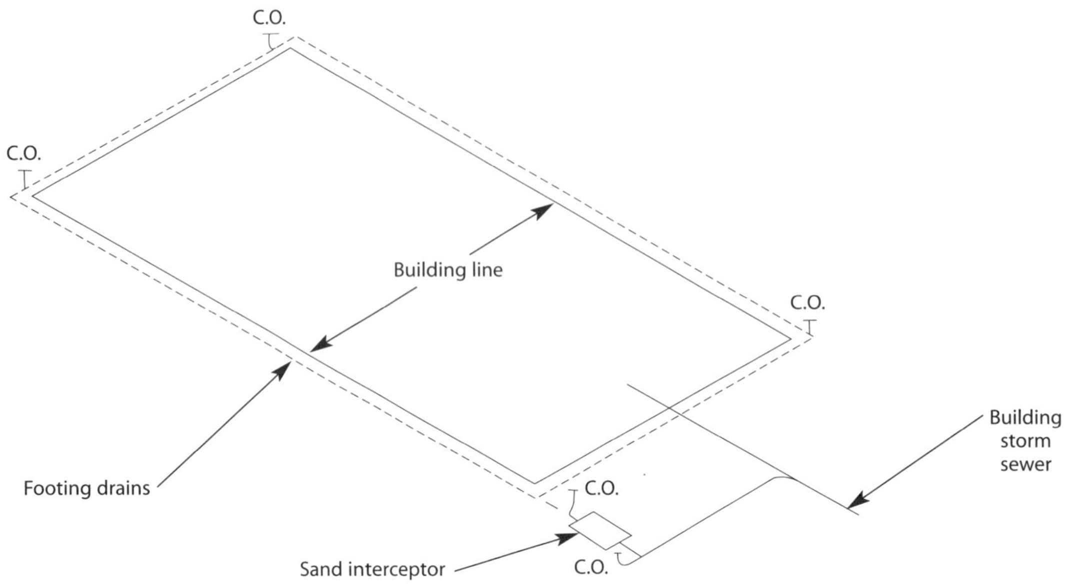


**Figure 7-18**
*Footing drains and a sand interceptor*

The piping itself is usually of clay or concrete. It may be open-jointed, horizontally split or perforated for seepage. You can usually get this piping in a 4-inch diameter and in lengths of 2 or 3 feet. After you install piping, use bituminous saturated paper or other approved materials at the openings to keep sand out. Since water always follows the least restrictive path, it will flow through the sub-soil drainage system to the point of discharge. Most local authorities will require you to have at least a $^1/_2$-inch slope per foot on a drainage pipe. Accumulated surface water must pass through an adequately-sized sand interceptor before it's discharged. Figure 7-18 shows a typical installation that most inspectors will accept.

## Other Miscellaneous Interceptors

*IPC* Section 1003.1 doesn't give every interceptor requirement. But it covers most by stating that unacceptable substances must not be allowed to enter a building drainage system. This gives local authorities considerable discretion. With my plumbing plans examiner background, I've highlighted a few locations where interceptors are a must.

There's a unique need for an interceptor in any bottling plant. Before liquid waste can discharge into the drainage system, it must be filtered through an interceptor to separate out broken glass or other solids. A properly-sized sand interceptor similar to the one in Figure 7-15 should work.

*IPC* Section 1003.8 requires interceptors or separators for all drains in slaughtering/dressing rooms. Figure 7-19 shows an acceptable slaughterhouse solids interceptor. Note the removable perforated sediment bucket for easy cleaning.

Figure 7-20 shows a typical hair interceptor with removable screen basket. It's possible that your code officials may require an interceptor like this one for barbershop sinks, beauty salon sinks and any fixtures used for bathing animals.

Where animals are confined, as in dog kennels and zoos, most local codes have special provisions before allowing waste to discharge to a legal point of disposal, whether private or public. In this type of system, the wash-down drain has a perforated or

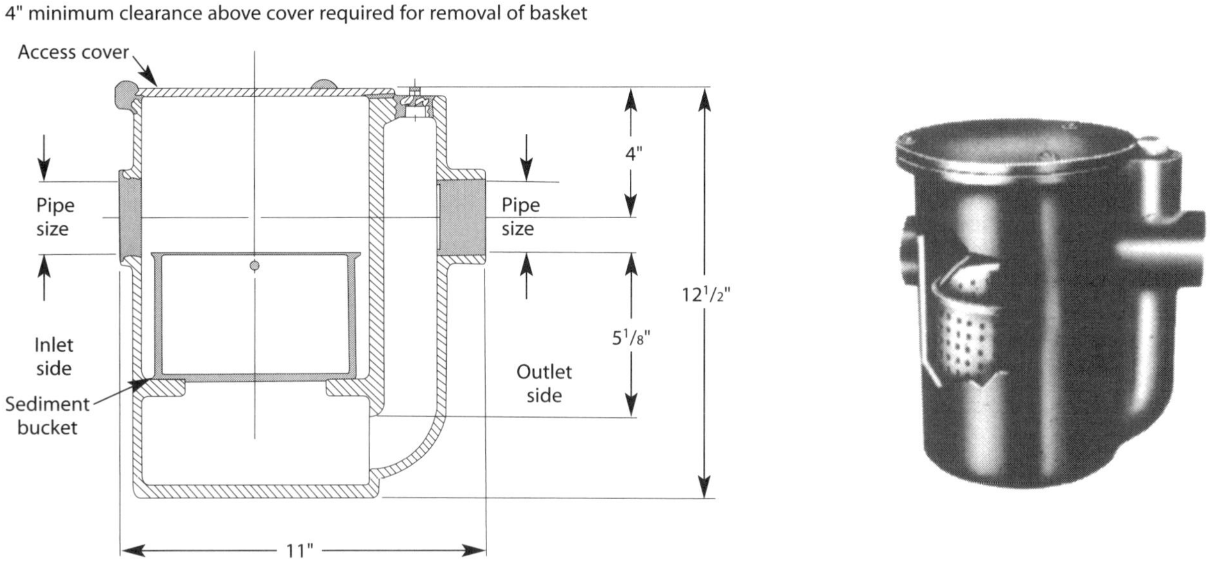


**Figure 7-19**
*Slaughterhouse solids interceptor*

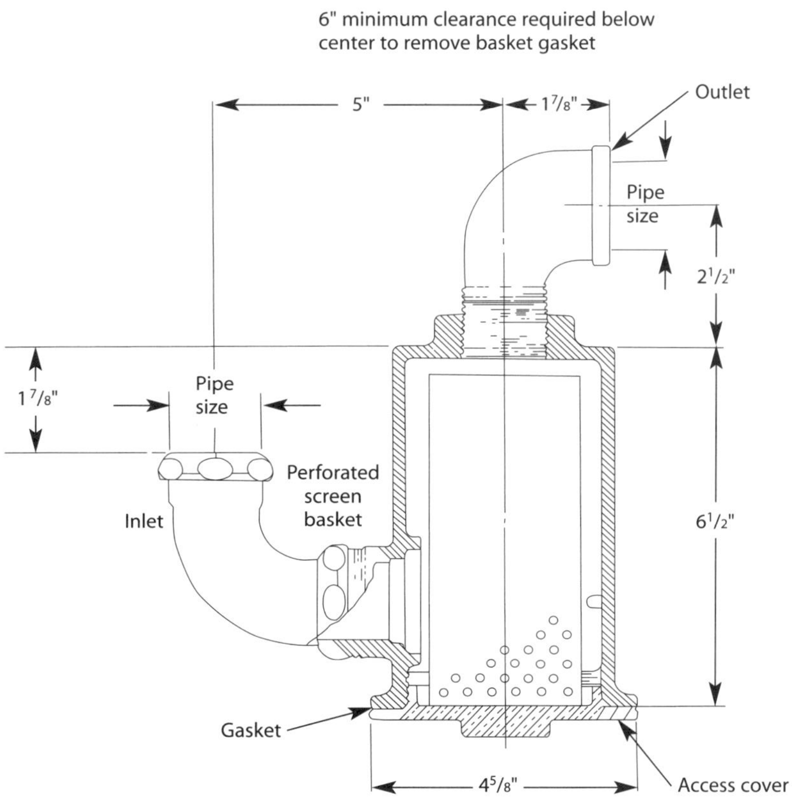


**Figure 7-20**
*Typical hair interceptor*

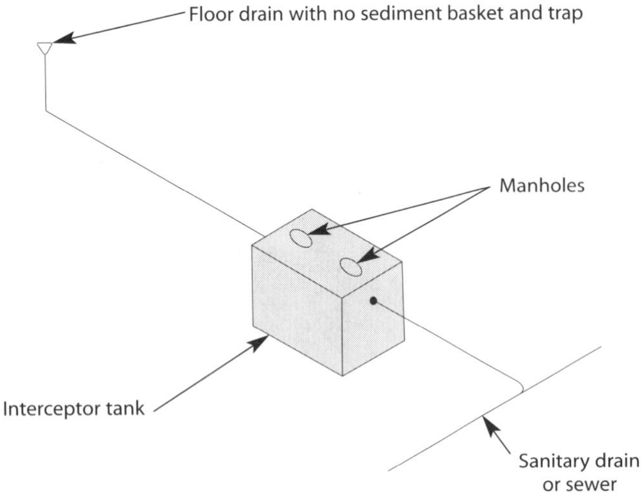


**Figure 7-21**
*Interceptor tank and component piping arrangements for an animal boarding establishment*

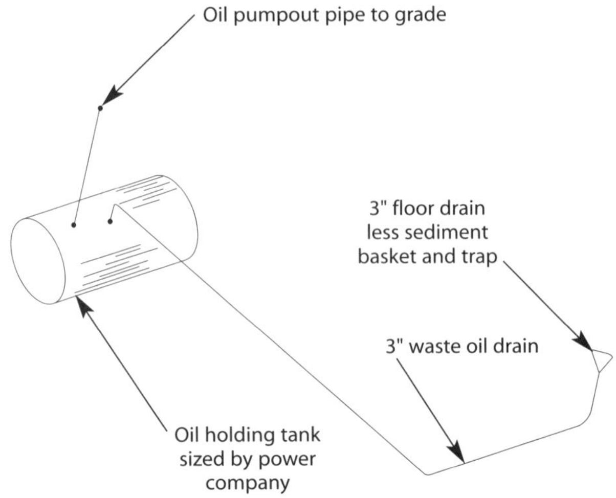


**Figure 7-22**
*Transformer oil spill holding tank and piping detail*

slotted grate, but no sediment basket. The floor drain isn't equipped with a trap. The drainage pipe empties into an interceptor tank, which may be designed like a septic tank. However, the bacterial process that's so effective in digesting human waste isn't very effective with animal waste. So animal solids have to be removed more frequently. You'll

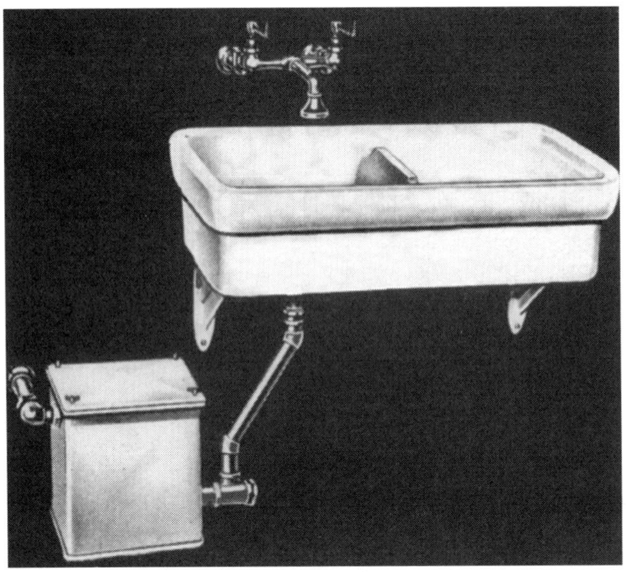

**Figure 7-23**
*Plaster work sink with integral right-hand tray and interceptor trap*

have to make sure there are enough manholes at grade to allow for easy cleaning if you design this type of waste system. See Figure 7-21.

Transformer vault rooms inside a building must have floor drainage. As a rule, you can use a 3-inch floor drain installed at floor level. There's no sediment basket or trap. The drainpipe that connects to an oil-spill holding tank is accepted as an indirect waste pipe and doesn't require a vent. The oil-spill holding tank must have a pump-out connection at grade. The local power company is your authority on sizing the tank, which must be able to hold the amount of oil contained in the transformers if a rupture occurs. Figure 7-22 shows a detailed drawing of a typical transformer oil-spill holding tank.

Local authorities will probably insist on an interceptor trap for dental and orthopedic sinks where plaster, wax or other objectionable substances can get into a building drainage system. Figure 7-23 shows a typical installation for this type of system.

## Neutralizing Tanks

Corrosive liquids, spent acids and other chemicals are dangerous and can damage or destroy a drainage system as well as create noxious or toxic fumes. They can't discharge into a building

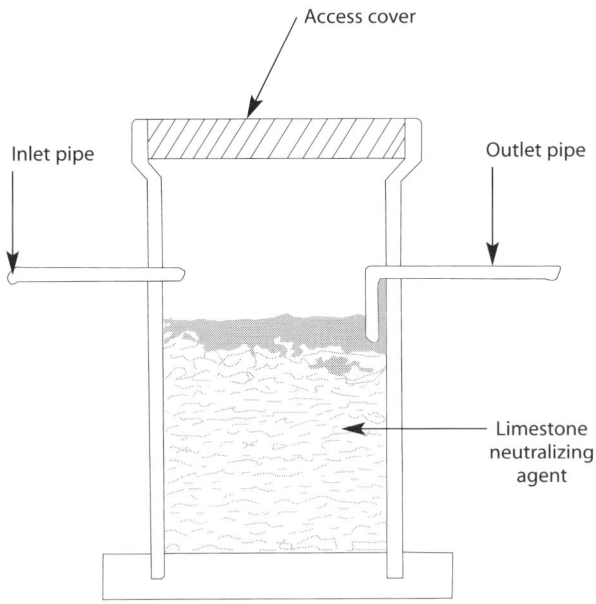


**Figure 7-24**
*Typical neutralizing tanks*

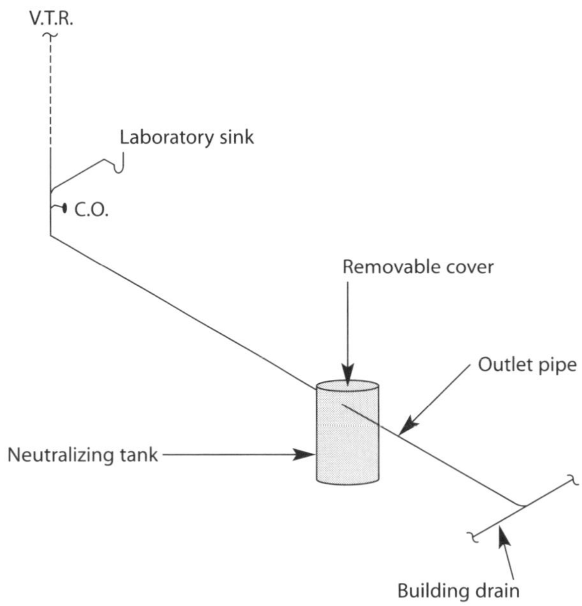


**Figure 7-25**
*Typical acid waste system with neutralizing tank*

drainage system until they've been diluted, neutralized or treated. That means the waste must discharge through a separate and independent system into a properly-designed dilution or neutralizing tank, as covered in *IPC* Sections 702.5, 803.3 and 901.3. The tank must automatically provide enough water or other agent to neutralize the toxic products before they discharge into a drainage system. See *IPC* Section 803.2. Figure 7-24 shows a typical neutralizing tank. Your local authority will want to know the type and amount of corrosive waste generated. It's up to them to approve the size and type of neutralizing tank you use. Figure 7-25 shows a typical acid waste system with neutralizing tank and piping. See "Acid waste" in the Glossary in this book for details.

# Chapter 7 — Questions and Answers

**7-1 Of the following, _______ need not be separated before discharging into a building's drainage system.**

(A) fecal matter

(B) oil

(C) grease

(D) sand

***Answer: A***

*Code response: Section 1003.2*
*This book, page 92*

**7-2 Interceptors and separators shall be designed and installed in accordance with the**

(A) code official's instructions.

(B) architect's instructions.

(C) job engineer's instructions.

(D) manufacturer's instructions.

***Answer: D***

*Code response: Section 1003.2*
*This book, page 91*

**7-3 According to the *IPC*, grease traps are not required in**

(A) commercial kitchens.

(B) bars.

(C) apartment buildings.

(D) garages.

***Answer: C***

*Code response: Section 1003.3.3*
*This book, page 92*

**7-4 If you connect a food waste grinder to a grease trap, the *IPC* requires the trap be**

(A) sized and rated for the grinder.

(B) approved by the health department.

(C) located on the outside of the building.

(D) installed in the area the food is handled.

***Answer: A***

*Code response: Section 1003.3.2*
*This book, page 92*

**7-5 According to the *IPC*, a grease interceptor must be equipped with**

(A) an easily-accessible, removable cover.

(B) a device to control the rate of water flow.

(C) at least a 3-inch vent.

(D) a device to control the water temperature.

***Answer: B***

*Code response: Section 1003.3.4.2*
*This book, page 92*

**7-6 The *IPC* requires that all commercial laundries be equipped with**

(A) a backflow preventer.

(B) an appropriately-sized vent.

(C) an interceptor.

(D) an indirect waste pipe.

***Answer: C***

*Code response: Section 1003.6*
*This book, page 96*

**7-7 According to the *IPC*, you don't have to install an oil separator in a(n)**

(A) commercial car storage garage.

(B) auto repair business.

(C) service station that does oil changes and lube jobs.

(D) car wash that cleans engines.

***Answer: A***

*Code response: Section 1003.4*
*This book, page 99*

**7-8 The minimum capacity of an oil separator installed to drain the first 100 square feet of a garage or full service gas station is**

(A) 3 cubic feet.

(B) 4 cubic feet.

(C) 5 cubic feet.

(D) 6 cubic feet.

***Answer: D***

*Code response: Section 1003.4.2.2*
*This book, page 99*

**7-9 According to the *IPC*, sand interceptors for commercial businesses must be designed and located**

(A) with a minimum 4-inch water seal.

(B) with a maximum 4-inch water seal.

(C) to allow for cleaning.

(D) no more than 10 feet from the drain it serves.

***Answer: C***

*Code response: Section 1003.5*
*This book, page 102*

**7-10 To drain a 400-square foot commercial auto repair garage, the *IPC* requires a _____ cubic foot oil interceptor.**

(A) 3

(B) 4

(C) 7

(D) 9

***Answer: D***

*Code response: Section 1003.4.2.2*
*This book, page 99*

◆ *Chapter 8* ◆

# Indirect Waste Piping for Special Wastes

Every commercial building generates what the code defines as special wastes. You must connect these wastes indirectly, rather than directly, to a drainage system. This type of drainage prevents sewage from backing up into a special fixture and contaminating its contents in case there's a stoppage somewhere. In a residence, the most common indirect drain fixtures are the ones on clothes washing machines and air conditioning units. Right now, let's look at some of the special fixtures in commercial buildings you have to drain indirectly because they have a drip or drainage outlet. These special fixtures or appliances include refrigerators, ice boxes, bar sinks, hand sinks, cooling or refrigerating coils, extractors, steam tables, egg boilers, coffee urns, stills, sterilizers, commercial dishwashers, water stations, water lifts, expansion tanks, cooling jackets, drip or overflow pans, air conditioning condensate drains, overflow drains, relief vents from the water supply systems, and similar applications. The *IPC* provides some guidelines for these in Section 802.

Overflow and relief pipes on a water supply system, and relief pipes on expansion tanks, sprinkler systems and cooling jackets must always connect indirectly to a sanitary drainage system. Note the separation between the drain pipe outlet and the top of the floor drain in Figure 8-1. This prevents any possibility of contaminating the potable water supply through cross connection. To avoid contamination, there must always be a positive separation by indirect means between the building sanitary drainage system and the waste outlet of hospital equipment and food-handling equipment. This means an indirect waste pipe outlet (one that doesn't connect directly with the drainage system) is permanently separated from the sanitary drainage system by an air gap or air break connection. This unique method of piping is very practical for fixtures with low discharge rates, since it's a one-pipe system approved for a certain type of installation and doesn't require a vent.

## Indirect Waste Pipes

In this chapter, we'll look at requirements for indirect waste pipes, including their materials, sizing, venting, location and cleanouts. Then we'll cover the appliances and equipment that require indirect drainage.

### *Piping Materials*

You can only use indirect waste piping that's made of approved, durable materials. Chapter 9 lists approved materials you can use for above- and below-ground drainage and vent pipes. The fittings you use must be compatible with, and have the same diameter as, the piping. When you use screw pipe, the fittings must be the recessed type, and tapped, to allow a fall of $^1/_4$ inch per foot.

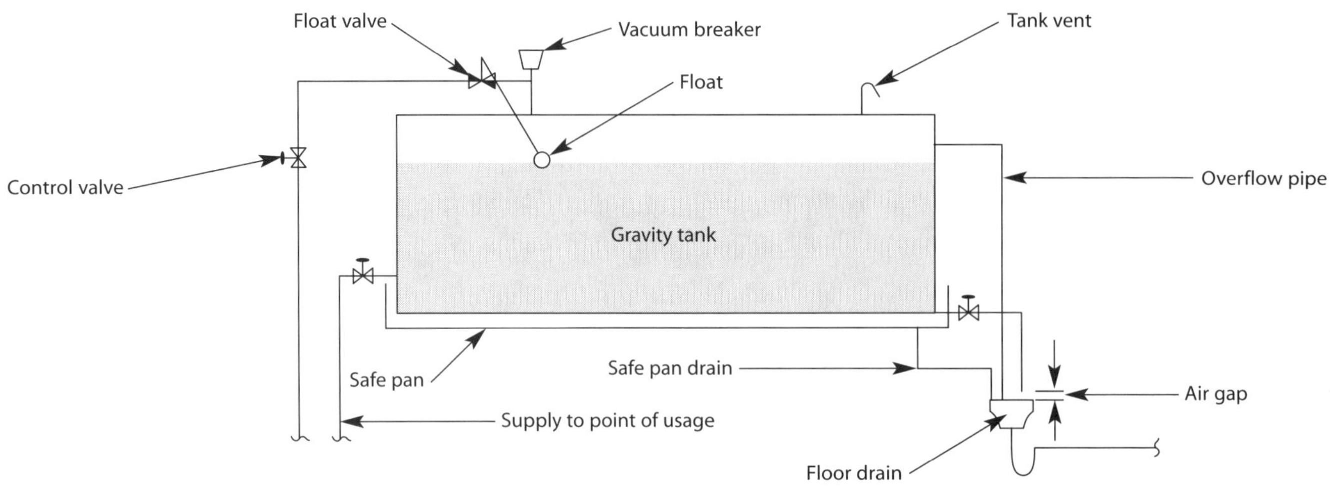


**Figure 8-1**
*Typical water gravity tank showing required air gap*

## *Sizing Indirect Waste Piping*

Sizing most indirect waste piping is fairly simple. The main thing to remember is that you can never use waste piping that's smaller than the drain outlet or tailpiece of the fixture, appliance or equipment it serves. No matter what, you can never use waste piping that's less than 1/2 inch. And don't forget that any fixture, appliance, or equipment you install can add dfu to the system.

Most of your work with indirect waste pipes will involve fixtures not listed in *IPC* Table 709.1 (see Figure 3-10 in Chapter 3). Size these according to *IPC* Table 709.2 (Figure 3-13 in Chapter 3). If you run into any devices with continuous or semi-continuous flow, *IPC* Section 709.3 says to figure dfu on a basis that 1 gpm of flow is equal to 2 dfu. Chapter 3 has an explanation of how to calculate and include these extra fixtures into a grand total when you size building drainage pipes.

## *When Traps and Cleanouts Are Required*

*IPC* Section 802.2 says you must install a trap in indirect waste pipes when the length of the piping ("x" on the hand sink in Figure 8-2) is more than 2 feet horizontally, or 4 feet in total developed length (the "x" distance plus the drop to the floor sink). You don't have to vent traps you locate on indirect waste piping.

Indirect waste pipes usually carry waste materials that don't have offensive odors to a receiving fixture (floor sink, floor drain, open pipe or pipe hub). Because the waste may move slowly in the pipe, slime deposits could build up in low places and eventually cause stoppages. It's important that you properly support indirect waste pipes to eliminate sags or low spots. This lets gravity do its job in preventing slime

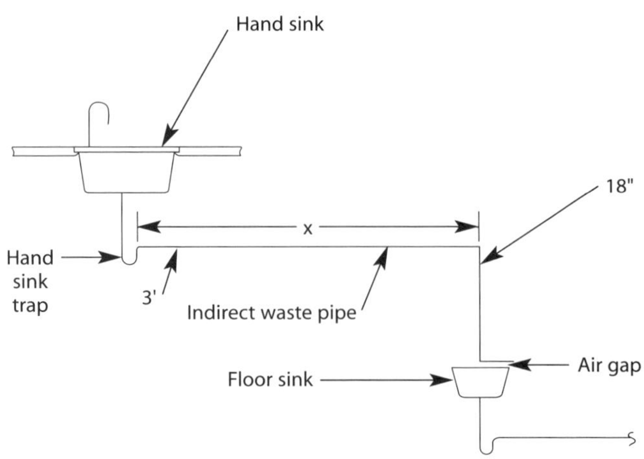


**Figure 8-2**
*Hand sink trap*

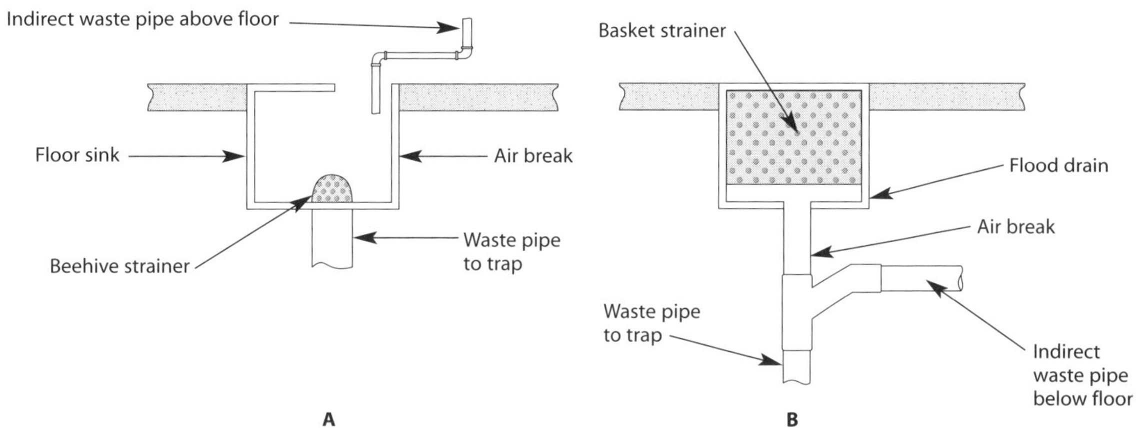


**Figure 8-3**
*Two indirect (air break) waste methods*

buildup. Place all receiving fixtures in well-ventilated areas to reduce the likelihood of slime buildup because good ventilation helps prevent slime buildup in the receiving fixture. The *IPC* doesn't talk about cleanout requirements, so you just need to install cleanouts that are accessible so the pipes can be cleaned and flushed.

## *Types of Indirect Waste Piping*

There are two types of indirect waste piping — air break and air gap. The *IPC* accepts both for certain types of installations. Let's briefly look at each one.

An air break drainage (see *IPC* Section 802.2.2) assembly usually uses a low inlet that discharges into an indirect waste receptor. There are two acceptable types of air break drainage systems — one that connects to a drainage system above a finished floor, shown in Figure 8-3 A and one that connects to a drainage system beneath a finished floor (the most common one), shown in Figure 8-3 B. The end of an indirect waste pipe can terminate below the overflow rim of the receiving fixture. If you're using the one above a finished floor, be sure to get your plumbing department's approval before installation.

An air gap drainage assembly uses an unobstructed vertical distance between a waste pipe outlet and the flood level rim at the top of a waste receptor. You must have at least 2 inches separation (or twice the drainpipe size, according to *IPC* Section 802.2.1) between a plumbing fixture, appliance or outlet and the rim of the receptor. See Figure 8-4. Figures 8-5 through 8-11 show air gaps for several types of installations.

When you install floor drains in walk-in refrigerators or freezers in food service establishments, you must connect them indirectly to the drainage system using an air gap. If you need to install a floor drain in an area subject to freezing, don't trap the waste line serving the cooler drain. The waste must

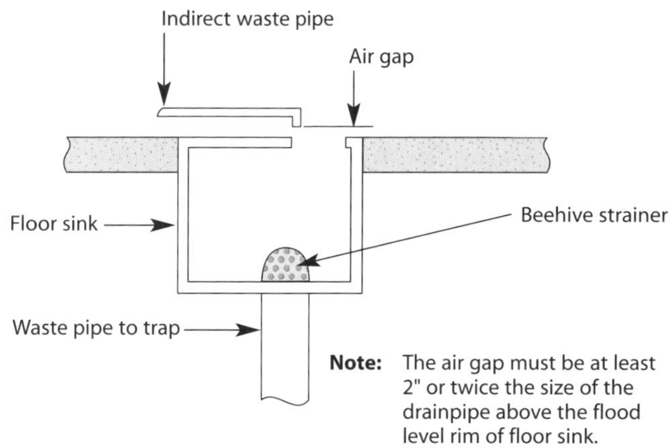


**Figure 8-4**
*Typical air gap indirect drainage method*

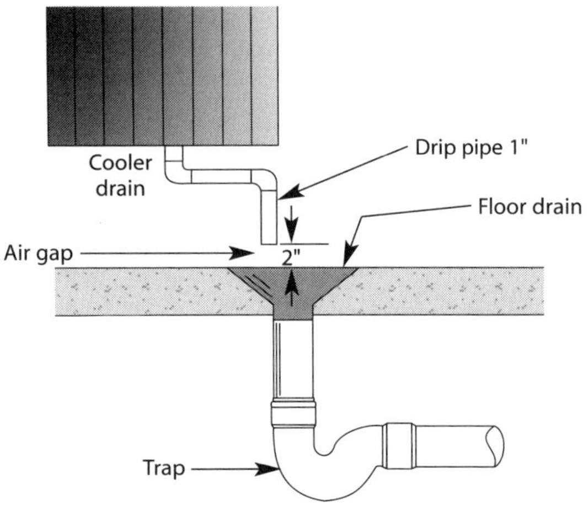


**Figure 8-5**
*Air gap between a drip pipe and a floor drain*

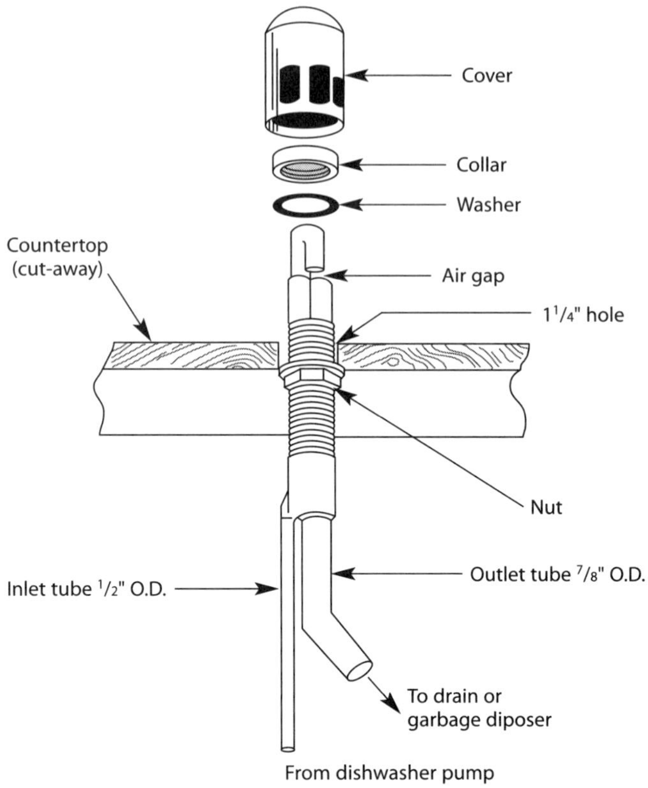


**Figure 8-6**
*Domestic dishwasher air gap device assembly*

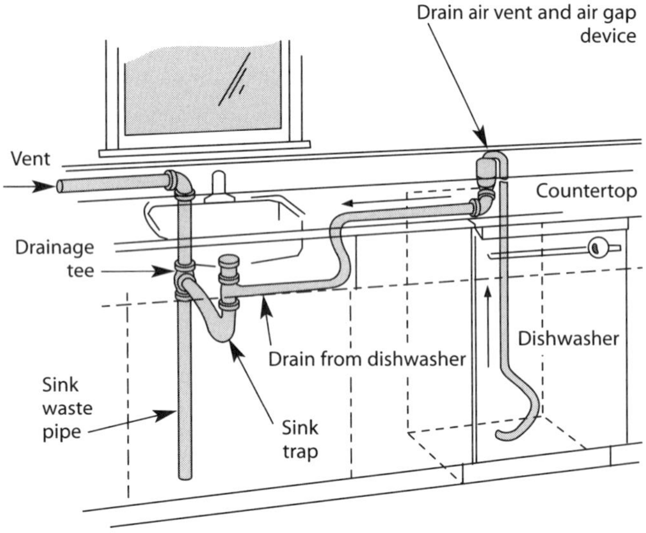


**Figure 8-7**
*Air gap device for a domestic dishwasher*

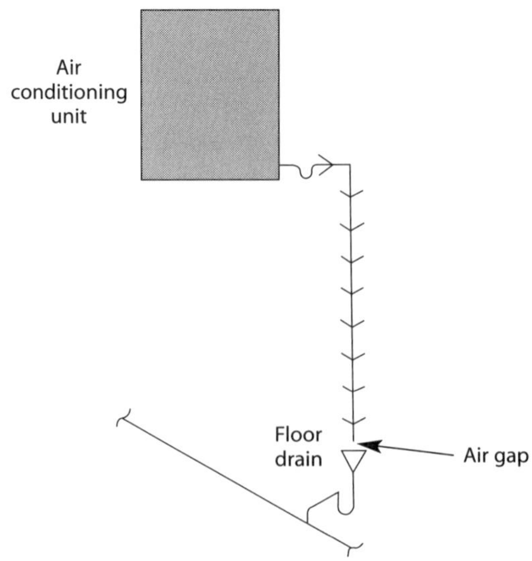


**Figure 8-8**
*Air gap isometric drawing*

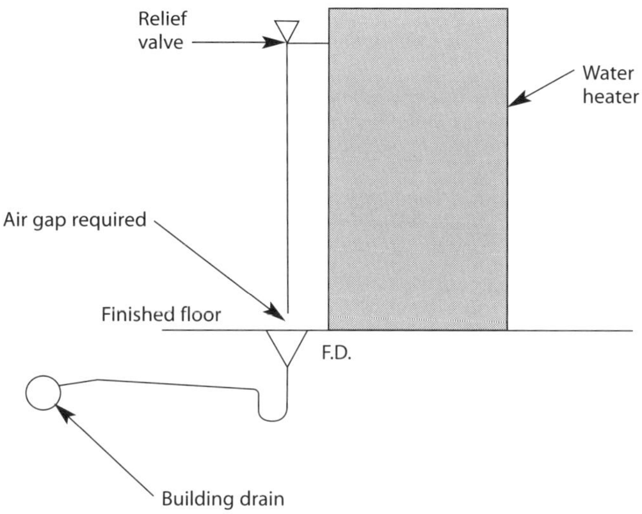


**Figure 8-9**

*Air gap for a relief valve on a water heater*

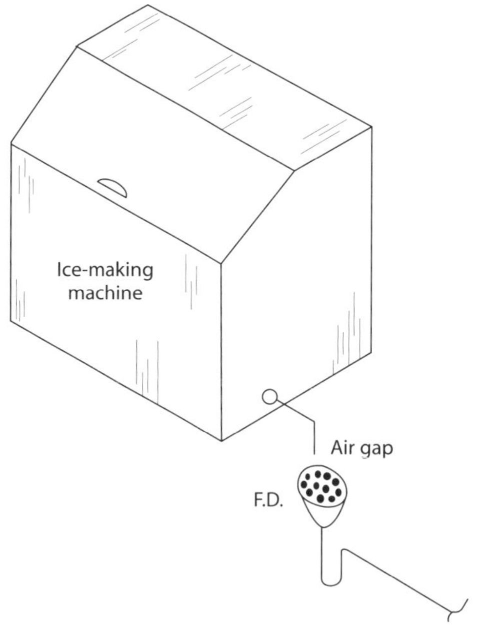


**Figure 8-10**

*Air gap for an ice-making machine*

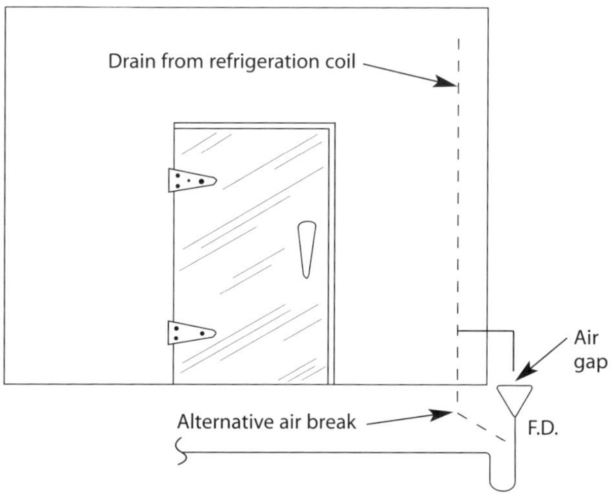


**Figure 8-11**

*Air gap and air break connection for a walk-in refrigerator*

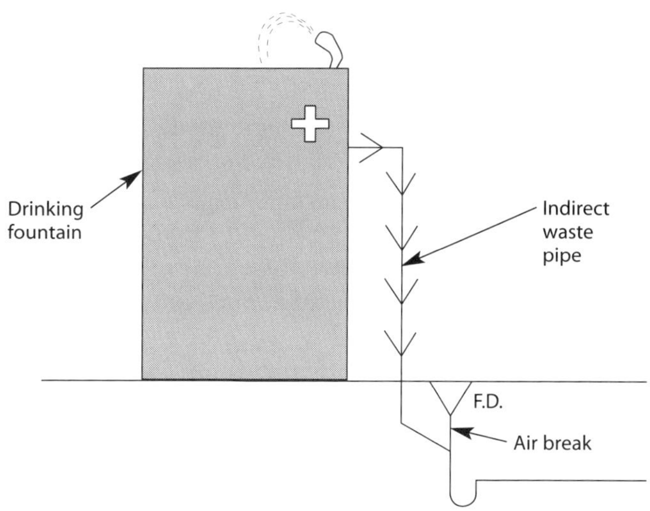


**Figure 8-12**

*Air break connection for a drinking fountain*

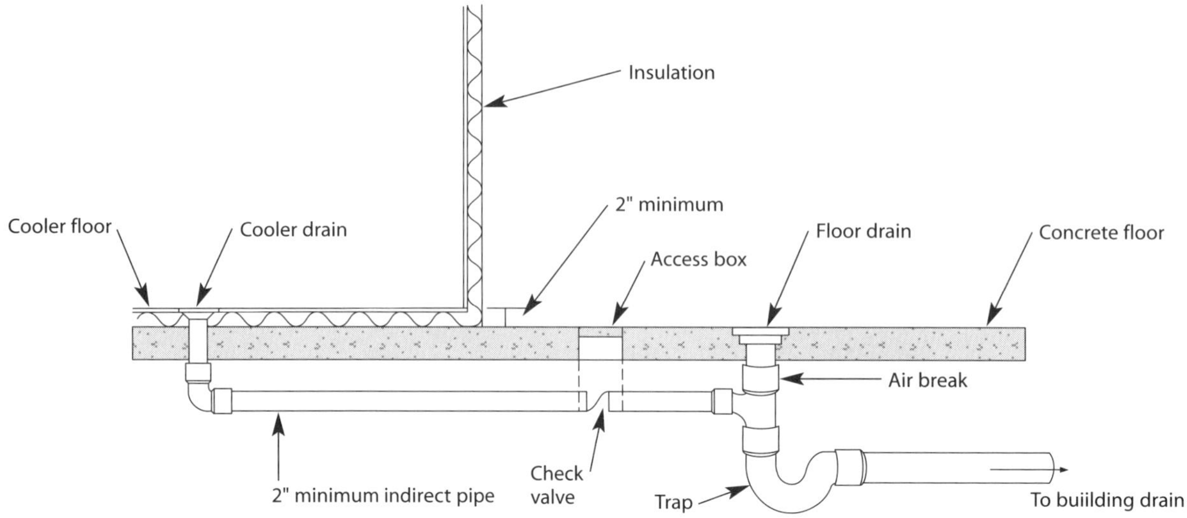


**Figure 8-13**

*Using air breaks for special cooler floor drainage*

go into a receptor located outside the area that freezes. You must install the receptor outside the unit. See Figure 8-11. One method typically acceptable to local authorities is shown in Figure 8-13. As a rule, local codes will accept a drip pipe from a walk-in refrigerator or a cooler room refrigerator coil with either an air gap or an air break connection to the drainage system.

## Waste Receptor Location and Design

Install waste receptors in accessible locations so they can be inspected and cleaned. Make sure workers can get to them without moving or disconnecting equipment or heavy objects. Don't put them in a bathroom, closet, cupboard or storeroom. Use only waste receptors that are approved and that have a removable strainer or basket. Always install any waste receptor in a well-ventilated area. Use receptors that are designed and sized to prevent splashing and flooding.

## Floor Sinks and Drains

Usually you use air gap indirect waste pipe with floor sinks, especially in restaurants or similar establishments. The top of the receptor is open to receive the indirect waste pipe. The type of floor sink you install depends on whether or not there will be any pedestrian traffic on the receptor. Use a type that has a grate in pedestrian traffic areas, as shown in Figure 8-14 A. If the floor sink isn't in a pedestrian traffic area and appearance isn't important, you can use a floor sink like the one shown in Figure 8-14 B.

Local codes generally require you to have a beehive strainer that's at least 4 inches high on the indirect waste receptor for a floor sink. But you don't need this strainer if the indirect waste receptor only receives clear water waste. The *IPC* also permits waste receptors to be in the form of a pipe hub or open pipe. It must extend at least 1 inch above a water-impervious floor. The code doesn't require it to have a grate. A floor drain is commonly used with an air break indirect waste pipe. It may also serve as a receptor to receive floor drainage. Look back at Figure 8-3.

The code says a floor drain is a plumbing fixture. And it's very specific about how to install and use floor drains. *IPC* Section 412 points out just a few simple requirements for floor drains:

- The traps must be accessible for cleaning.

**A** Floor sink with grate for a pedestrian traffic area

**B** Floor sink for an area with no pedestrian traffic

**Figure 8-14**
*Floor sinks*

- The area of the strainer must equal the area of the drain outlet.
- The drain outlet for a floor drain must be at least 2 inches in diameter.
- A drain outlet in a public laundry or a central washing facility must be at least 3 inches.
- The floor drain for any public washing facility must be located so it will readily drain the entire floor area.

### *Standpipes, Sumps and Drinking Fountains*

The standpipe on an automatic clothes washer is usually considered to be an indirect waste receptor. Here are some *IPC* standpipe requirements:

- A standpipe must be at least 2 inches in diameter.
- A standpipe must be installed above floor level.
- A standpipe must be at least 18, but not more than 42, inches above the trap.
- A standpipe must provide access to the trap and drain for rodding purposes.

If gravity won't drain a fixture, appliance or piece of equipment, you have to use a sump. A sump pump can empty clear water waste from a basement floor, subsoil drain, air conditioning condensate drain, or similar applications. The pump will lift the liquid and carry it to an approved place of disposal. See Figure 8-15 for a typical installation.

The code classifies the waste discharge from a drinking fountain as a low rate flow. The dfu of a fountain is $^1/_2$. Fountain waste contains little or no organic matter. If there are no partitions to conceal the waste and vent pipe, you can discharge fountain waste indirectly into a receptor (floor drain) and then to the building drainage system. Figure 8-12 shows an air break installation on a drinking fountain.

## Special Wastes

Let's look at some of the special wastes you'll probably run into. Where the *IPC* hasn't fleshed out the skeleton, we'll look at the general requirements usually accepted by local authorities.

### *Swimming Pool Waste*

The *IPC* briefly discusses swimming pool wastewater in Section 802.1.4. Their guiding principle is that you must install an air gap if the wastewater discharges into the building drainage system. This gives you and your code authorities many opportunities for decision-making. Since there are variations in various local requirements, it's best to ask your questions before installation.

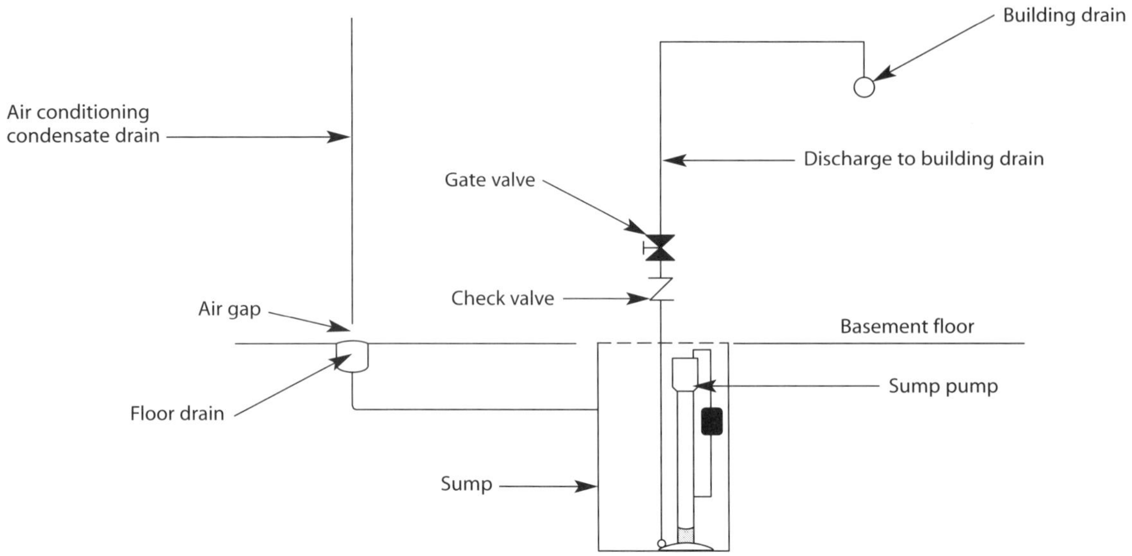


**Note:** The sump need not be vented.

**Figure 8-15**

*Clear water waste sump and piping detail*

Commercial swimming pools, wading pools, spas, backwash from filters and pool deck drains all contain objectionable wastes. Many local codes won't allow this waste to discharge into public sewers. Why? Because ordinary sewers and treatment plants simply can't handle huge amounts of waste that suddenly discharge into the system. If your approved plans aren't clear about how to dispose of commercial swimming pool wastewater, here are a few installations you can discuss with your plumbing department. With their advance approval, you can use any of these to dispose of backwash water and to empty a pool.

- An indirect air gap connection to the building sanitary or storm water system
- A disposal well
- An open waterway, bay or ocean
- A sprinkler system used for irrigation purposes
- A soakage pit or drainfield designed for this purpose. See Figure 8-16 for specific design criteria usually accepted by most authorities.

Note that your local authorities may treat private pool wastes differently. You may be able to puddle these wastes onto private property. If so, the following conditions usually apply:

- The disposal area must be large enough, and graded, to keep the wastewater within the confines of the property.
- The percolation rate of the disposal area must be high enough so water won't remain standing more than one hour after discharge.
- The disposal area must be at least 50 feet from any water supply well.

Local authorities will never let you discharge pool wastes into a sanitary drainage system connected to a septic tank. There are several good reasons for this. The waste could, and possibly would, damage both the septic tank and the subsoil leach field. Sudden flooding can create a reverse flow of wastes into the building drainage pipes. And the chemicals in pool water can destroy the bacterial process a septic tank needs to function properly.

| Pool Capacity (gallons) | Diameter S & G filter | Soil percolation rates (minutes/inch) | | | | | | | | | | | |
|---|---|---|---|---|---|---|---|---|---|---|---|---|---|
| | | 1 | | 2 | | 3 | | 4 | | 5 | | 6 | |
| | | Sq. ft. | Gallons | Sq. ft. | Gallons | Sq. ft. | Gallons | Sq. ft. | Gallons | Sq. ft. | Gallons | Sq. ft. | Gallons |
| 17,000 | 24" | 53.5 | 2,000 | 96 | 3,590 | 130 | 4,860 | 158 | 5,910 | 182 | 6,800 | 202 | 7,560 |
| 17,000 to 26,000 | 30" | 83 | 3,100 | 149 | 5,560 | 200 | 7,550 | 247 | 9,240 | 280 | 10,500 | 315 | 11,780 |
| 26,000 to 38,000 | 36" | 120 | 4,490 | 215 | 8,050 | 292 | 10,910 | 358 | 13,400 | 408 | 15,290 | 452 | 16,900 |
| 38,000 to 52,000 | 42" | 163 | 6,100 | 293 | 10,970 | 400 | 14,980 | 485 | 18,150 | 555 | 20,800 | 618 | 23,100 |

**Note:** Effective depth of soakage pits is 5'0".
Square feet refers to the area of the bottom of the pit.

**Figure 8-16**

*Minimum area and volume of soakage pits for swimming pool soil percolation rates*

## High Temperature Wastewater

*IPC* Section 803.1 says you can't let any steam or water that's 140 degrees F or higher discharge into any part of a drainage system. Pipes that carry steam must first discharge into an approved intercepting sump or boiler blow-off tank. *IPC* Section 701.7 and Figure 8-17 show that you need to have an air gap in the drainage system. Vent any closed tanks from the top. Extend the vent separately, full size, above the roof.

You have to make sure pipes that discharge water above 140 degrees F discharge into an approved receptor connected to the drainage system. I assume you would use an air-gap method, but the code doesn't say specifically which method to use.

## Air Conditioning Condensate Drains

Air conditioning equipment, air washers, air-cooling coils, fuel-burning condensing appliances, the overflow from evaporative coolers and similar water supplied equipment or similar air conditioning equipment generate special waste. This waste must be collected and discharged to an approved receptor or other approved disposal area. I can't find anything in the *IPC* to rely on that relates to the waste from air conditioning units. However, this is a very important part of a plumber's work, so I'll tell you about some installations that most local authorities will accept. Be sure to get their input before you begin any job.

- If air conditioning units are located on a roof, you may discharge the waste onto the roof and into the roof drain. You can do this only if the leader pipe (the vertical pipe of a storm water drainage system) discharges into a storm sewer, soakage pit or adequately-sized pervious area. But you can't do this if the leader pipe discharges onto any paved or non-pervious area where wastes can become a health hazard or nuisance.
- When an air conditioning unit is centrally located below the roof of a building, you may connect it indirectly to a rain leader pipe. Figure 8-18 shows a method that's accepted by most local authorities. But you can't use this method — ever — on a vertical sanitary drainage, waste or vent pipe.

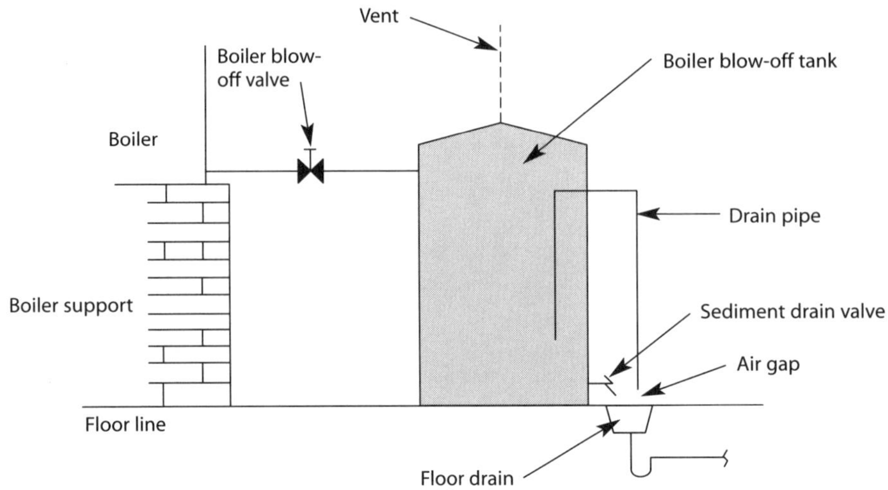


**Figure 8-17**
*Typical boiler blow-off and blow-off tank installation*

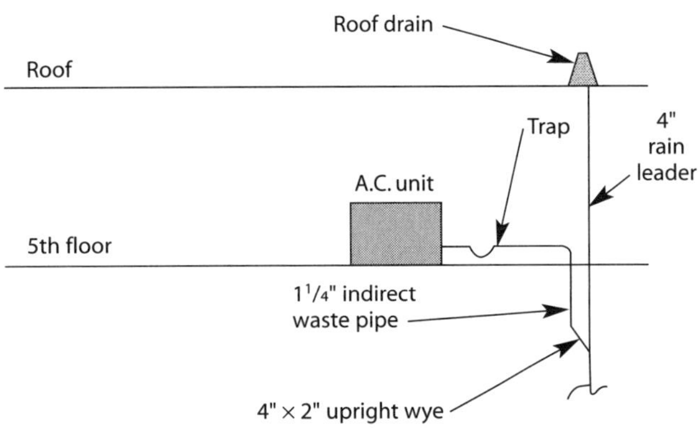


**Figure 8-18**
*Indirect waste connection to an inside rain leader*

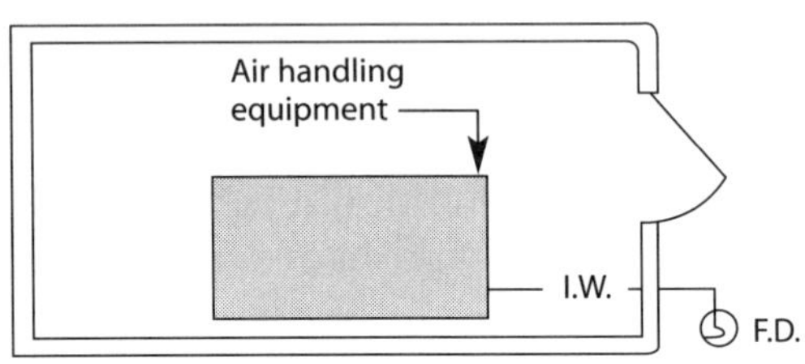


**Figure 8-19**
*Indirect waste pipe connected to a floor drain located outside a room*

- Use an air gap or air break installation to drain an air handling equipment room. Always locate the receptor (floor drain) outside of the equipment room, as shown in Figure 8-19.
- The main condensate drain line in a multistory building may receive discharge from many air conditioning units, but each individual unit must be trapped. Some local authorities require a vent in the main drain line above the highest connection. The vent may extend up through the roof independently, or you can equip it with a perforated cap 2 feet above the highest connection. The condensate drain may discharge indirectly into the building sanitary drainage system, storm water drainage system, or a soakage pit. Figure 8-20 shows a typical connection.
- The following pipe sizes are considered adequate for a combination of multiple air conditioning units. For larger units, the local authority will need to approve pipe size.

1. Up to 3-ton units, pipe size must be 3/4 inch.

2. Up to 20-ton units, pipe size must be 1 inch.
3. Up to 90-ton units, pipe size must be $1^1/_4$ inches.
4. Up to 125-ton units, pipe size must be $1^1/_2$ inches.
5. Up to 250-ton units, pipe size must be 2 inches.

- If air conditioning waste discharges into a drainage system, include this waste when you size the system using the fact that 1 gpm of flow is equivalent to 2 dfu. In large buildings, this waste can make a difference in the drainage pipe size you can use.
- Never use condensate waste from air conditioning unit(s) to reseal the trap of a floor drain since air conditioning condensate waste is not a dependable source of liquid to reseal floor drain traps, especially in winter.
- Usually you can't slope horizontal condensate drainpipes more than $^1/_8$-inch per foot. Get approval before using a steeper slope.
- Use air conditioning waste pipe that's made of materials included in the list in Chapter 9.
- Usually, the local authority will let you use a dry well, leach pit, or in some cases, the tailpiece of certain plumbing fixtures, to dispose of condensate.

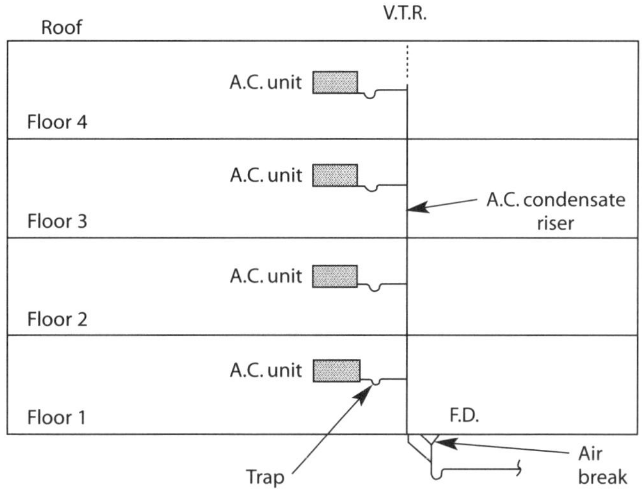


**Figure 8-20**
*Typical air conditioning riser condensate drain connected to a building drain by air break method*

## Storm Water Drainage Systems

Rainwater that's not properly collected can be a nuisance — even a health hazard. Storm water from paved parking lots and large residential and commercial buildings can collect into stagnant pools of water that can be ideal breeding grounds for mosquitoes and offensive odors. The *IPC* also recognizes the importance of collecting and disposing of storm water from commercial properties. Many older cities near lakes, rivers or the ocean have used a dual sewer system that carries both storm water and partially-treated or untreated sewage. These "designs" usually ended in outfall lines that dumped the waste into the water far from the city itself. But the urban situation is improving. Most such cities are trying to correct these environmental assaults by establishing very strict controls.

You'll need advance approval to connect any new construction into an old dual system. And you probably won't get it. But in case you do, *IPC* Section 1108.1 tells you how to convert square feet of rainwater drainage into dfu for sizing purposes. We'll go into this later in this chapter. Today, most local authorities require two separate systems, one for sewage and one for storm water collection.

Install only approved durable materials for storm drainage systems. There are several approved materials available. Except for rainwater pipes on the exterior of a building (we'll delve into this later in the "Protecting Leaders or Conductors" section in this chapter), you can use the same materials, installation methods, protection and supports approved for sanitary drainage, waste and vent systems.

### Strainers

*IPC* Section 1105.1 says you must install a roof drain wherever a roof surface drains into the inlet pipe of an inside leader (the vertical pipe of a storm water drainage system located on the inside of a building that extends to a roof drain located somewhere on the building's roof). You have to make sure the strainer portion extends at least 4 inches above the roof surface. The strainer itself must have an available inlet that's at least $1^1/_2$ times the area of the leader or conductor the drain connects to.

On sun decks, parking decks and other public areas, *IPC* Section 1105.2 says you have to use flat deck strainers. This is so pedestrians won't trip on them and nobody will get their automobile tires damaged. The available inlet area has to be at least 2 times the area of the leader or conductor the drain is connected to. Figure 8-21 shows both types of drains.

## Roof Drains and Special Drains

Most roof drains are made of cast iron. The code accepts other approved materials, but they must be manufactured of high-quality corrosion-resistant materials. *IPC* Section 1105.3 requires that you (or your roofer) install drain flashings on roof drains that connect to interior leaders that penetrate a building roof. To help make the roof penetration watertight, use approved flashing material.

Sometimes you need to install a special drain to carry liquid waste to a legal point of disposal. The *IPC* doesn't give any particular advice on this, but leaves it to local authorities. So here are a few suggestions I believe will be acceptable in most places in the United States.

### *Areaway Drain*

Sometimes an architect will place a floor drain at the base of a stairwell. The stairwell may open to the outside, making a great opportunity for rainwater to collect there. The plumber must know how to install a drain and dispose of this liquid waste. *IPC* Section 1104.3 says you can't connect a floor drain to a storm water drain. What should you do? In this case, you install a floor drain for the sole purpose of carrying the rainwater to a legal point of disposal. If you connect it to the sanitary drainage system, you'll need to install a backwater valve on the drainpipe. This will prevent backflow into the stairwell through the floor drain. Make sure the valve is accessible for inspection and maintenance. See Figure 8-22 and the glossary for details.

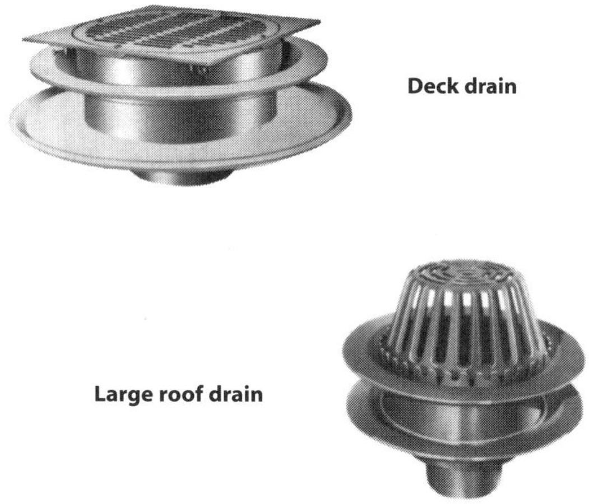


**Figure 8-21**
*Two types of roof drains*

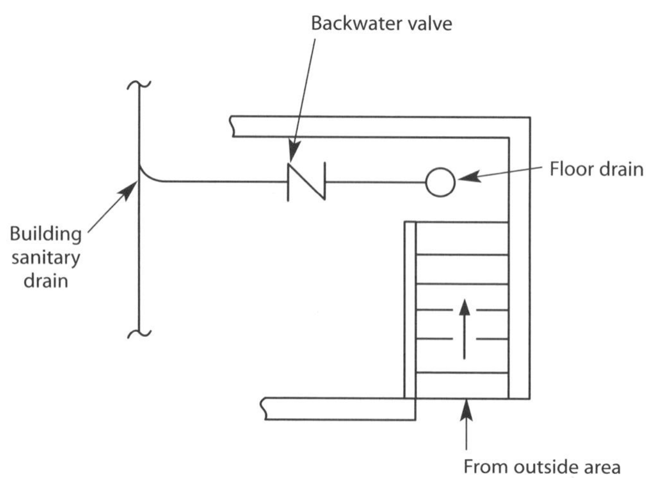


**Figure 8-22**
*Areaway drain connection to a sanitary drain*

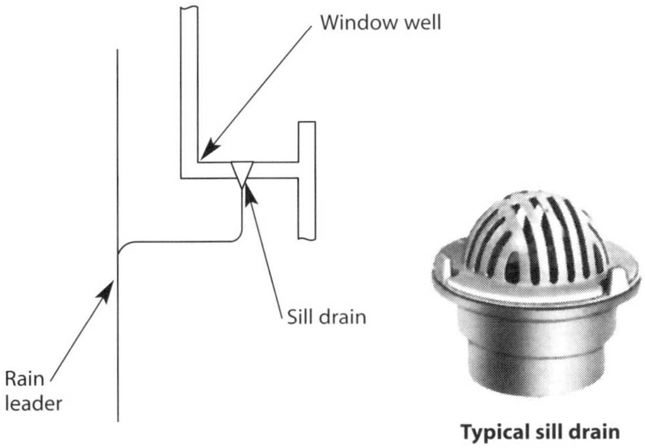


**Figure 8-23**
*Sill drain connection to rain leader*

**Figure 8-24**
*Typical planter drain*

### *Window Well Drains*

Sill drains are very similar to roof drains in design and are constructed of similar materials. Since they drain much less water, however, they are smaller than roof drains. These are required mostly in commercial buildings where the building is designed with a flat surface between a window and an outside wall. Rainwater has to be drained away to keep it from damaging the building. You can usually connect wastes from these drains to an inside rain leader or to a storm water drain, as shown in Figure 8-23.

### *Planter Drains*

You may not often work with indoor planter areas or decorative pools, but they do have wastewater that has to be drained. Most often you find them in commercial buildings. If there's a possibility of sand entering the drainage system, your local authority will require you to install a sand interceptor. And you have to make sure the waste goes to a legal point of disposal. Most local authorities will accept a storm drainage system, an adequately-sized dry well, a soakage pit, or a drainfield. Figure 8-24 shows a typical planter drain.

## *Trap Requirements*

Leaders and storm drains that connect to building sanitary drainage pipes must keep sewer gases and objectionable odors from escaping. There are two ways to do this and stay within the intent of the *IPC*. First, install a trap in the horizontal branch drain from each leader before you connect the leader to the building sanitary drainage system. Use a trap that's the same size as the horizontal drain it serves, as stated in *IPC* Section 1103.1. Figure 8-25 shows such an arrangement. Second, for a separate storm water collection system, install a trap on the main storm drain that connects to the sanitary drain, as shown in Figure 8-26.

To avoid flow interference when combining the two separate systems (the building drainage system and the building storm water drainage system), connect the storm drain in the same horizontal

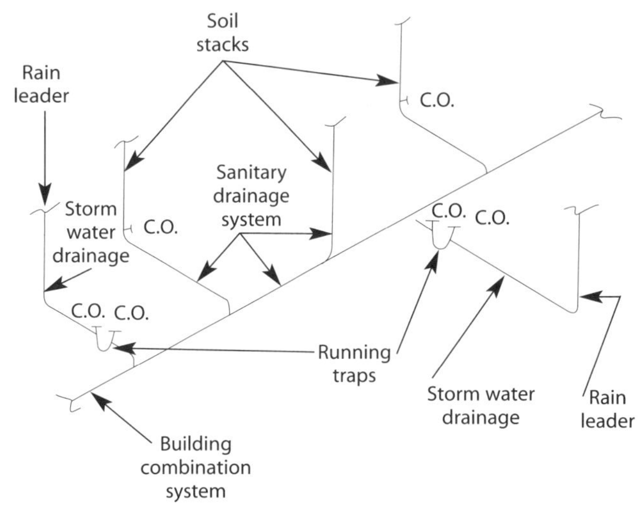


**Figure 8-25**
*Each leader served by a single trap*

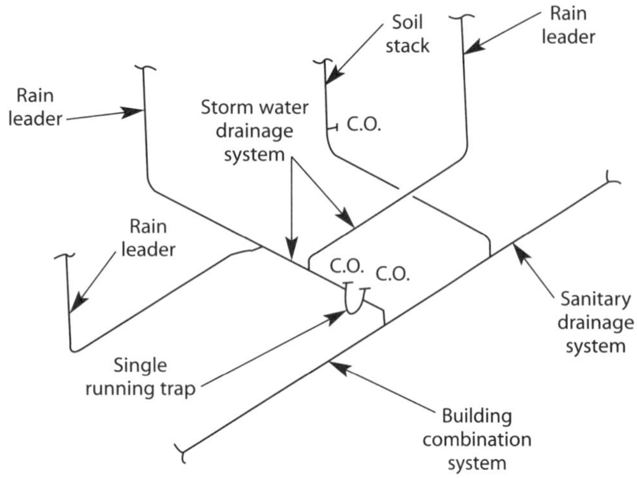


**Figure 8-26**
*Single trap installed in a main storm drain*

plane through a single "wye" fitting to the combined drain or sewer. Make the connection at least 10 feet downstream from any soil stack.

You don't need to install traps in regular storm water drains connected to a storm drainage system.

### *Cleanouts*

*IPC* Sections 1101.8 and 1103.4 state the requirements for cleanouts. Cleanout location and installation requirements are the same as those for sanitary drainage systems. See Chapter 6. Also, you must install an accessible cleanout on the building side of a trap.

Don't forget that you can't use leader pipes as soil, waste or vent pipes, and you can't use soil, waste or vent pipes as leaders. Each must be independent of the other.

### *Subsoil Drains*

Subsoil drains are sometimes called footing drains. They're installed to stop surface water before it reaches a building's foundation wall or footings. Some local authorities require subsoil drains for buildings with basements, rooms, cellars or crawl spaces below grade. You can install them inside or outside of the footings. Look back to Figure 7-17 in Chapter 7. Liquid in subsoil drains must pass through a sand interceptor before it can be pumped or discharged into a storm water system. See Figure 7-18 for a gravity subsoil system.

Subsoil drains only collect sub-surface water or seepage. If a liquid can't be drained by gravity, it must go into a sump or receiving tank. From there you have to make sure it's pumped to a legal point of disposal. You don't need a vent or a cover on a subsoil sump.

### *Protecting Leaders or Conductors*

Be sure to protect all exposed rainwater leaders that you install in parking garages, alleys, driveways or anywhere they could be damaged. Most local authorities will accept one of the following methods:

- Use a 3-inch galvanized steel pipe supported in a concrete base placed on each side of the leader.
- Install a cast iron boot on the lower 5-foot portion of the leader to protect fragile materials (sheet metal conductors or plastics).
- Recess the leader into the building wall.

## Sizing Storm Water Drainage Systems

A mechanical engineer is usually responsible for figuring out the size and disposal methods of a storm water drainage system. His calculations are based on the square footage of impervious areas that have to be drained (roofs, parking lots and similar surfaces) and the maximum inches of rainfall anticipated in an hour. So why should you be concerned with sizing commercial storm drainage and disposal systems if an engineer will do the calculations? First, it's the plumber, not the engineer, who actually installs this kind of work. If you install a system that's wrong, even if you can show it's the engineer's fault, not yours, you're going to look bad and possibly have to correct it at your own expense. After all, you're a licensed plumber — you should have known. The second reason is that you'll have to interpret storm water tables and understand regulations to pass the journeyman and master plumber's exams.

There are big differences in local code storm water tables used to size storm water drains because the maximum anticipated rainfall per hour varies across the country. Naturally, less rainfall means you can use smaller pipe. And more rainfall means larger pipe is required. For local jurisdictions that use the *IPC*, the information you need to figure out leader or conductor sizes and the number of roof drains to use are found in *IPC* Table 1106.2, shown in Figure 8-27, or *IPC* Table 1106.6, shown in Figure 8-28. You need to install enough roof drains, using good spacing, to eliminate puddling on roof surfaces. If you don't space roof drains properly, you won't pass inspection.

The first step is to look up the maximum rainfall for your locality. Each local code will have maximum anticipated rainfall listed in a table like the one in Figure 8-29. Be sure to use your own local code book rainfall tables to figure out the size of leaders and horizontal drainpipes. Every local code book has rainfall tables that give the diameter of vertical leaders and horizontal piping based on the square footage of the area that has to be drained. I'll use *IPC* tables for my calculations. I'll explain the steps and you can follow along in your own code book even if your tables are different. *IPC* Table 1106.2 shows vertical leaders and *IPC* Table 1106.3, shown in Figure 8-30, shows horizontal piping.

First you need to find the total square feet of roof area you have to drain. Next, check the plans for any vertical walls that will let storm water drain onto the roof area. *IPC* Section 1106.4 says you need to add to the total square footage of the roof area to be drained one-half of the area of a vertical wall that sends water to a roof. When you add these vertical wall areas, you can see a big increase in the size of leaders and horizontal pipes you need.

Figure 8-31 shows one way to arrange roof drains and size storm water pipes for a typical roof with a vertical wall. You can follow this example for draining any size roof. In this case, the main roof is 5,000 square feet (100' × 50' = 5,000 SF), with an attached 900 square foot vertical wall (100' × 9' = 900 SF). So we'll add 450 square feet (one half of the vertical wall area) to the 5,000 square feet, which totals 5,450 square feet to be drained. Notice that two roof drains (each one with a drain capacity of 2,725 SF, as just calculated) will do an adequate job. *IPC* Section 1110.4 says that you must install at least two roof drains in any roof area of 10,000 square feet or less. Our example meets code requirements.

Use *IPC* Table 1106.2 shown in figure 8-27, to size rain leaders. The first column shows the leader pipe size in inches. Columns 2 through 12 show the maximum number of square feet that each pipe size can drain. The bold number at the top of each column is the rainfall rate in inches per hour. Let's suppose our example roof is in Montgomery, Alabama. According to the maximum rainfall rates in Figure 8-29, Montgomery has a maximum anticipated rainfall of 3.8 inches per hour. Now look at the 4-inch rainfall rate column in *IPC* Table 1106.2. As you move down the column, you'll see that you need a 4-inch vertical leader pipe that can drain 4,600 square feet to drain each 2,725 square-foot area.

Use *IPC* Table 1106.3, shown in Figure 8-30, to do the horizontal rainwater pipes. This table is divided into three sections for slopes of 1/8-inch, 1/4-inch and 1/2-inch. The greater the slope, the larger the area each pipe size can drain. The left column shows pipe size in inches. The bold number at the top of each column is the rainfall rate. The line immediately below the rainfall rate is the slope per foot. The other columns show the maximum area that can be drained by that size pipe during a rainfall of 1 to 6 inches per hour.

The slope shown in Figure 8-31 is 1/8-inch per foot, so you'll use the top section of *IPC* Table 1106.3 to figure out the horizontal pipe size you need. Note that the horizontal pipe in Figure 8-31 is divided into two sections, A and B. Section B carries the rainwater flow for both leaders (5,450 SF). At 4 inches of maximum rainfall, you'll need to install a 5-inch pipe that can drain up to 3,340 square feet for

**TABLE 1106.2**
**SIZE OF VERTICAL CONDUCTORS AND LEADERS**

| DIAMETER OF LEADER (inches)[a] | HORIZONTALLY PROJECTED ROOF AREA (square feet) Rainfall rate (inches per hour) | | | | | | | | | | | |
|---|---|---|---|---|---|---|---|---|---|---|---|---|
| | 1 | 2 | 3 | 4 | 5 | 6 | 7 | 8 | 9 | 10 | 11 | 12 |
| 2 | 2,880 | 1,440 | 960 | 720 | 575 | 480 | 410 | 360 | 320 | 290 | 260 | 240 |
| 3 | 8,800 | 4,400 | 2,930 | 2,200 | 1,760 | 1,470 | 1,260 | 1,100 | 980 | 880 | 800 | 730 |
| 4 | 18,400 | 9,200 | 6,130 | 4,600 | 3,680 | 3,070 | 2,630 | 2,300 | 2,045 | 1,840 | 1,675 | 1,530 |
| 5 | 34,600 | 17,300 | 11,530 | 8,650 | 6,920 | 5,765 | 4,945 | 4,325 | 3,845 | 3,460 | 3,145 | 2,880 |
| 6 | 54,000 | 27,000 | 17,995 | 13,500 | 10,800 | 9,000 | 7,715 | 6,750 | 6,000 | 5,400 | 4,910 | 4,500 |
| 8 | 116,000 | 58,000 | 38,660 | 29,000 | 23,200 | 19,315 | 16,570 | 14,500 | 12,890 | 11,600 | 10,545 | 9,600 |

For SI: 1 inch = 25.4 mm, 1 square foot = 0.0929 $m^2$.

a. Sizes indicated are the diameter of circular piping. This table is applicable to piping of other shapes provided the cross-sectional shape fully encloses a circle of the diameter indicated in this table.

From the *International Plumbing Code* ©2003, with permission of the International Code Council, Inc.

**Figure 8-27**

*Size of vertical conductors and leaders*

**TABLE 1106.6**
**SIZE OF SEMICIRCULAR ROOF GUTTERS**

| DIAMETER OF GUTTERS (inches) | HORIZONTALLY PROJECTED ROOF AREA (square feet) RAINFALL RATE (inches per hour) | | | | | |
|---|---|---|---|---|---|---|
| | 1 | 2 | 3 | 4 | 5 | 6 |
| $^1/_{16}$ unit vertical in 12 units horizontal (0.5-percent slope) | | | | | | |
| 3 | 680 | 340 | 226 | 170 | 136 | 113 |
| 4 | 1,440 | 720 | 480 | 360 | 288 | 240 |
| 5 | 2,500 | 1,250 | 834 | 625 | 500 | 416 |
| 6 | 3,840 | 1,920 | 1,280 | 960 | 768 | 640 |
| 7 | 5,520 | 2,760 | 1,840 | 1,380 | 1,100 | 918 |
| 8 | 7,960 | 3,980 | 2,655 | 1,990 | 1,590 | 1,325 |
| 10 | 14,400 | 7,200 | 4,800 | 3,600 | 2,880 | 2,400 |
| $^1/_8$ unit vertical in 12 units horizontal (1-percent slope) | | | | | | |
| 3 | 960 | 480 | 320 | 240 | 192 | 160 |
| 4 | 2,040 | 1,020 | 681 | 510 | 408 | 340 |
| 5 | 3,520 | 1,760 | 1,172 | 880 | 704 | 587 |
| 6 | 5,440 | 2,720 | 1,815 | 1,360 | 1,085 | 905 |
| 7 | 7,800 | 3,900 | 2,600 | 1,950 | 1,560 | 1,300 |
| 8 | 11,200 | 5,600 | 3,740 | 2,800 | 2,240 | 1,870 |
| 10 | 20,400 | 10,200 | 6,800 | 5,100 | 4,080 | 3,400 |
| $^1/_4$ unit vertical in 12 units horizontal (2-percent slope) | | | | | | |
| 3 | 1,360 | 680 | 454 | 340 | 272 | 226 |
| 4 | 2,880 | 1,440 | 960 | 720 | 576 | 480 |
| 5 | 5,000 | 2,500 | 1,668 | 1,250 | 1,000 | 834 |
| 6 | 7,680 | 3,840 | 2,560 | 1,920 | 1,536 | 1,280 |
| 7 | 11,040 | 5,520 | 3,860 | 2,760 | 2,205 | 1,840 |
| 8 | 15,920 | 7,960 | 5,310 | 3,980 | 3,180 | 2,655 |
| 10 | 28,800 | 14,400 | 9,600 | 7,200 | 5,750 | 4,800 |
| $^1/_2$ unit vertical in 12 units horizontal (4-percent slope) | | | | | | |
| 3 | 1,920 | 960 | 640 | 480 | 384 | 320 |
| 4 | 4,080 | 2,040 | 1,360 | 1,020 | 816 | 680 |
| 5 | 7,080 | 3,540 | 2,360 | 1,770 | 1,415 | 1,180 |
| 6 | 11,080 | 5,540 | 3,695 | 2,770 | 2,220 | 1,850 |
| 7 | 15,600 | 7,800 | 5,200 | 3,900 | 3,120 | 2,600 |
| 8 | 22,400 | 11,200 | 7,460 | 5,600 | 4,480 | 3,730 |
| 10 | 40,000 | 20,000 | 13,330 | 10,000 | 8,000 | 6,660 |

For SI: 1 inch = 25.4 mm, 1 square foot = 0.0929 $m^2$.

From the *International Plumbing Code* ©2003, with permission of the International Code Council, Inc.

**Figure 8-28**

*Size of semicircular roof gutters*

| State | City | Rate |
|---|---|---|
| Alabama | Birmingham | 3.7 |
| | Mobile | 4.5 |
| | Montgomery | 3.8 |
| Alaska | Fairbanks | 1.0 |
| | Juneau | 0.6 |
| Arizona | Phoenix | 2.2 |
| Arkansas | Ft. Smith | 3.9 |
| | Little Rock | 3.7 |
| California | Eureka | 1.5 |
| | Lake Tahoe | 1.3 |
| | Los Angeles | 2.0 |
| | Lucerne Valley | 2.5 |
| | San Diego | 1.5 |
| | San Francisco | 1.5 |
| | San Luis Obispo | 1.5 |
| Colorado | Denver | 2.2 |
| | Durango | 1.8 |
| | Stratton | 3.0 |
| Connecticut | Hartford | 2.8 |
| | New Haven | 3.0 |
| District of Columbia | Washington | 4.0 |
| Florida | Daytona Beach | 4.0 |
| | Jacksonville | 4.3 |
| | Melbourne | 4.0 |
| | Miami | 4.5 |
| | Tampa | 4.2 |
| Georgia | Atlanta | 3.5 |
| | Brunswick | 4.0 |
| | Macon | 3.7 |
| | Savannah | 4.0 |
| | Thomasville | 4.0 |
| Hawaii | Honolulu | Consult local data |
| Idaho | Boise | 1.0 |
| | Lewiston | 1.0 |
| | Twin Falls | 1.1 |
| Illinois | Chicago | 2.7 |
| | Peoria | 2.9 |
| | Springfield | 3.0 |
| Indiana | Evansville | 3.0 |
| | Indianapolis | 2.8 |
| | South Bend | 2.7 |
| Iowa | Davenport | 3.0 |
| | Des Moines | 3.4 |
| | Sioux City | 3.6 |
| Kansas | Topeka | 3.8 |
| | Wichita | 3.9 |
| Kentucky | Lexington | 2.9 |
| | Lousville | 2.8 |
| Louisiana | New Orleans | 4.5 |
| | Shreveport | 4.0 |
| Maine | Eastport | 2.0 |
| | Portland | 2.4 |
| Maryland | Baltimore | 3.6 |
| Massachusetts | Boston | 2.7 |
| | Nantucket | 2.7 |
| Michigan | Detroit | 2.5 |
| | Grand Rapids | 2.6 |
| | Port Huron | 2.7 |
| Minnesota | Duluth | 2.6 |
| | Minneapolis | 3.0 |
| Mississippi | Meridian | 4.2 |
| | Vicksburg | 4.5 |
| Missouri | Kansas City | 3.9 |
| | St. Louis | 3.2 |
| | Springfield | 3.7 |
| Montana | Havre | 2.0 |
| | Kalispell | 1.5 |
| | Missoula | 1.3 |
| Nebraska | Lincoln | 2.7 |
| | North Platte | 3.5 |
| | Omaha | 3.7 |
| Nevada | Reno | 1.2 |
| | Tonopah | 1.3 |
| | Winnemucca | 1.0 |
| New Hampshire | Concord | 2.5 |
| New Jersey | Atlantic City | 3.4 |
| | Trenton | 3.2 |
| New Mexico | Albuquerque | 2.0 |
| | Roswell | 2.6 |
| | Santa Fe | 2.1 |
| New York | Binghamton | 2.4 |
| | Buffalo | 2.3 |
| | New York | 3.1 |
| | Syracuse | 2.4 |
| North Carolina | Asheville | 3.2 |
| | Charlotte | 3.4 |
| | Raleigh | 4.0 |
| | Wilmington | 4.4 |
| North Dakota | Bismarck | 2.7 |
| | Devils Lake | 2.8 |
| | Williston | 2.5 |
| Ohio | Cincinnati | 2.8 |
| | Cleveland | 2.4 |
| | Columbus | 2.7 |
| | Toledo | 2.4 |
| Oklahoma | Oklahoma City | 4.1 |
| Oregon | Baker | 1.5 |
| | Portland | 1.3 |
| Pennsylvania | Erie | 2.4 |
| | Harrisburg | 2.9 |
| | Philadelphia | 3.2 |
| | Pittsburgh | 2.5 |
| | Scranton | 2.8 |
| Puerto Rico | San Juan | 5.7 |
| Rhode Island | Block Island | 3.0 |
| | Providence | 2.9 |
| South Carolina | Charleston | 4.1 |
| | Columbia | 3.5 |
| | Greenville | 3.3 |
| South Dakota | Huron | 2.5 |
| | Rapid City | 2.7 |
| | Yankton | 2.0 |
| Tennessee | Chattanooga | 3.2 |
| | Knoxville | 3.1 |
| | Memphis | 3.5 |
| | Nashville | 3.0 |
| Texas | Austin | 4.5 |
| | Corpus Christi | 4.6 |
| | Dallas | 4.2 |
| | El Paso | 2.0 |
| | Galveston | 4.7 |
| | Houston | 4.6 |
| | San Antonio | 4.4 |
| Utah | Modena | 2.1 |
| | Salt Lake City | 1.3 |
| Vermont | Burlington | 2.3 |
| | Northfield | 2.5 |
| Virginia | Lynchburg | 3.3 |
| | Norfolk | 4.0 |
| | Richmond | 4.0 |
| Washington | Seattle | 1.0 |
| | Spokane | 1.0 |
| | Walla Walla | 1.0 |
| West Virginia | Elkins | 2.9 |
| | Parkersburg | 3.0 |
| Wisconsin | Green Bay | 2.5 |
| | LaCrosse | 2.9 |
| | Milwaukee | 2.7 |
| Wyoming | Casper | 1.9 |
| | Cheyenne | 2.5 |
| | Yellowstone Park | 2.5 |

Rates given in this table are based on U.S. Weather Bureau Technical Paper No. 40, Chart 14/ 100-year one-hour rainfall (inches)

**Figure 8-29**

*Maximum rainfall rates in inches per hour*

**TABLE 1106.3**
**SIZE OF HORIZONTAL STORM DRAINAGE PIPING**

| SIZE OF HORIZONTAL PIPING (inches) | HORIZONTALLY PROJECTED ROOF AREA (square feet) Rainfall rate (inches per hour) 1 | 2 | 3 | 4 | 5 | 6 |
|---|---|---|---|---|---|---|
| $^1/_8$ unit vertical in 12 units horizontal (1-percent slope) | | | | | | |
| 3 | 3,288 | 1,644 | 1,096 | 822 | 657 | 548 |
| 4 | 7,520 | 3,760 | 2,506 | 1,800 | 1,504 | 1,253 |
| 5 | 13,360 | 6,680 | 4,453 | 3,340 | 2,672 | 2,227 |
| 6 | 21,400 | 10,700 | 7,133 | 5,350 | 4,280 | 3,566 |
| 8 | 46,000 | 23,000 | 15,330 | 11,500 | 9,200 | 7,600 |
| 10 | 82,800 | 41,400 | 27,600 | 20,700 | 16,580 | 13,800 |
| 12 | 133,200 | 66,600 | 44,400 | 33,300 | 26,650 | 22,200 |
| 15 | 218,000 | 109,000 | 72,800 | 59,500 | 47,600 | 39,650 |
| $^1/_4$ unit vertical in 12 units horizontal (2-percent slope) | | | | | | |
| 3 | 4,640 | 2,320 | 1,546 | 1,160 | 928 | 773 |
| 4 | 10,600 | 5,300 | 3,533 | 2,650 | 2,120 | 1,766 |
| 5 | 18,880 | 9,440 | 6,293 | 4,720 | 3,776 | 3,146 |
| 6 | 30,200 | 15,100 | 10,066 | 7,550 | 6,040 | 5,033 |
| 8 | 65,200 | 32,600 | 21,733 | 16,300 | 13,040 | 10,866 |
| 10 | 116,800 | 58,400 | 38,950 | 29,200 | 23,350 | 19,450 |
| 12 | 188,000 | 94,000 | 62,600 | 47,000 | 37,600 | 31,350 |
| 15 | 336,000 | 168,000 | 112,000 | 84,000 | 67,250 | 56,000 |
| $^1/_2$ unit vertical in 12 units horizontal (4-percent slope) | | | | | | |
| 3 | 6,576 | 3,288 | 2,295 | 1,644 | 1,310 | 1,096 |
| 4 | 15,040 | 7,520 | 5,010 | 3,760 | 3,010 | 2,500 |
| 5 | 26,720 | 13,360 | 8,900 | 6,680 | 5,320 | 4,450 |
| 6 | 42,800 | 21,400 | 13,700 | 10,700 | 8,580 | 7,140 |
| 8 | 92,000 | 46,000 | 30,650 | 23,000 | 18,400 | 15,320 |
| 10 | 171,600 | 85,800 | 55,200 | 41,400 | 33,150 | 27,600 |
| 12 | 266,400 | 133,200 | 88,800 | 66,600 | 53,200 | 44,400 |
| 15 | 476,000 | 238,000 | 158,800 | 119,000 | 95,300 | 79,250 |

For SI: 1 inch = 25.4 mm, 1 square foot = 0.0929 $m^2$.

From the *International Plumbing Code* ©2003, with permission of the International Code Council, Inc.

**Figure 8-30**

*Size of horizontal storm drainage piping*

Section A and an 8-inch pipe that can drain up to 11,500 square feet for Section B.

Again, the greater the slope, the larger the area each pipe size can drain. This means that each pipe size may be smaller. Now that we know what we're doing, let's run through the other slopes just for practice. From the drawing in Figure 8-31, let's size Sections A and B, using the other two slopes listed in *IPC* Table 1106.3 and the same rainfall rate of 4 inches per hour. At $^1/_4$-inch slope per foot, you still need a 5-inch pipe for Section A (an area of 2,725 SF). But for Section B (an area of 5,450 SF) you'll need a 6-inch pipe, which can handle up to 7,550 square feet. At a $^1/_2$-inch slope per foot, you can use a 4-inch pipe to drain Section A and a 5-inch pipe to drain Section B.

As a rule, you'll find that tables in your local code book for sizing storm water drains and leaders are already calculated for the particular geographical area you work in. But what if there's more rainfall than IPC Table 1106.3 shows? What do you do? Let's use a method from one of my other books, Plumber's Handbook, that usually gets the OK to adjust for other amounts of rainfall that aren't listed in IPC Table 1106.3. Here's how it works.

In the 1-inch rainfall column, adjust the figures by dividing the square feet in the inch/hour column by the rainfall rate you want in inches/hour.

Let's do some examples. Using each of the three acceptable horizontal pipe slopes, we'll figure the number of square feet a 4-inch pipe will carry in a geographic area with a maximum 8 inches/hour rainfall rate.

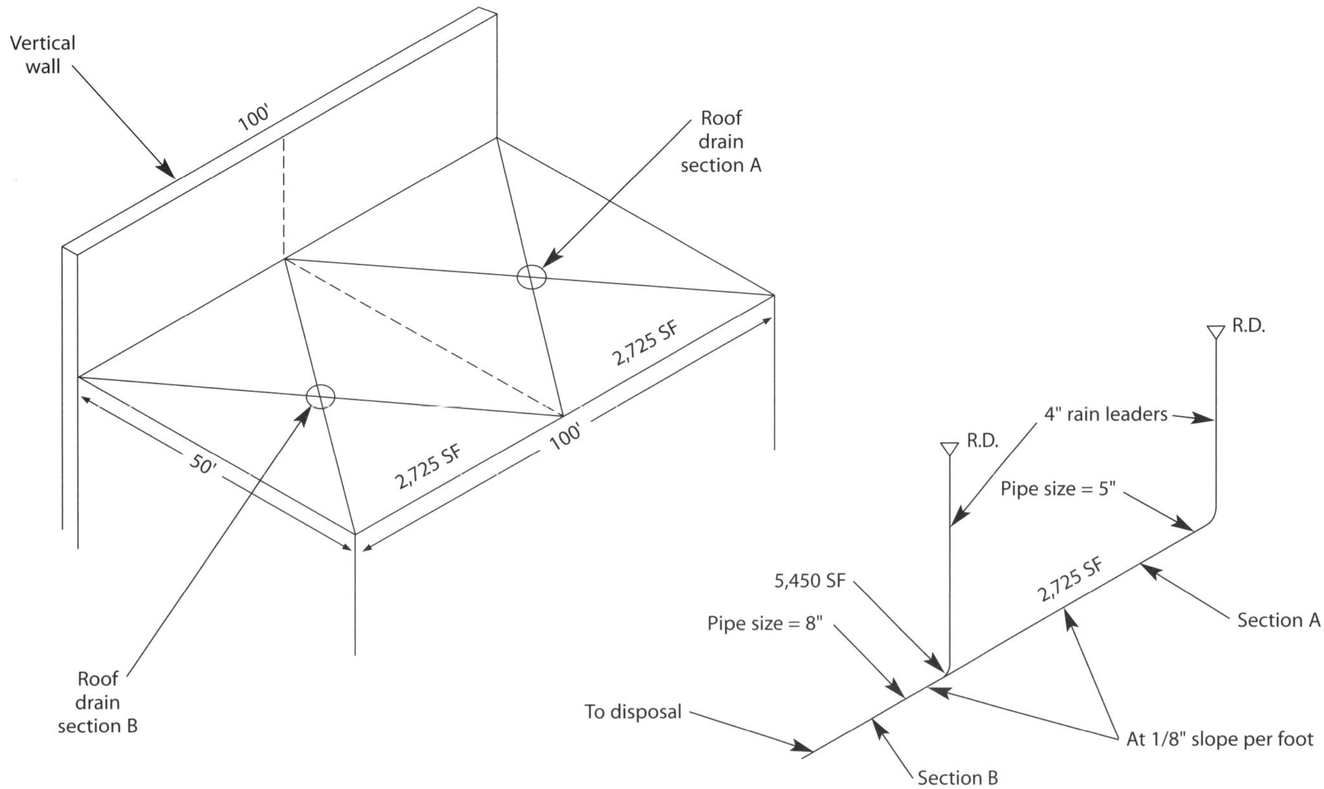


**Figure 8-31**

*Storm water drainage system*

For a 1/8-inch slope, in *IPC* Table 1106.3, the 1-inch rainfall column says the 4-inch pipe can drain 7,520 SF. So we'll divide 7,520 SF by 8 to get 940 SF. This is the area the 4-inch pipe can drain in a location that has a maximum 8 inches/hour rainfall rate.

For a 1/4-inch slope, in *IPC* Table 1106.3, the 1-inch rainfall column says the 4-inch pipe can drain 10,600 SF. So we'll divide 10,600 SF by 8 to get 1,325 SF. This is the area the 4-inch pipe can drain in a location that has a maximum 8 inches/hour rainfall rate.

For a 1/2-inch slope, in *IPC* Table 1106.3, the 1-inch rainfall column says the 4-inch pipe can drain 15,040 SF. So we'll divide 15,040 SF by 8 to get 1,880 SF. This is the area the 4-inch pipe can drain in a location that has a maximum 8 inches/hour rainfall rate.

## Testing Storm Water Drainage Piping

Your testing must be done in accordance with *IPC* Section 312.8. It's the same test that's required for sanitary drainage, waste and vent pipes. If you use an air test, the standard is 5 pounds per square inch held for 15 minutes. If you use a water test, you must have at least a 10-foot head of water held without leaking for 15 minutes.

## Scuppers and Gutters

Although *IPC* Table 1106.6 shows gutter size and Section 1106.5 the requirements for overflow scuppers, these almost never fall under the jurisdiction of a plumbing contractor. It's the general contractor who arranges, sizes, designs and locates scuppers. And it's the roofing contractor or sheet metal worker who designs and installs gutters. You may have to reference your code book if a gutter question is thrown your way on an exam. But that's probably the only time you'll need to be concerned with scuppers and gutters.

## Combined Sanitary and Storm Water Systems

You'll probably never be required to size a combined system (as discussed earlier) because most local jurisdictions won't permit one in the first place. But it's best to be prepared just in case. So let's take a look.

First, you'll need a formula to size the sanitary drain when a storm building drain connects to a sanitary building drain. Generally, the flow in a storm building drain during maximum rainfall periods far exceeds the flow in a sanitary building drain at peak load. So you have to size the combined building drain based on the storm drainpipe sizes listed in *IPC* Table 1106.3. Then convert the sanitary dfu into the equivalent square feet of drained area.

*IPC* Section 1108 has the key to change dfu into an equivalent projected roof or other impervious area. It works like this: For a total sanitary fixture load of 256 dfu or less, use an equivalent drainage area of 4,000 square feet. If the fixture load is more than 256 dfu, add 15.6 square feet for each extra dfu. The *IPC* formula is based on a rainfall rate of 1 inch per hour. To determine the load on the combined system, add the total equivalent drained area, based on the dfu, to the actual drained area. Then just select the appropriate size of combined drain from *IPC* Table 1106.3.

Let's do another example. Assume 1 inch of rainfall per hour, and a building with 300 dfu and 20,000 square feet of roof area. The pipe has a $^{1}/_{8}$-inch slope. First, add the equivalent square feet to drain:

| | |
|---|---|
| Original drainage area | 20,000 SF |
| SF equivalent to 256 dfu | 4,000 SF |
| SF equivalent to extra 44 dfu | 686.4 SF |
| Total SF to drain | 24,686.4 SF |

Now look at *IPC* Table 1106.3. For a 1-inch rainfall, you need an 8-inch pipe to carry the 24,686.4 square feet that has to be drained. A 6-inch pipe can only handle 21,400 square feet.

# *Chapter 8 — Questions and Answers*

**8-1 Of the following devices, appurtenances, or apparatus serving a special function, the only one that the *IPC* does not require to have indirect waste piping or connection is/are**

(A) sterilizing equipment.

(B) residential broiler.

(C) steam tables.

(D) distillation equipment.

***Answer:* *B***

*Code response: Section 801.2*
*This book, page 111*

**8-2 The *IPC* requires that indirect waste pipe longer than ______ feet in total developed length, measured horizontally, be trapped.**

(A) 4

(B) 6

(C) 10

(D) 20

***Answer:* *A***

*Code response: Section 802.2*
*This book, page 112*

**8-3 The air gap between a 1-inch indirect waste pipe and the rim of a floor sink shall be**

(A) 1 inch.

(B) $1^1/_2$ inches.

(C) 2 inches.

(D) $2^1/_2$ inches.

***Answer:* *C***

*Code response: Section 802.2.1*
*This book, page 113*

**8-4 To drain water from a swimming pool into the public drain system, the *IPC* requires you to**

(A) install a back water valve.

(B) make an air break connection.

(C) make a direct connection.

(D) install an air gap.

***Answer:* *D***

*Code response: Section 802.1.4*
*This book, page 118 and Figure 8-15*

**8-5 The *IPC* never allows you to discharge waste water that's more than _____ degrees F. into a building's drainage system.**

(A) 140

(B) 160

(C) 175

(D) 180

***Answer: A***

*Code response: Section 803.1*
*This book, page 119*

**8-6 The *IPC* states that roof drain strainers must extend at least _____ inches above the roof surface.**

(A) 3

(B) 4

(C) 5

(D) 6

***Answer: B***

*Code response: Section 1105.1*
*This book, page 122*

**8-7 When installing a roof drain strainer on a parking deck, the *IPC* requires you to**

(A) use a flat type strainer.

(B) use a recessed type strainer.

(C) install a cast iron protective grate.

(D) use a cone-type strainer.

***Answer: A***

*Code response: Section 1105.2*
*This book, page 122*

**8-8 When figuring the total roof area you need to provide drainage for, you must add ______ of the area of any vertical wall that may add rain water to the total roof area.**

(A) 1/4

(B) 1/3

(C) 1/2

(D) 3/4

***Answer: C***

*Code response: Section 1106.4*
*This book, page 125*

**8-9 If you install a 3-inch leader on a flat deck, your available inlet area has to be at least**

(A) 3 square inches.

(B) 4 square inches.

(C) 5 square inches.

(D) 6 square inches.

***Answer: D***

*Code response: Section 1105.2*
*This book, page 122*

**8-10 According to the *IPC*, you must install at least _______ roof drain(s) for a roof area that's 10,000 square feet or smaller.**

(A) 1

(B) 2

(C) 3

(D) 4

***Answer: B***

*Code response: Section 1110.4*
*This book, page 125*

◆ Chapter 9 ◆

# Installing Drainage and Vent Pipes

In this chapter we'll cover the *IPC* requirements for materials and installation for drain, waste and vent systems. Local authorities, however, may have other requirements. We'll also look at the importance the code places on change in direction in drainage pipes, what type of fittings you can use, where each can be used in drainage and vent systems, and fittings you *can't* use.

## Plumbing Materials Standards

The *IPC* always emphasizes protection. But even if your plumbing system has an excellent design and installation, it may not do its job — unless your materials provide satisfactory service. Materials must be free from defects and meet all code and local standards. All pipe, fittings and fixtures have to be listed or labeled by one or more of the organizations listed in Figure 9-1.

To gain approval, a material must meet one or more of the standards cited in Figure 9-2. You must decipher the standards table in the *IPC*. If you can't get the materials that are listed, you can substitute materials if the plumbing department gives you a green light. Most likely, your local plumbing supplier will stock materials that meet code standards.

You need to be familiar with the abbreviations in Figure 9-1 and the standards in Figure 9-2. There's a distinct possibility that your plumbing exam will use them.

### *Permitted Uses for Materials*

The most common materials used for drainage, waste and vent systems are listed in Figure 9-3. This chart covers sanitary and storm drainage, vent piping, and chemical and acid system materials. Figure 9-3 shows that there are some materials you can use *only* for building sewers; other materials *only* for underground use within buildings; and certain materials *only* for aboveground use within buildings. There are other materials you can use in all three locations. Pay attention to the footnotes which say that the use of other materials is limited, restricted or requires special protection.

You may wonder, after reviewing Figure 9-3, "How do I decide?" Well, several things should come to mind. First look at the actual job situation — the building type, its use and location, what type of fill material the contractor plans to use, whether there will be traffic over piping, and of course, the type of waste conveyed. There's a steady stream of new materials in the plumbing market, so standards can change. Check regularly with your local plumbing department about new materials and code changes. Don't take it for granted that all materials stocked by your local supplier are approved for immediate use. Always buy the updated code supplement sheets, when available, and put them in your code book. These sheets come out several times a year, though the actual code book is printed only about once every three years.

**American National Standards Institute (ANSI)**
11 W. 42nd Street
New York, NY 10036

**Air-Conditioning & Refrigeration Institute (ARI)**
4301 N. Fairfax Drive, Suite 425
Arlington, VA 22203

**American Society of Mechanical Engineers (ASME)**
Three Park Avenue
New York, NY 10016-5900

**American Society of Sanitary Engineering (ASSE)**
28901 Clemens Road
Westlake, OH 44145

**American Society for Testing and Materials (ASTM)**
100 Bar Harbor Drive
West Conshohocken, PA 19428-2959

**American Welding Society (AWS)**
550 N.W. LeJeune Road
Miami, FL 33135

**American Water Works Association (AWWA)**
6666 W. Quincy Avenue
Denver, CO 80235

**Canadian Standards Association (CSA)**
178 Rexdale Blvd.
Rexdale (Toronto), Ontario, Canada M9W 1R3

**Cast Iron Soil Pipe Institute (CISPI)**
5959 Shallowford Road, Suite 419
Chattanooga, TN 37421

**Federal Specification (FS)**
General Service Administration
7th and D Streets
407 E. L'enfant Plaza, S.W., Suite 8100
Washington, DC 20024-2124

**International Code Council (ICC)**
5203 Leesburg Pike, Suite 600
Falls Church, VA 22041-3401

**National Fire Protection Association (NFPA)**
1 Batterymarch Park
Quincy, MA 02269

**National Sanitation Foundation (NSF)**
3475 Plymouth Road
P. O. Box 130140
Ann Arbor, MI 48113-0140

**Plumbing and Drainage Institute (PDI)**
1106 W. 77th Street, South Drive
Indianapolis, IN 46260-3318

**Figure 9-1**

*Addresses and abbreviations for plumbing-related organizations*

| Materials | Standard |
|---|---|
| **Ferrous pipe**<br>Cast-iron | ASTM A 74; ASTM A 888; CISPI 301 |
| **Ferrous fittings**<br>Cast-iron | ASTM A 74; ASTM A 888; ASME B 16.4;<br>ASME B 16.12; CISPI 301 |
| Gray iron and ductile iron | AWWA C 110 |
| **Nonferrous pipe**<br>Brass pipe | ASTM B 43 |
| Copper or copper-alloy tubing (type K, L, M or DWV) | ASTM B 75; ASTM B 88; ASTM 251; ASTM B 306 |
| Stainless steel drainage systems, types 304 and 3161 | ASME A 112.3.1; ANSI A 112.3.1 |
| **Nonferrous fittings**<br>Copper or copper alloy | ASTM B 16.15; ASTM B 16.22; ASTM B 16.23;<br>ASTM B 16.26; ASTM B 16.29; ASTM B 16.32 |
| **Glass**<br>Glass pipe and fittings | ASTM C 1053 |
| **Nonmetallic pipe**<br>Asbestos-cement pipe | ASTM C 428 |
| Concrete pipe | ASTM C 14; ASTM C 76; CSA A 257.1; CAN/CSA A 257.2 |
| Vitrified clay pipe | ASTM C 4; ASTM C 700 |
| **Plastic pipe**<br>ABS plastic pipe | ASTM D 2661; ASTM F 628; ASTM D 2751; CSA B 181.1 |
| ABS DWV sch 40 IPS pipe (solid) | ASTM F 1488 |
| ABS DWV sch 40 IPS pipe (cellular core) | ASTM F 1488 |
| PVC DWV sch 40 IPS pipe (solid) | ASTM F 1488 |
| PVC DWV sch 40 IPS pipe (cellular core) | ASTM F 1488 |
| PVC IPS-DR, PSI 140, PS 200, DWV | ASTM F 1488 |
| PVC plastic pipe (type DWV) | ASTM D 2665; ASTM D 2949; ASTM F 891; ASTM F 1488; CSA B 181.2 |
| Polyolefin pipe | CSA B 181.2 |
| ABS sewer and drain DR-Ps in PS 35, PS 50, PS 100, PS 140, PS 200 | ASTM F 1488 |
| (PVC) plastic pipe (Type DWV, SDR 26, SDR 35, SDR 41, PS 50, PS 100) | ASTM D 2665; ASTM D 2948; ASTM D 3034; ASTM F 891;<br>CSA B 182.2; CAN/CSA B 182.4 |
| Galvanized steel pipe | ASTM A 53 |
| **Fittings**<br>ABS plastic | ASTM D 3311; ASTM D 2661; CSA B 181.1 |
| ABS DWV sch 40 IPS pipe (solid or cellular core) | ASTM D 2661; ASTM D 3311; ASTM F 628 |
| PVC DWV sch 40 pipe IPS-DR, PS 140, PS 200 (solid or cellular core) | ASTM D 2665; ASTM D 3311; ASTM F 891 |
| ABS sewer and drain DR-PS in PS 35, PS 50, PS 100, PS 140, PS 200 | ASTM D 2751 |
| PVC sewer and drain DR-PS in PS 35, PS 50, PS 100, PS 140, PS 200 | ASTM D 3034 |
| Malleable iron | ASME B 16.3 |
| (PVC) plastic | ASTM D 3311; ASTM D 2665 |
| Steel | ASME B 16.9; ASME B 16.11; ASME B 16.28 |
| **Note**: Although these are the standards according to the *IPC*, you may find that pipe, tube or fittings tables may vary in other codes. | |

**Figure 9-2**

*Plumbing materials standards for drainage systems*

| Pipe material | Building sewer | Underground within buildings | Aboveground within buildings |
|---|---|---|---|
| **Sanitary drainage and waste systems** | | | |
| Asbestos-cement sewer | X | | |
| ABS Schedule 40 | X | X | X |
| Borosilicate glass | | | X |
| Brass | | | X |
| Cast iron soil | X | X | X |
| Concrete drain | X | | |
| Copper, DWV, K,L,M | X | X | X |
| Copper, Type K, L | X | X | X |
| Galvanized steel | | | X |
| Wrought iron | | | X |
| PVC Schedule 40 | X | X | X |
| Vitrified clay, standard | X | | |
| Vitrified clay, extra strength | X | X | |
| **Vent pipes for drainage and waste systems** | | | |
| ABS Schedule 40 | N/A | X | X |
| Borosilicate glass | N/A | | X |
| Brass | N/A | | X |
| Cast iron soil | N/A | X | X |
| Asbestos - cement pipe | N/A | X | |
| Copper, Type K, L, M or DWV | N/A | X | X |
| Galvanized steel | N/A | | X |
| PVC Schedule 40 | N/A | X | X |
| Stainless steel drainage | N/A | X | X |
| **Storm drainage systems** | | | |
| Asbestos-cement pipe | X | | |
| ABS Schedule 40 | X | X | X |
| Brass | | | X |
| Cast iron soil | X | X | X |
| Concrete drain | X | | |
| Copper, DWV | X | X | X |
| Copper, Type K, L, M | X | X | X |
| Galvanized steel | | | X |
| Stainless steel drainage pipe | X | X | X |
| PVC Schedule 40 | X | X | X |
| Vitrified clay, extra strength | X | X | |
| Wrought iron | | | X |
| **Chemical waste and acid systems** | | | |
| ABS Schedule 40 | N/A | X | X |
| Borosilicate glass | N/A | | X |
| High silicon content iron | N/A | X | X |
| Plastic lined | N/A | X | X |
| PVC Schedule 40 | N/A | X | X |
| Lead | N/A | | X |
| Vitrified clay, extra strength | N/A | X | |

**Notes:** Other codes may list materials that don't appear in above charts. Some codes limit plastic pipe to first three floors of a building. Fragile sewer piping materials, like asbestos, plastics, concrete and clay, generally require special protection. See Figure 9-23.

**Figure 9-3**

*Common materials and locations to use them*

## Drainage and Vent Materials

### *Building Sewers*

Here is information on some common building sewer materials.

*Cast iron pipe* — The most common cast iron is tar-coated centrifugal-spun service weight pipe and fittings. It comes in 5- and 10-foot lengths, with and without hubs. Some architects may still specify extra-heavy building sewer pipe and fittings. If they do, it's OK to use it, though it might not be easy to find.

*Vitrified clay pipe* — You can use standard or extra-strength vitrified clay pipe and fittings.

*Plastic pipe* — You can use ABS or PVC Schedule 40 plastic pipe and fittings. If plastic sewer pipe is under a heavy traffic area, your local authorities may require you to use the stronger Schedule 80 plastic pipe. Check it out before installation to make sure.

*Concrete drainpipe* — *IPC* Table 702.3 says you can use concrete drainpipe and fittings. However, since concrete pipe is easily corroded by acid and sewer gas, it may not be acceptable for ordinary building sewers. Check with your plumbing department before installation.

*Asbestos-cement sewer pipe* — Contrary to some people's opinion, most codes still approve asbestos-cement pipe. See *IPC* Table 702.3. But it's often restricted to drainage systems located outside of buildings.

*Bituminous fiber pipe* — This pipe isn't listed in the *IPC* standards or most other plumbing codes. However, the government housing authority (HUD) often specifies it for their projects. It's much cheaper than most other sewer materials. Its life span is questionable and chances are, you'll never have to install it. But since you're a plumber, you should know about it.

*Copper pipe* — There are several types (grades) of copper. But the *IPC* only allows sewer piping in Type K or L. Because of its cost and fragility, you'll seldom have to install it.

*Stainless steel* — Though the *IPC* lists stainless steel piping for building sewers, this would have to be for a very special situation since it's so expensive and requires special protection. If specified for a particular job, follow the installation and protection for this pipe from the approved job set of plans.

### *Underground and Aboveground Piping*

The *IPC* list of acceptable materials for DWV underground installation includes ABS or PVC Schedule 40 plastic pipe, cast iron pipe, copper (Type K, L, M,) or DWV pipe, stainless steel drainage pipe, polyolefin pipe and asbestos-cement pipe. You'd be wise to check with your plumbing department before you install certain pipes (such as stainless steel and asbestos-cement) not commonly used for underground building DWV systems.

The *IPC* list of acceptable materials for aboveground installation in DWV systems includes ABS or PVC Schedule 40 plastic pipe, brass pipe, cast iron pipe, copper or copper-alloy pipe or tubing (Type K, L, M) or DWV, galvanized pipe, glass pipe, polyolefin pipe and stainless steel pipe. Make sure pipe fittings in a drainage system are compatible with the pipe you use. *Don't mix different types of plastics in a system.* Make sure the drainage fittings have no ledges, shoulders or reductions that could restrict or obstruct flow. Check your threaded fittings to see that they're the recessed drainage type.

### *Chemical Waste Piping*

The *IPC* doesn't list pipe and fittings materials approved for chemical or acid systems. But it does say in Sections 702.5, 803.2 and 803.3 that you must give this waste special attention before discharging it into a building or public drainage system. Most local authorities will require that piping carrying this waste be made of corrosion-resistant material. Some codes may allow such materials as borosilicate glass, high silicon content cast iron, ABS and PVC plastic, plastic-lined pipe, lead pipe and vitrified clay pipe to carry chemical waste. See Figure 9-3. And *always* match the fittings to the material and pipe you're using.

### Indirect Waste Piping

At this writing the *IPC* doesn't specify any particular indirect waste piping materials. It's pretty safe to assume that you can use any materials approved for potable water (see Chapter 13), or for drainage systems. The most common materials are cast iron, galvanized steel, wrought iron, brass, copper (Type K, L, M,) or DWV pipe or tubing, ABS or PVC plastic and lead pipe. Figure 9-3 shows where you can use each type material, above or below ground.

### Footing and Subsoil Drain Piping

Whenever the design of a building calls for subsoil drains under a cellar or basement floor, or around the outer walls of a building, your subsoil drainpipe must be at least 4 inches in diameter. *IPC* Section 1102.5 says that all subsoil drains must be open-jointed, horizontally split or perforated pipe.

### Storm Drainage Pipe and Fittings

Materials recognized by the *IPC* for storm drainage systems are listed in Figure 9-3. This is a concise, dependable source when you're working on a building storm sewer, or on an under and above-ground storm drainage system. For a building storm sewer, you can use asbestos-cement pipe, ABS and PVC Schedule 40 pipe, cast iron pipe, concrete drain pipe, copper (Type K, L, M,) or DWV pipe and vitrified clay pipe.

For underground storm drainage piping within a building, choose from ABS and PVC Schedule 40 plastic pipe, asbestos-cement pipe, cast iron pipe, copper (Type K, L, M) or DWV pipe, vitrified clay pipe and stainless steel drainage pipe.

Acceptable materials for aboveground storm drains and leaders within a building are ABS and PVC Schedule 40 plastic pipe, brass pipe, cast iron pipe, copper (Type K, L, M) or DWV pipe, galvanized steel pipe, polyolefin pipe and stainless steel drainage pipe.

Plan to protect all exposed rainwater leaders you install in parking garages, alleys, driveways or any place subject to vehicular damage. The conventional protection is a 3-inch steel pipe supported in a concrete base, placed in front of the leader. To protect fragile materials (sheet metal conductors or plastics) install a cast iron pipe boot on the lower 5-foot portion of the leaders. Again, be sure you use fittings that conform to the materials and pipe you're using, and don't mix different types of plastic materials in any plumbing system. For example, don't link ABS with PVC.

## Joints and Connections

Joints and connections in a DWV system must be gastight and watertight at the pressure required for testing, or for their intended use. The code makes an exception for the parts of a drainage system that use perforated or open-joint piping designed to collect and carry underground water to a legal point of disposal.

Now let's look at the different type joints you may have to deal with in a DWV system. The code is very strict on how and where you can use these joints. Obviously, you can't use plastic pipe to support the weight of any plumbing fixture.

### Types of Joints

*Cast iron caulked* — Every lead-caulked joint for cast iron bell-and-spigot soil pipe must be firmly packed with oakum or hemp and filled with molten lead at least 1 inch deep. See Figure 9-4. The lead can't extend more than $^1/_8$ inch below the rim of the hub. You should run the lead in one pouring and caulk it tight. Never use paint, varnish or other coatings on the jointing material until after the joint's been tested and approved. Acid-resistant rope and acid-proof cement are permitted when necessary. See *IPC* Section 705.5.1.

*Cast iron compression* — Compress neoprene rubber compression gaskets for bell-and-spigot cast iron pipe and fittings when you insert the spigot into the hub of the pipe. You can use this kind of joint as an alternate for a lead and oakum joint. See *IPC* Section 705.5.2 and Figure 9-5.

*Cast iron hubless* — Use mechanical joint couplings to join hubless cast iron pipe and fittings for DWV plumbing systems. Use stainless steel couplings

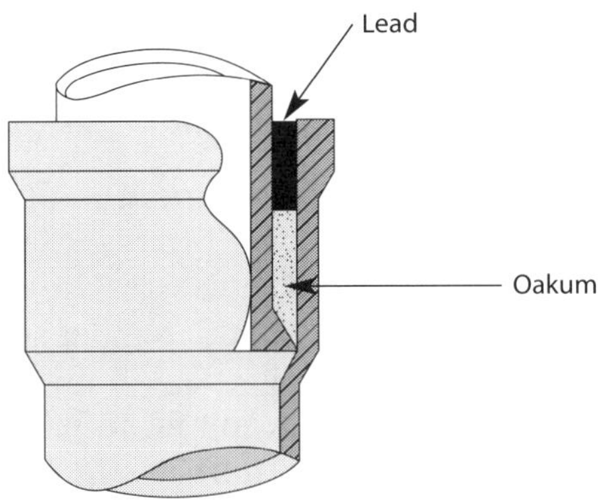


**Figure 9-4**
*Lead and oakum joint*

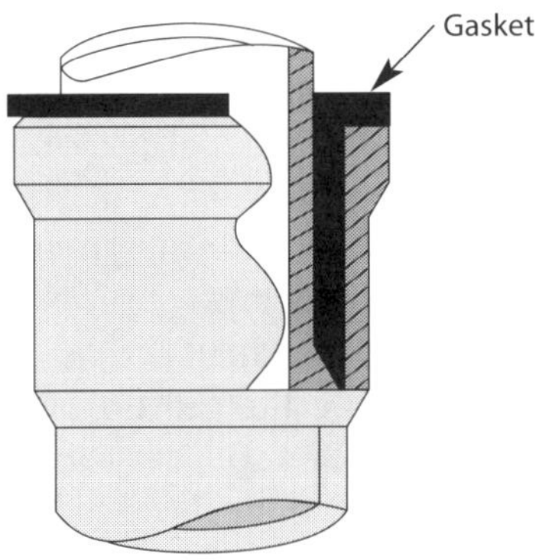


**Figure 9-5**
*Compression joint*

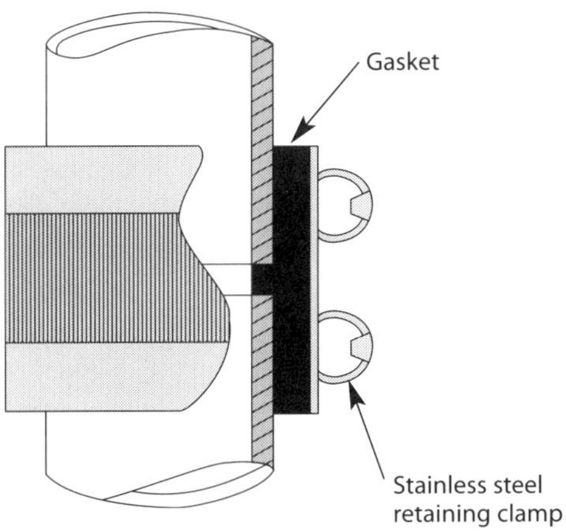


**Figure 9-6**
*Hubless joint*

that meet the Cast Iron Soil Pipe Institute (CISPI) standards. Use an elastomeric sealing sleeve with a center stop. Follow the manufacturer's installation instructions when you install mechanical joint couplings. See *IPC* Section 705.5.3 and Figure 9-6. According to the *IPC*, you'll need prior approval from your local authority to install mechanical type joints on drainage systems above ground. See *IPC* Sections 705.2.1, 705.7.1, 705.8.1 and 705.14.1.

*Asbestos-cement* — Make all joints in asbestos-cement pipe with sleeve couplings made of the same composition as the pipe, sealed with elastomeric preformed rings. See *IPC* Section 705.3 and Figure 9-7.

*ABS plastic* — Make sure both surfaces are free of grime and moisture when you connect ABS plastic pipe Schedule 40 to plastic fittings. Apply solvent cement to all surfaces. Make joints quickly while the cement is wet or you'll have to begin again, using new materials. Solvent-cement joints are permitted above or below ground. See *IPC* Section 705.2.2. You can also use mechanical joints to connect ABS plastic pipe Schedule 40 to plastic fittings, according to *IPC* Section 705.2.1.

ABS plastic pipe Schedule 80 may be threaded if necessary for use in a threaded piping system. Use dies specifically designed to thread plastic pipe. Use only lubricant or tape approved for plastic pipe and fittings. Apply only to male threads. See *IPC* Section 705.2.3

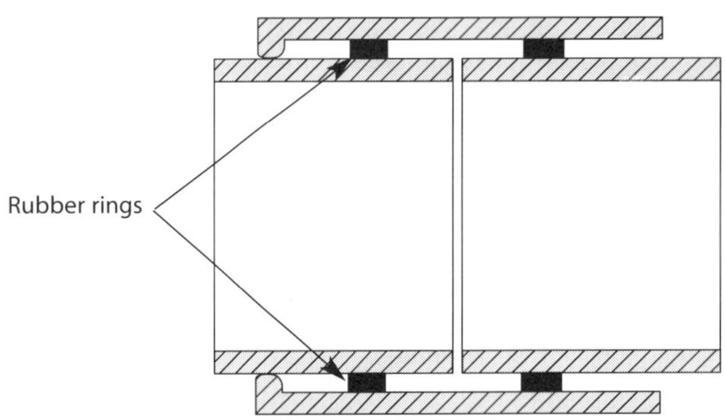


**Figure 9-7**
*Asbestos-cement pipe coupling*

*ABS coextruded composite* — According to *IPC* Section 705.7 and 705.7.1, it's OK to use mechanical joints between coextruded composite pipe having an ABS outer layer or ABS fittings.

*ABS coextruded solvent cemented* — You can install this type of joint both above and below ground. Clean grime and moisture off plastic pipe and fittings before you apply any cement to any surface. Create joints while cement is still wet.

*PVC plastic pipe* — You can use solvent cement for joints connecting PVC plastic pipe and fittings. As always, clean off all grime and moisture. Notice that for PVC pipe and fittings you must apply a primer (purple in color). Then, apply a solvent-cement (not a purple color) to all surfaces. Again, make the joint while the cement is wet. You can use these joints above or below ground. See *IPC* Section 705.14.2.

As with ABS, you can thread PVC Schedule 80 pipe for use in a threaded piping system. Use dies designed for threading plastic pipe. Apply approved lubricant or tape for plastic threads to male threads only. Never use regular pipe dope or compound on plastic threaded joints. See *IPC* Section 705.14.3.

*Coextruded PVC composite pipe* — You can use mechanical joints between coextruded composite pipe having a PVC outer layer or PVC fittings. You must use an elastomeric seal. Install joints according to the manufacturer's recommendations. Unless first approved by your plumbing department, don't install them above ground. See *IPC* Section 705.8 and 705.8.1.

You can use solvent cement joints between coextruded PVC pipe and fittings. Clean all joint surfaces of grime and moisture. Use a primer (purple in color). Then apply the regular solvent cement to all surfaces and make the joint while the cement is still wet. Use these joints both above and below ground. See *IPC* Section 705.8.2 for details.

*Plastic pipe hub and spigot end* — The *IPC* doesn't mention this type of joint for plastic pipe. But it's a common joint for building sewers. Make this joint by pushing the spigot end of the pipe or fittings into the pipe or fitting hub which contains the gasket. See Figure 9-8 for an example.

*Bituminous fiber pipe* — This type of pipe isn't mentioned in the *IPC* or most other codes. Use it only if the architect approves it. This joint will most likely be used in low-income government housing. To make joints in this type of pipe, use tapered couplings and fittings made of the same material as the pipe. When you need to cut the pipe, you must use a specially-designed tapering tool to taper the pipe end, as shown in Figure 9-9. Otherwise you'll have a bad joint that the plumbing inspector won't approve.

*Vitrified clay sewer pipe* — Use a neoprene preformed elastomeric ring to seal the joints between vitrified clay pipe and fittings with bell-and-spigot

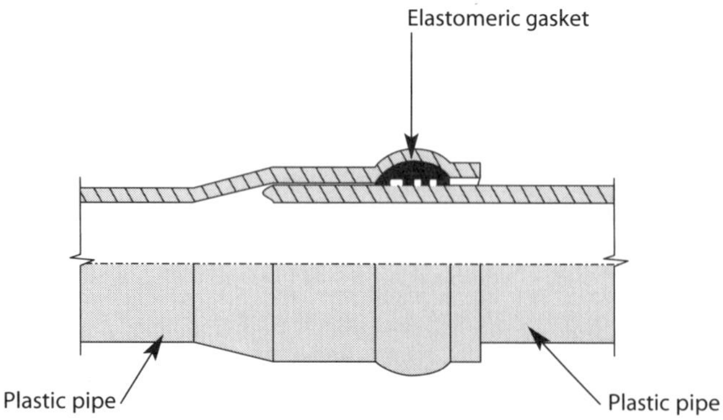


**Figure 9-8**

*A typical push-on elastomeric gasket joint for plastic pipe with hub and spigot end*

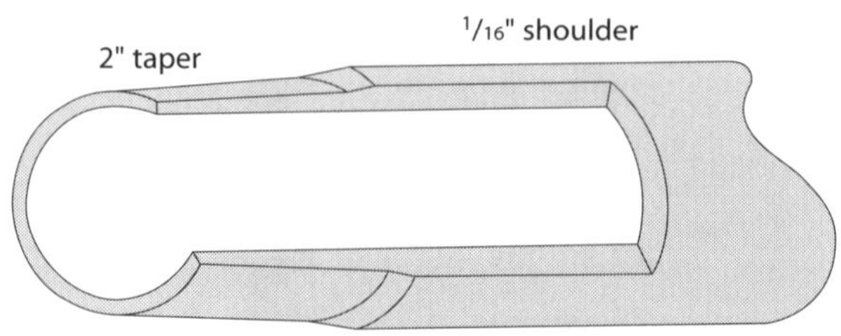


**Figure 9-9**

*Tapered pipe end*

connections similar to those in Figure 9-10. Join plain end pipe connections with mechanical joint couplings. See *IPC* Section 705.15 and Figure 9-6.

Although the *IPC* prohibits it, your local code may let you use this method to seal vitrified clay pipe and fittings with bell-and-spigot joints. Here's how to make the joint. Use hot poured bitumastic compound with a bond strength of at least 100 psi in shear. Fill approximately 25 percent of the joint space at the base of the hub (bell) with jute or hemp. Pour each joint in one operation until the joint is filled. Don't test any joint until about an hour after pouring, to allow the compound to cool. This joint is made similarly to a lead and oakum joint (Figure 9-4), but with less hemp or jute and with a bitumastic compound. Vitrified clay sewer pipe joints made with bitumastic compound provide flexibility to avoid breakage. Remember, don't use this method unless you know your local code allows it.

*Concrete sewer pipe* — For new construction, your local code may prohibit installing concrete pipe and fittings with cement mortar joints. In most cases, the code will let you use mortar joints and connections for repairs or for connections to existing lines that have mortar joints. The *IPC* limits joints for concrete pipe and fittings to those containing an elastomeric seal. That's simply an O-ring located in the bell, which makes a tight joint when the spigot end of the pipe or fitting is forced into the hub. See Figure 9-10.

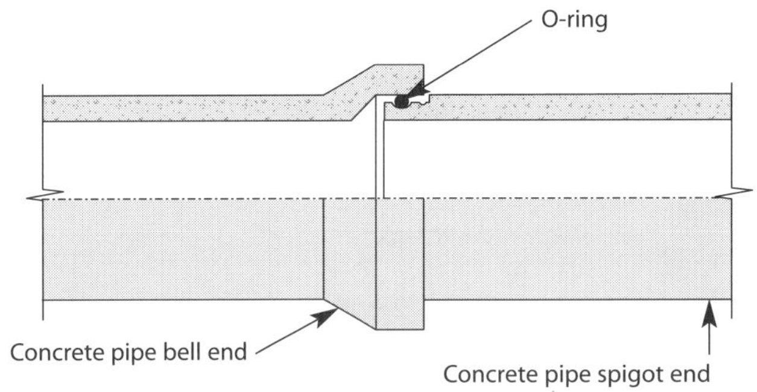


**Figure 9-10**
*Concrete pipe joint made with O-ring*

*Welded* — *IPC* Sections 705.4.4 and 705.9.5 mention this only briefly, saying that joined surfaces must be clean, and joints welded with an approved filler metal. Your local plumbing department has the details you'll need.

*Burned* — Lap burned (welded) lead joints. Fuse the sections together uniformly to make one continuous piece. Make the weld as thick as the lead you join to. Use the same material for the weld as the material you're joining. See *IPC* Section 705.13.1.

*Brass brazed* — *IPC* Section 705.4.1 is pretty concise when it comes to brazed joints. After you clean all surfaces, use an approved flux. Then braze the joint with filler metal conforming to AWS A5.8. If you come up against a piping system requiring brass pipe that needs brazed joints, check with your local authorities for the type of flux and filler material they'll accept.

*Copper brazed* — *IPC* Section 705.9.1 says you can braze these joints. It's extremely important that you clean all surfaces and use an approved flux. The filler metal you use for brazing is usually made of silver and is very expensive. It must conform to AWS A5.8.

*Wiped* — These joints have pretty much disappeared today. But in my early days of plumbing, they were often used to join a 4-inch lead pipe to a brass ferrule for use as a water closet stub (connection). We also had to wipe one of these joints to pass the examination.

Here are the code requirements for these connections: lead pipe or fittings with lead, and lead pipe with brass or copper (includes ferrules, solder nipples or traps). Be sure any wiped joint has an exposed surface on each side of the joint no less than $3/4$ inch that's at least as thick as the material being joined. At the center of a joint, at its thickest point, the joint must be at least 0.325 inch. See *IPC* Section 705.13.2.

*Copper soldered* — The *IPC* requires that all copper piping must be cut square, using a tubing cutter. *IPC* Sections 705.9.3 and 705.10.3 say you have to ream the ends to the full diameter of the pipe. Clean joining surfaces well and use approved flux.

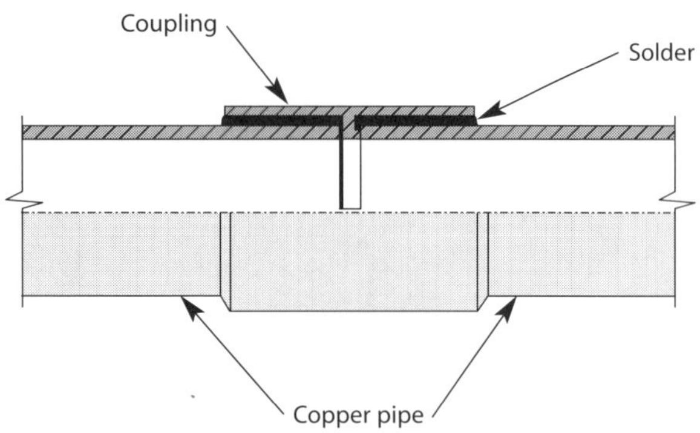


**Figure 9-11**
*Typical copper soldered joint*

When you join copper pipe in a DWV system, always use approved cast brass or wrought copper fittings. Make your joints with approved solder containing no more than 0.2 percent lead. See Figure 9-11.

*Steel pipe threaded joints* — Use only approved standard pipe threads on threaded joints in a DWV system. See Chapter 12. Always use recessed drainage fittings with a smooth interior waterway. The manufacturer taps the threads to allow 1/4 inch fall per foot. Apply pipe dope (compound) or tape to male threads only.

The *IPC* permits using mechanical joints with approved elastomeric seals in a galvanized steel system. See *IPC* Section 705.12. Install these joints according to the manufacturer's instructions or requirements of your local controlling authorities.

*Borosilicate glass* — Make glass-to-glass connections with a bolted compression-type stainless steel coupling that has a contoured acid-resistant elastomer compression ring and fluorocarbon polymer inner seal ring. See *IPC* Section 705.11.

## Joints Between Different Materials

*Borosilicate glass* — Make glass joints as you would cast iron caulked joints, except you must use acid-resistant oakum or hemp rope and acid-proof cement. Use approved adapters having a TFE seal to join borosilicate glass joints to other types of piping material. Follow the manufacturer's recommendations. See local regulations and *IPC* Section 705.16.6.

*Copper to cast iron* — *IPC* Section 705.16.1 explains how to join copper pipe to cast iron hub pipe. Make the joint by soldering the copper pipe to a brass ferrule. Caulk the ferrule into the hub of the cast iron pipe, using lead. It's OK to make the connection with a compression gasket, as defined in *IPC* Section 705.5.2.

*Copper to galvanized steel* — To connect copper pipe or tubing to galvanized steel pipe, use a brass converter fitting or a dielectric fitting. This prevents corrosion caused by electrolytic action. Solder the copper tubing to the sweat side of the fitting and the screw side of the fitting screwed to the threaded pipe. See *IPC* Section 705.16.2.

*Cast iron to galvanized steel or brass* — Caulk joints between cast iron pipe and galvanized steel or brass pipe as stated in *IPC* Section 705.5.1. Use an approved adapter fitting to join pipes. Check with your plumbing inspector for the adapter that's approved in your area.

*Plastic to other material* — Use only approved transition fittings to join plastic pipe and any other piping material. The manufacturer will probably recommend the type of fitting and installation method. *IPC* Section 705.16.4 says that joints between plastic pipe and cast iron hub pipe must be made by a caulked joint (Figure 9-4) or a mechanical compression joint (Figure 9-12). Remember, don't mix ABS or PVC plastic pipe and fittings in the same system without transition fittings. If you do, be ready to tangle with your plumbing inspector.

*Lead to other material* — Use a wiped joint to join lead pipe with other piping materials. The details are in *IPC* Section 705.13.2. Acceptable transition fittings for wiping joints are caulking ferrules, soldering nipples, bushings or an approved adapter fitting. See *IPC* Section 705.16.5.

*Stainless steel to other materials* — Use approved mechanical couplings to join stainless steel and other type piping materials.

*Trap slip joints* — For fixture traps, you can use connectors (nuts and washers) on both sides of the trap and in the trap seal. If the trap is concealed,

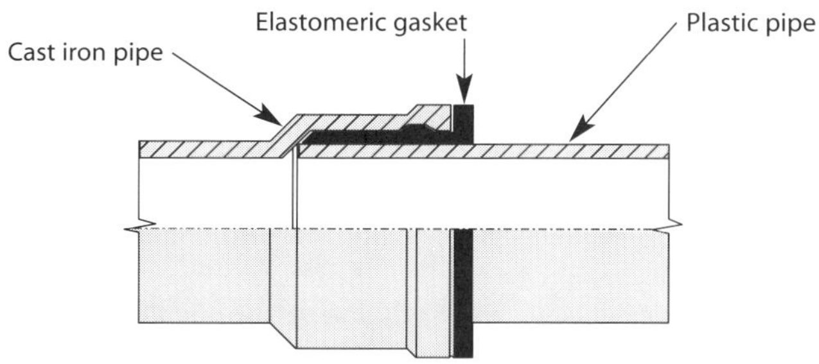


**Figure 9-12**

*Typical transition mechanical compression joint — plastic to cast iron pipe*

TABLE 705.18
CAULKING FERRULE SPECIFICATIONS

| PIPE SIZES (inches) | INSIDE DIAMETER (inches) | LENGTH (inches) | MINIMUM WEIGHT EACH |
|---|---|---|---|
| 2 | $2^1/_4$ | $4^1/_2$ | 1 pound |
| 3 | $3^1/_4$ | $4^1/_2$ | 1 pound 12 ounces |
| 4 | $4^1/_4$ | $4^1/_2$ | 2 pounds 8 ounces |

For SI: 1 inch = 25.4 mm, 1 ounce = 28.35 g, 1 pound = 0.454 kg.

From the *IPC*® ©2003, with permission of the ICC.

**Figure 9-13**

*Caulking ferrule specifications*

*IPC* Section 405.8 states that you must provide access for service or repair.

*Caulking ferrules* — Lead pipe is usually wiped to caulking ferrules so it can be joined to another type of material. Ferrules must be made of red brass. *IPC* Table 705.18, shown in Figure 9-13, gives information on caulking ferrule specifications.

*Soldering bushings* — Soldering bushings must be constructed of red brass. Figure 9-14 shows the specifications for soldering bushings from *IPC* Table 705.19, used most often to join a copper fixture drain to a cast iron drainage system.

TABLE 705.19
SOLDERING BUSHING SPECIFICATIONS

| PIPE SIZES (inches) | MINIMUM WEIGHT EACH |
|---|---|
| $1^1/_4$ | 6 ounces |
| $1^1/_2$ | 8 ounces |
| 2 | 14 ounces |
| $2^1/_2$ | 1 pound 6 ounces |
| 3 | 2 pounds |
| 4 | 3 pounds 8 ounces |

For SI: 1 inch = 25.4 mm, 1 ounce = 28.35 g, 1 pound = 0.454 kg.

From the *IPC*® ©2003, with permission of the ICC.

**Figure 9-14**

*Soldering bushing specifications*

## Prohibited Joints and Connections

Never use a double sanitary tee to receive the discharge of back-to-back water closets or other fixtures and appliances that have pumping action discharge. See *IPC* Section 706.3. When you connect different-size pipes or different-size pipes and fittings, take care to select the proper size increaser, reducer or reducing fitting. Never use fittings or connections (such as an enlarged chamber, a ledge, shoulder or reduced pipe area) that could obstruct the drain flow.

Don't use double hub fittings on drainage lines, as shown in Figure 9-15. And don't use fittings with a drainage pattern in the direction opposite to flow, as shown in Figure 9-16. You can only use these fittings in a vent system. An inexperienced plumber is likely to make this mistake when using check valves, no-hub cast iron, copper or plastic drainage fittings. Figure 9-17 shows a simple isometric of a drainage system with correct drainage patterns.

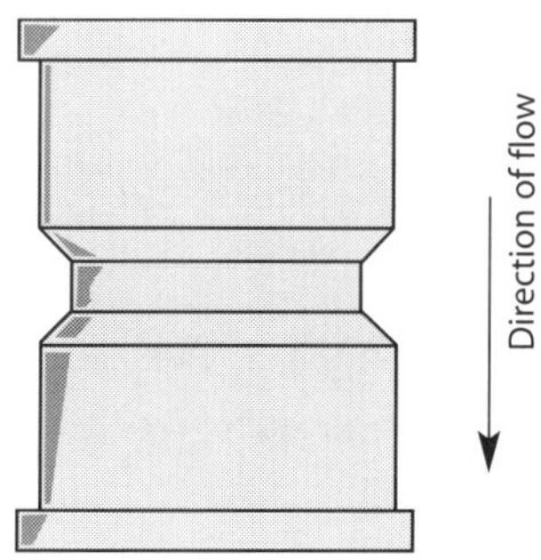


**Figure 9-15**

*Double hub fitting with hub opposite to flow (prohibited)*

Most codes don't permit drilling or tapping any drainage or vent piping. Drilling or tapping can create fractures that will let raw sewage seep out of the pipe. In the case of a vent pipe, sewer gases could escape into the building. See Figure 9-18.

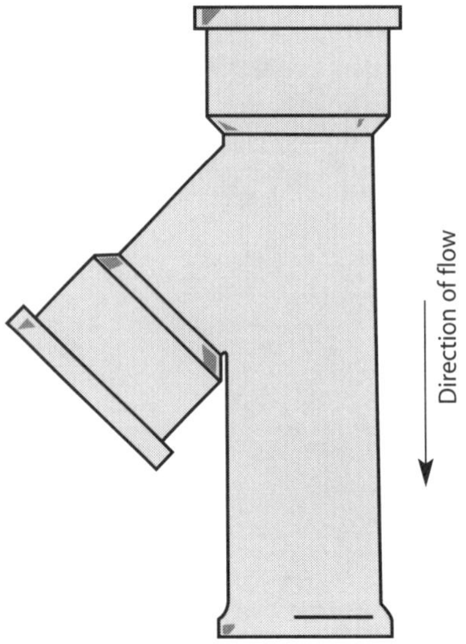


**Figure 9-16**

*Inverted Y fitting with branch drainage pattern opposite to flow (prohibited)*

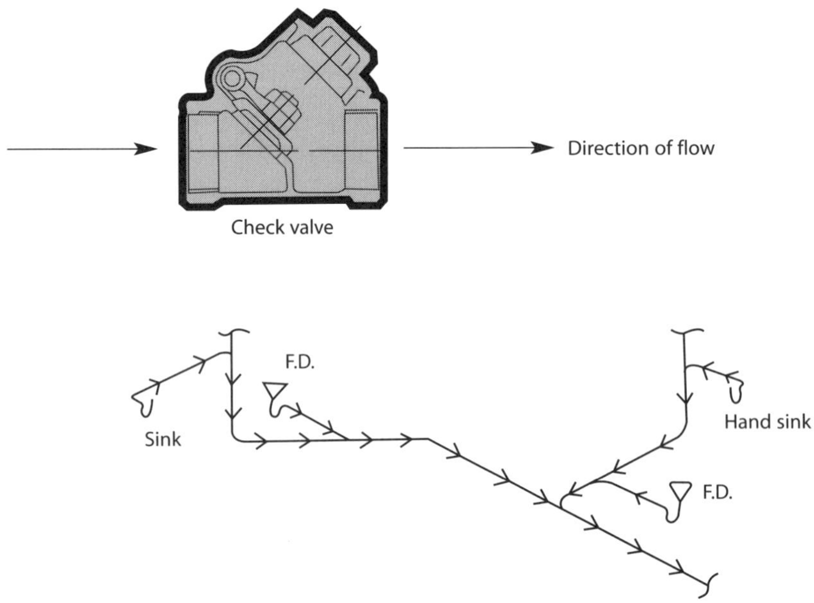


**Figure 9-17**

*Pipe and fittings showing correct direction of flow*

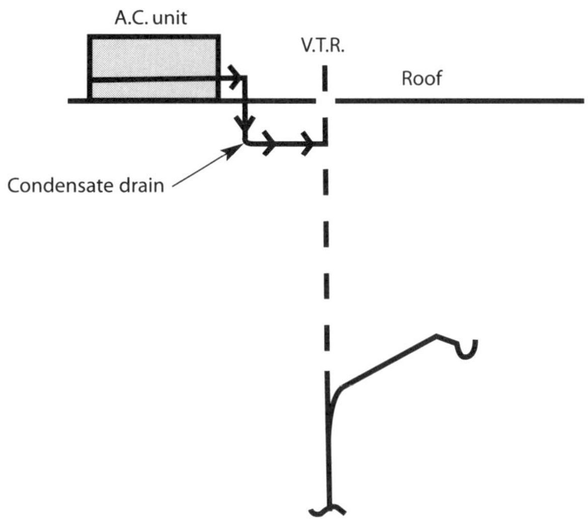


**Figure 9-18**
*Prohibited connection*

The *IPC* says the following joints and connections are unacceptable when you connect pipe with fittings. See *IPC* Section 707.

- cement or concrete joints
- mastic or hot-pour bituminous joints
- joints made with unapproved fittings for a specific installation (see earlier illustrations)
- joints that connect pipes of different outside diameters with elastomeric rolling O-rings
- solvent-cement joints between different types of plastic pipe and fittings
- saddle-type fittings

## Changes in Direction

Proper fittings guide sewage and waste in the direction of flow. Figure 9-19 has the more-common DWV fittings showing various flow directions. Here are the code-approved fittings for change in direction:

A. Horizontal to horizontal fittings — You can use $^1/_{16}$ bends, $^1/_8$ bends, $^1/_6$ bends, $^1/_4$ bends and short sweeps (only on 2-inch and smaller fixture drains), long sweeps, wyes and combination wyes and $^1/_8$ bends.

B. Vertical to horizontal fittings — You can use $^1/_{16}$ bends, $^1/_8$ bends, $^1/_6$ bends, $^1/_4$ bends, short sweeps, long sweeps, wyes, combination wyes and $^1/_8$ bends. You can only use the $^1/_4$ bend and the short sweep on 2-inch or smaller fixture drains. But you can use the short sweep on pipe that's 3 inches or larger.

C. Horizontal to vertical fittings — You can use $^1/_{16}$ bends, $^1/_8$ bends, $^1/_6$ bends, $^1/_4$ bends, short sweeps, long sweeps, sanitary wye or combination wye and $^1/_8$ bends.

## Building Sewers

*IPC* Section 701.2 says you must connect any plumbing fixtures you install in a building to a public sewer, if one is available, or to an approved private disposal system (septic tank). Any building with plumbing fixtures intended for people to use that's located on a street, alley or easement must also be connected separately to a public sewer, if one is available. Figure 9-20 shows the correct and incorrect way to make this type of installation. If you have an accessory building with plumbing fixtures located on the same lot as a primary building, you can connect to a common building sewer, as shown in Figure 9-21. See *IPC* Section 701.3. If a building sewer and storm water sewer are installed on the same lot, you can place them side by side in the same trench. See *IPC* Section 703.3.

If you're working on an old building that's being torn down or relocated, you must make sure that the existing building sewer and/or drainage pipes meet code requirements. See *IPC* Section 703.4. Don't use an existing drainage system for a new building if the old material doesn't conform to new material requirements. When renovating the plumbing in an existing building, you can try to get approval from the code official if you have a

**Figure 9-19**

*Changes of direction*

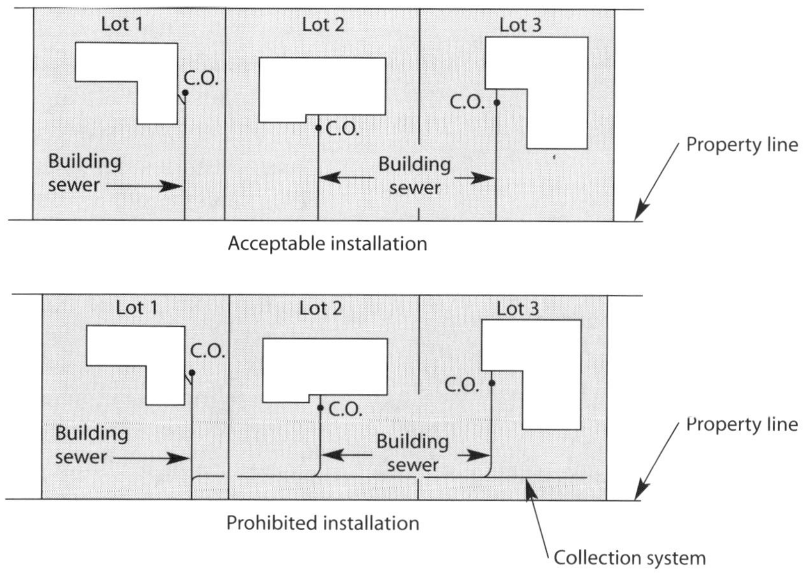


**Figure 9-20**

*Each building's drainage system must be independent*

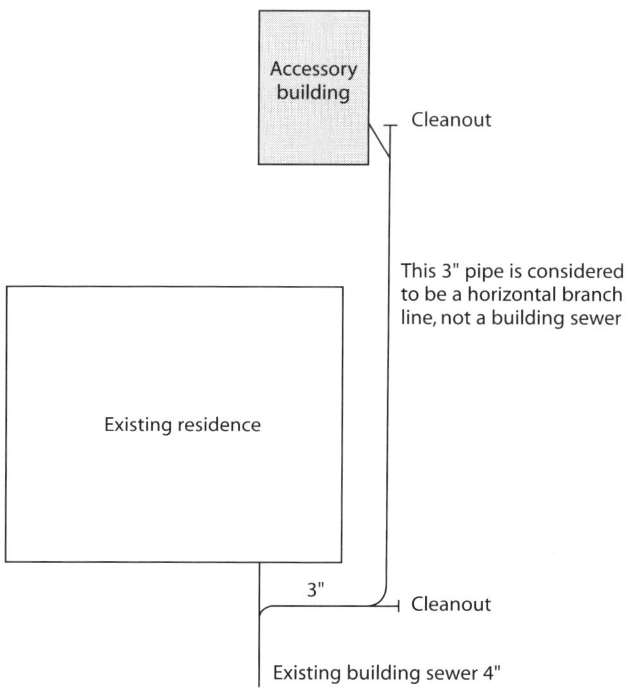


**Figure 9-21**

*Sewer line for an accessory building*

problem following code requirements. Be sure your changes won't cause a health problem.

## Installing a Building Sewer

Here are a few code tips for installing cast iron sewer pipe. Make sure the barrel of the pipe is in firm contact with the trench bottom. Excavate for the hub (bell). This will allow even distribution of the weight (backfill) along the full length of the pipe barrel. See Figure 9-22. Excavation depth isn't as important with cast iron pipe as it is with fragile pipe. Cast iron has a high resistance to trench loads. Nevertheless, you don't want to backfill the trench with large boulders, rocks, cinder and the like. This could damage or corrode the pipe. Always backfill with clean material.

Vitrified clay, plastic, asbestos-cement, copper and bituminous fiber pipe are considered fragile sewer materials. When you install these pipes, you must use special installation methods. Here are some precautions the code or manufacturer will require. Place fine material in the trench in 6-inch layers and tamp it in place. Be sure it extends at least 12

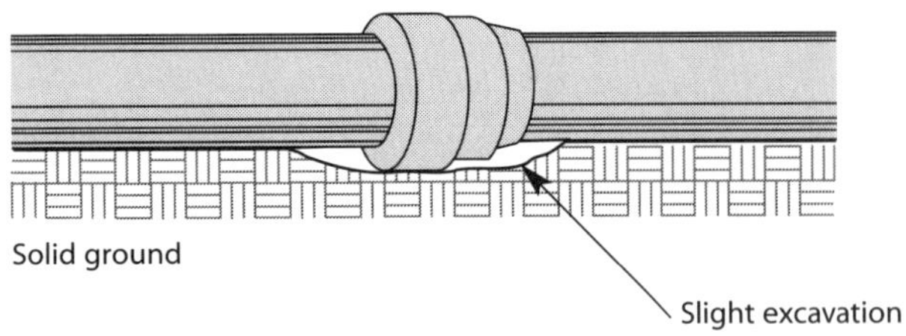


**Figure 9-22**
*Excavation for piping hub projection*

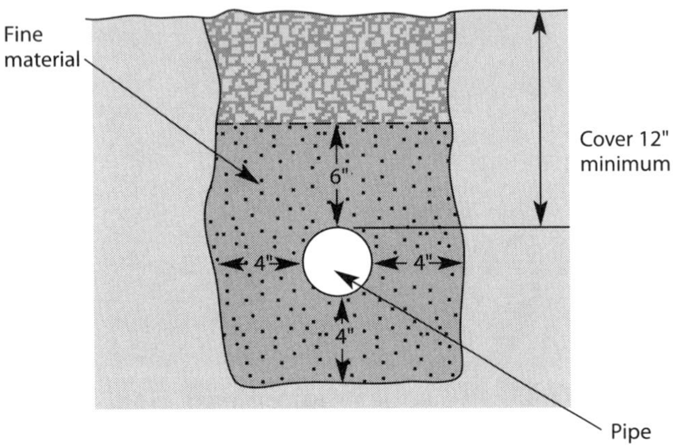


**Figure 9-23**
*Protecting fragile piping materials*

inches above the crown of the pipe. Compact the 4-inch backfill under and beside the pipe for support and to keep the pipe aligned. See Figure 9-23. Excavate for piping hub projections and couplings so the weight is evenly distributed along the pipe barrel. See *IPC* Section 306.3. Never support a sewer pipe on blocks or other loose type materials. Make sewer pipe trenches wide enough to provide adequate space for placing and joining the pipe.

Sometimes trenches are overexcavated and the bottom of the trench won't form a bed for the pipe. When this happens, backfill the trench bottom with sand or fine gravel to the installation level of the bottom of the pipe. Place it in 6-inch layers and compact it after each placement. See *IPC* Section 306.2.1.

When you dig sewer trenches in certain geographical locations, you'll run into rock. Miami-Dade in Florida is such an area because it's filled with coral rock. The code requires that you remove the rock at least 3 inches below installation level of the bottom of a sewer pipe. Remember, you must use fine sand tamped into place and brought up to the installation bottom of the sewer pipe. The code advises that no part of the pipe or fittings be in contact with the rock. See *IPC* Section 306.2.2.

Sometimes you have to install sewers in poor load-bearing soil. The code says you must remove at least two pipe diameters of this poor soil below the installation level of the bottom of a pipe. See *IPC* Section 306.2.3. Then replace the poor soil with fine gravel (crushed stone) or a concrete foundation. Use gravel, unless you're on a government job that can afford a concrete foundation for the sewer pipe. The new backfill material must provide a solid foundation for the barrel of the pipe between joints. This should pretty much take care of most sewer installations.

## Code Restrictions and Requirements

Whether you're installing underground or aboveground pipes (drainage or vent) there's a great variety of piping materials to choose from. But you'll find code restrictions when you use certain pipe and fitting materials. Here are some that are very important.

- Use the same installation methods and backfill protection to install underground piping as for sewer pipes (discussed a few paragraphs back).
- Use approved hangers or masonry supports to keep pipe in alignment and prevent sagging, for all underground or aboveground piping installations.
- Support the base of stacks with concrete, brick laid in cement mortar, metal brackets attached to a building, or other approved methods. See *IPC* Section 308.9.

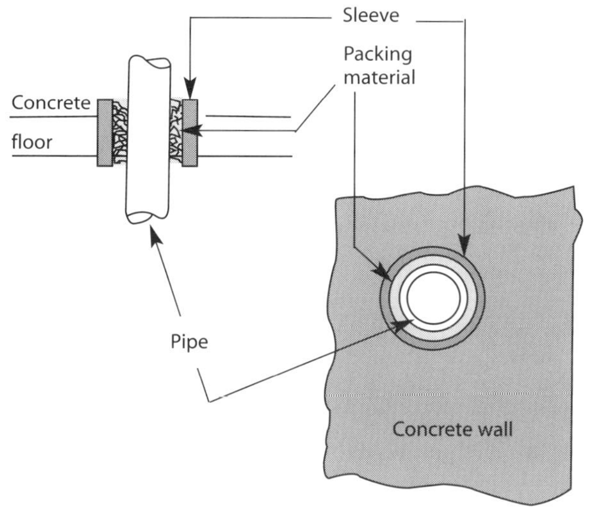


**Figure 9-24**
*Protecting pipes in poured-in-place concrete*

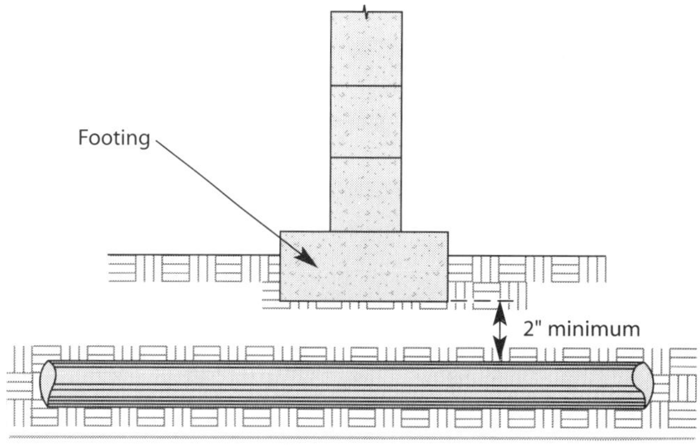


**Figure 9-25**
*Clearance for building drains beneath foundations*

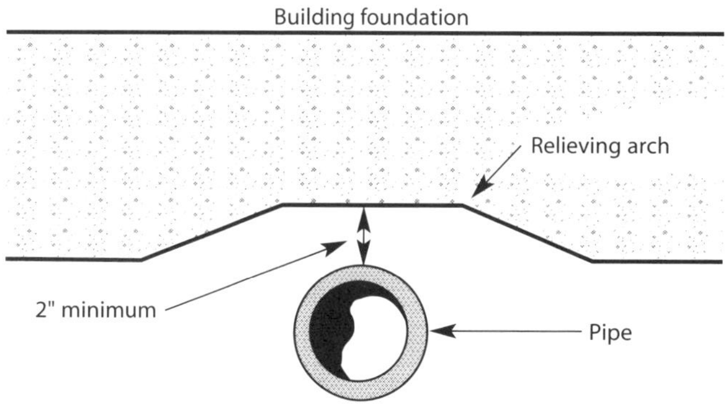


**Figure 9-26**
*Relieving arch*

- When you install drainage pipes through concrete floors or walls, use sleeves that create at least 1/2 inch of annular space around the entire circumference of the pipe. This prevents external corrosion on the pipe. Fill the annular space between the pipe and sleeve with coal tar or asphaltum compound, lead or other approved material. See Figure 9-24. If a pipe has to penetrate a firewall, fill or tightly caulk the annular space between the sleeve and the pipe. See *IPC* Section 305.4.

- To prevent corrosion to drainage or vent pipes that come in contact with cinders, concrete or other corrosive materials, it's important to use sleeves if possible. Sometimes pipes are encased in a protective sheathing or wrapping. *IPC* Section 305.1 says that the material you use to wrap the piping must be at least 0.025 inches thick.

- *IPC* Section 305.2 says you must protect pipes from breakage, but isn't specific about how to accomplish this. If the code's not specific, here's what most local authorities will accept:

If a pipe passes under a building foundation, be sure there's at least a 2-inch clearance between the top of the pipe and the bottom of the foundation. See Figure 9-25. Sometimes this isn't possible. If the architect knows this, he should design and draw on his plan a relieving arch in the building foundation, as shown in Figure 9-26. If neither is possible, *IPC* Section 305.5 will permit you to use an iron pipe sleeve two pipe sizes larger than the drainage pipe you're using. The building department will require the general contractor to use additional steel and concrete to strengthen the building foundation at this location. Some buildings may have poured-in-place stem or outer walls with no provision for hiding drainage pipes that must pass up these vertical

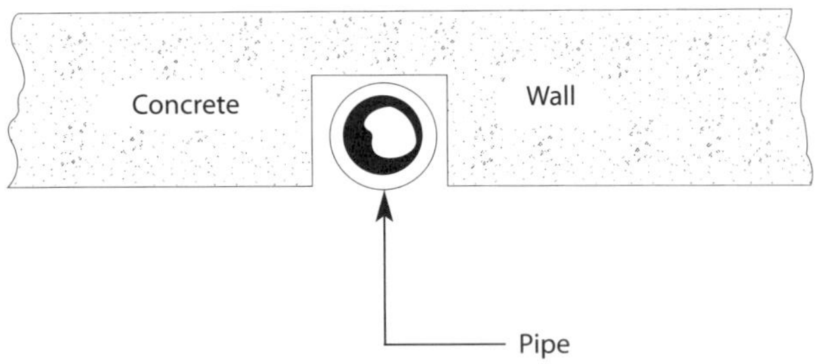


**Figure 9-27**
*One type of approved installation method*

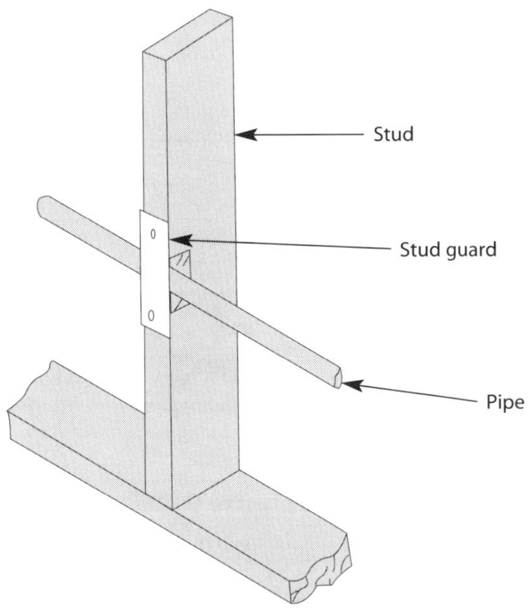


**Figure 9-28**
*Stud guard protects pipe and reinforces notched studs*

walls. Every code prohibits piping from being directly embedded in concrete walls. Figure 9-27 shows one acceptable way of boxing out a chase to leave a space for the pipe installation, with the cooperation of the general contractor.

One other thought about stress and strain on pipes. Don't expect pipes to support fixtures or any type of equipment. Actually, you must support them with approved brackets, hangers or other means directly secured to the building.

- What about freezing? See *IPC* Section 305.6. Don't install DWV pipes outside a building, in an attic or crawl space, concealed in outside walls, or any other place subject to freezing temperatures unless you provide adequate insulation or heat (or both) to protect the pipes from freezing.
- Close any openings for pipes through walls, floors or ceilings by fastening approved metal collars securely to the structure.
- Never install used drainage piping that doesn't conform to your local code's standards.
- Make all joints and connections gastight and watertight to withstand any required pressure testing.
- Don't install fittings or connections that present any abnormal obstruction to the flow of waste. That's a *bona fide* "no-no."
- If you're installing work with a difficult-to-find legal point of disposal or difficult-to-locate cleanout (for rodding) don't drill or tap the first convenient waste or vent pipe. It's against the code, and the law!
- *IPC* Section 305.6.1 says that sewer depth is between you and your code authorities.
- When you install pipes that penetrate an outside wall, make certain the opening around the pipe is properly sealed against the intrusion of rainwater. See *IPC* Section 305.7.
- *IPC* Section 305.8 says you must use shield plates (stud guards) to protect pipes (other than cast iron or galvanized) installed through holes or notches in studs, joists, rafters or similar members less than 1.5 inches from either edge of the member. Use 0.062-inch-thick plates that extend 2 inches above and below the notched area. See Figure 9-28.
- Unless you get approval for a special design from the building department, don't install any drainage pipes running parallel to a building within the 45-degree bearing plane of the footing. In other words, don't dig a

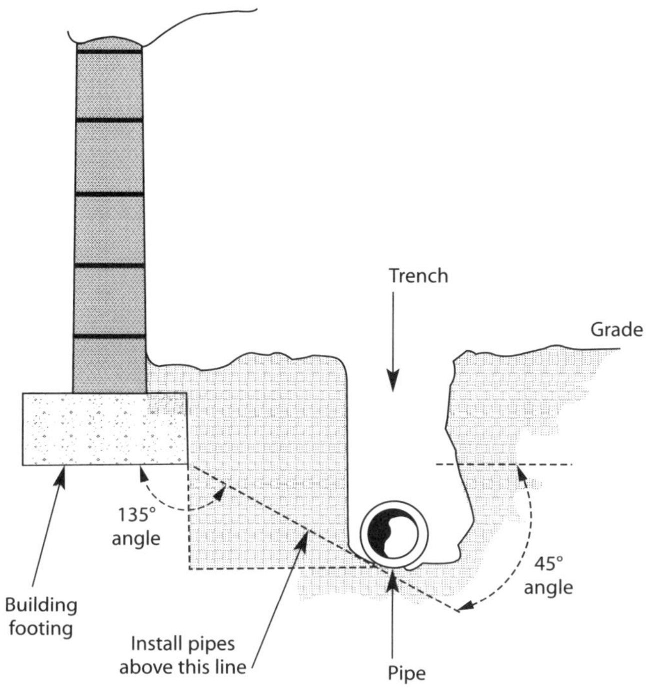


**Figure 9-29**
*45-degree angle for pipe parallel with building foundation*

drainage pipe trench parallel to a building and deeper than the building's footing that's inside of a 45-degree right triangle extending down and outward from the base of the footing. See Figure 9-29.

**TABLE 308.5**
**HANGER SPACING**

| PIPING MATERIAL | MAXIMUM HORIZONTAL SPACING (feet) | MAXIMUM VERTICAL SPACING (feet) |
|---|---|---|
| ABS pipe | 4 | 10[b] |
| Aluminum tubing | 10 | 15 |
| Brass pipe | 10 | 10 |
| Cast-iron pipe[a] | 5[a] | 15 |
| Copper or copper-alloy pipe | 12 | 10 |
| Copper or copper-alloy tubing, $1^1/_4$-inch diameter and smaller | 6 | 10 |
| Copper or copper-alloy tubing, $1^1/_2$-inch diameter and larger | 10 | 10 |
| Cross-linked polyethylene (PEX) pipe | 2.67 (32 inches) | 10[b] |
| Cross-linked polyethylene/aluminum/crosslinked polyethylene (PEX-AL-PEX) pipe | 2.67 (32 inches) | 4[b] |
| CPVC pipe or tubing, 1 inch or smaller | 3 | 10[b] |
| CPVC pipe or tubing, $1^1/_4$ inches or larger | 4 | 10[b] |
| Steel pipe | 12 | 15 |
| Lead pipe | Continuous | 4 |
| PB pipe or tubing | 2.67 (32 inches) | 4 |
| Polyethylene/aluminum/polyethylene (PE-AL-PE) pipe | 2.67 (32 inches) | 4[b] |
| PVC pipe | 4 | 10[b] |
| Stainless steel drainage systems | 10 | 10[b] |

For SI: 1 inch = 25.4 mm, 1 foot = 304.8 mm.

a. The maximum horizontal spacing of cast-iron pipe hangers shall be increased to 10 feet where 10-foot lengths of pipe are installed.

b. Midstory guide for sizes 2 inches and smaller.

From the *IPC*® ©2003, with permission of the ICC.

**Figure 9-30**
*Hanger spacing*

## *Piping Hangers and Supports*

It's important to properly support drainage and vent piping above grade to prevent sags and keep piping aligned. Earthquake areas are especially at risk. You can find information on piping supports designed for seismic forces in the *International Building Code*. Your architect should specify on the working plans all that you (a plumber) will need on your job. See *IPC* Section 308.2.

Use hangers, anchors and supports made of materials that don't promote galvanic action. See *IPC* Section 308.3. Follow regulations in *IPC* Section 308.4 for attaching supports to a building. Use *IPC* Table 308.5 shown in Figure 9-30 to figure out how to space hangers horizontally and vertically for various piping materials. Figure 9-31 shows five common plumbing pipe supports and hangers. Figure 9-32 shows hanger rod sizing. This information isn't in the *IPC*, but the values in the figure will in most cases be accepted by code officials.

*IPC* Section 703.1 says that when you install a sewer within 5 feet of building water service, you must use either cast iron pipe, copper or copper-alloy tubing, ABS plastic pipe, or PVC plastic pipe. The same guidelines apply in *IPC* Section 703.2 when

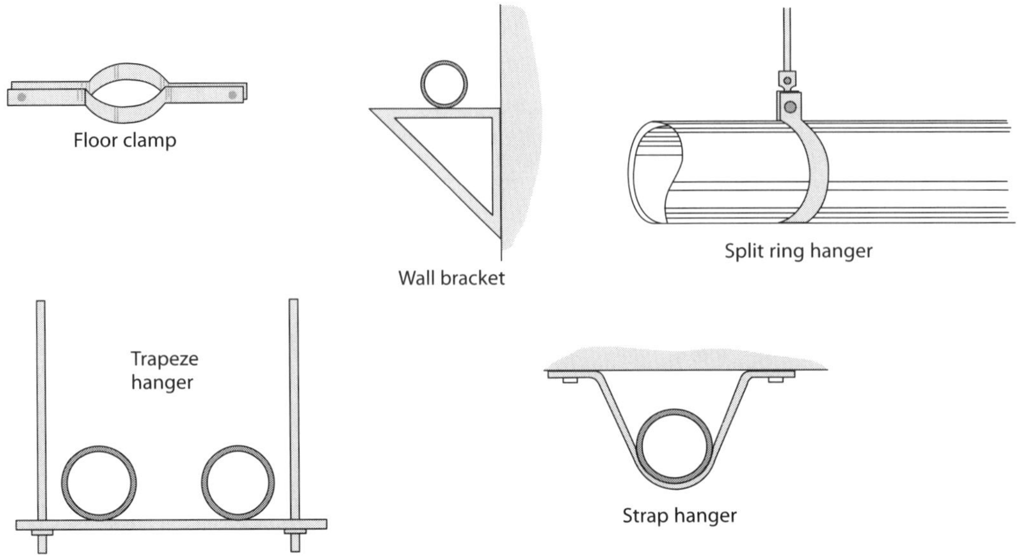


**Figure 9-31**
*Five common horizontal and vertical pipe supports*

| Pipe and tube size (in) | Rod size (in) |
|---|---|
| 1/2 to 4 | 3/8 |
| 5 to 8 | 1/2 |
| 10 to 12 | 5/8 |

**Figure 9-32**
*Hanger rod sizes*

you install building sewer or drainage pipes in filled or unstable ground.

*IPC* Section 704.5 says that when you install or remove any part of a drainage system, you can't leave any dead ends. This means that it's not acceptable for you to extend a drainage system using a blind plug to close up the pipe end having no future useful purpose. But it is acceptable if you create a dead end for some future plumbing installation, as shown in Figure 3-20, or for a cleanout extension, as shown in Figure 9-33.

## Indirect Waste Piping

When installing indirect waste piping, you can use any material approved for potable water or sanitary drainage. The most common materials are cast iron, copper (Type K, L, M), ABS or PVC plastic pipe and fittings. Figure 9-3 shows where you can use each material, above or below ground.

Always size and install indirect waste piping to accommodate the outlet drainage of any fixture or appliance (generally the same size). Though you don't need to vent indirect waste piping that doesn't have its own trap, you must never connect it directly to the sanitary drainage system.

Some indirect waste piping installed on certain equipment (meat and milk coolers) carries offensive odors to the receiving fixture floor sink or floor drain. The liquid content, flowing slowly, may create slime deposits. Make sure you install piping without sags to eliminate the potential for slime buildup and eventually stoppages. Always install a slight slope toward the receiving fixture so the piping will drain dry. Locate any receiving fixture in a

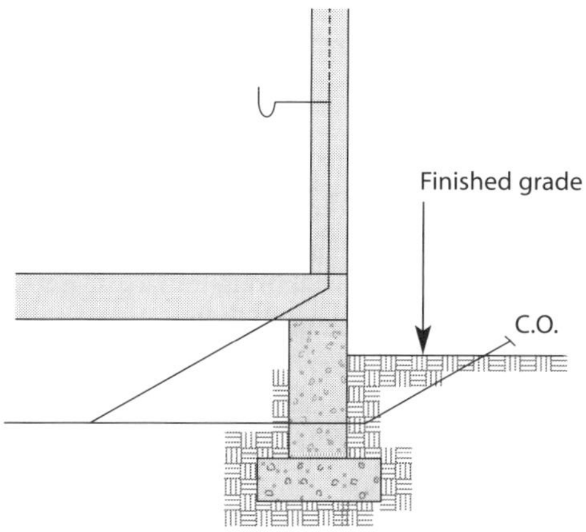


**Figure 9-33**
*Dead end created with cleanout extension*

well-ventilated area and provide accessible cleanouts for cleaning and flushing.

Whenever possible, install indirect waste piping (air break type) below the floor. Connect it to the receiving fixture above the water seal of the trap. The receiving end of the indirect waste piping should turn up through the floor as close as possible to the fixture or appliance it serves. See Figure 8-13.

When you install an indirect waste pipe above the floor, your local authority may require you to locate the indirect waste piping at least 3 inches above the floor. This will permit proper cleaning of the floor area. Terminate the pipe outlet so there's an air gap of at least 2 inches (or two pipe sizes) above the receiving fixture. See Figure 8-5. If you have a job coming up that has indirect waste piping, review the particulars in Chapter 8.

*IPC* Section 802.1.7 states that commercial dishwashers must discharge through an air gap or air break into a standpipe or waste receptor.

Remember you should never install a receiving fixture (floor drain or floor sink) for indirect waste pipe in a toilet room.

## Air Conditioning Condensate Drains

Materials used for condensate drains are generally made of Schedule 40 plastic pipe and fittings. Most any drainage pipe listed in Figure 9-3 is acceptable and can be used below or aboveground. When installing indirect waste piping for air conditioning units, locate the pipe at least 2 inches below the bottom of the building floor slab. Be sure to wait until the fill is well compacted before you begin work. After you lay the pipe on a firm base (its entire length), backfill the trench with sand. Always protect the risers passing through the floor slab with sleeves or by wrapping. You may want to review Chapter 8 and read the additional requirements for indirect waste piping before you begin an actual installation.

## Vent Pipes

In Chapter 4 we discussed vent systems. Now we'll point out some code requirements about installing vent piping. You can find approved materials for vent pipes in Figure 9-2 and where they can be used in Figure 9-3. The most commonly-used materials are plastic Schedule 40, cast iron, copper and galvanized pipe. You can use plastic, copper and cast iron pipe and fittings belowground and aboveground. Galvanized steel pipe and fittings are restricted to aboveground use only. Let's look at some other code requirements covering installation of vent piping and fittings.

## Vent Connections and Grades

You must connect all vent pipes that serve individual fixtures — or serve as a branch or circuit vent for a number of fixtures — to a vent stack, stack vent, or air admittance valve. Or you can extend them separately above the building roof to the open air. See *IPC* Section 905.1.

You must slope horizontal vent piping to drain back by gravity to the vertical vent pipe connection. See *IPC* Section 905.2. The reason for this is simple. Improper grading or sags in the pipe promote accumulation of condensation in low places. That restricts air circulation and reduces venting capacity. You should connect vertical vent piping to a

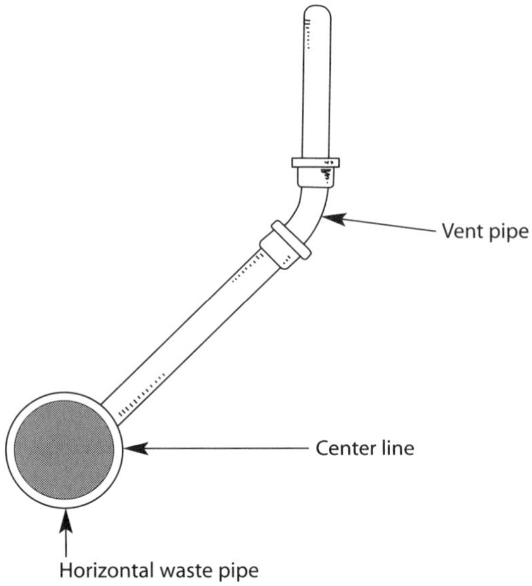


**Figure 9-34**
*Proper connection of vent pipe to horizontal waste pipe*

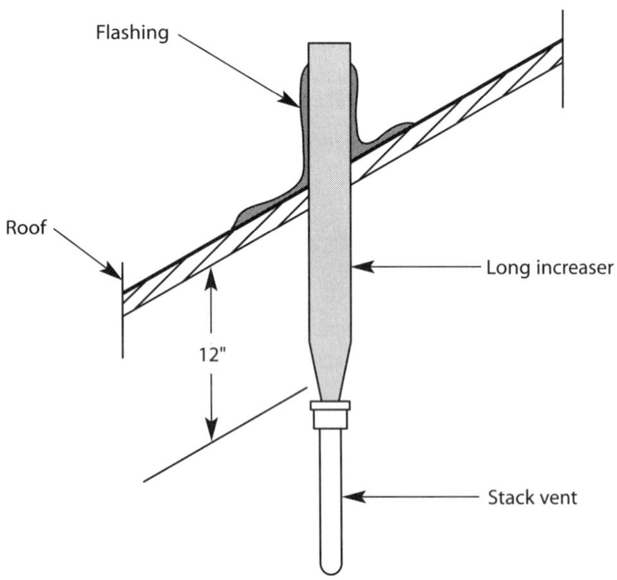


**Figure 9-35**
*Cold-climate vent terminals*

stack vent in an upward slope. That way, warmer air in the system is free to circulate upward into the stack. Trapped moist air can accelerate pipe corrosion and greatly reduce a pipe's usable life.

Don't connect dry vent pipes to the side or bottom of horizontal drainage pipes. See *IPC* Section 905.3. Instead, connect the vent pipe above the center line of the drainage pipe, as shown in Figure 9-34. The vent pipe should rise at a 45-degree angle and then straight up from the horizontal drainage pipe.

*IPC* Section 906.1 says that you have to protect each fixture trap by a vent that's located within the requirements shown in *IPC* Table 906.1. See Figure 4-32. The slope and the developed length are measured from the trap weir to the vent fitting. Connect the vent for a fixture drain above the weir of the fixture trap being vented (except fixtures with built-in integral traps such as a water closet). Review Chapter 5 for information on trap weirs.

## *Vent Terminals*

Most codes require vent pipe extensions to terminate at least 6 inches above the roof. The *IPC* doesn't state a specific height. If the roof is used for purposes such as a sun deck or play area, the vent must extend at least 7 feet above the roof. See *IPC* Section 904.1. Don't use a vent pipe as a flagpole, television aerial, clothesline or anything similar, unless it's been anchored properly and approved by your plumbing department. See *IPC* Section 904.4.

Be careful with pipes that penetrate the roof of a building. Use lead, copper, galvanized steel, aluminum, plastic or other flashing materials approved by your local authority to make them watertight. See *IPC* Section 305.7. If you use sheet copper, it must weigh 8 ounces per square foot. If you use sheet lead flashings made at the job site, it must weigh 3 pounds per square foot. If you use manufactured flashings, they must weigh 2.5 pounds per square foot. See *IPC* Section 902.3.

In climates where temperatures can drop to less than 0 degrees F, frost is likely to form in a vent stack and obstruct the free flow of air in the plumbing system. To eliminate this possibility, make each vent through the roof or wall at least 3 inches in diameter. Start your increase in pipe size at least 12 inches inside the building. See *IPC* Section 904.2. See Figure 9-35 for a long increaser fitting used to enlarge vent diameter. The long increaser fitting may have either a spigot end for caulking into a hub or threads for joining to a steel pipe threaded vent stack. Protect (using insulation, heat, or both) any vent pipes installed on outside walls of a building. See *IPC* Section 904.7.

Never install a vent terminal for a sanitary system directly beneath any door, window, or air intake opening of a building. Actually, a vent terminal has to be 10 feet horizontally from any such opening. If it's closer than that, you'll have to extend it at least 2 feet above the opening. See *IPC* Section 904.5 and Figure 9-36.

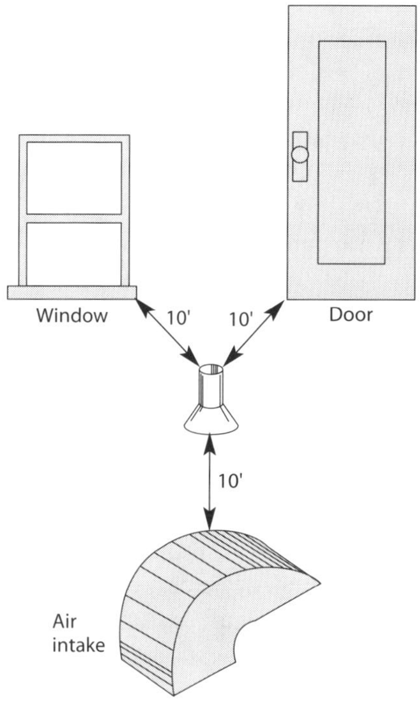


**Figure 9-36**
*Proper separation from a vent terminal*

Any vent extension through a building sidewall must terminate at least 10 feet from any lot line and be at least 10 feet above ground level. Don't terminate a vent pipe under any overhang of a building having soffit vents. Be sure to place a screen on sidewall vent terminals to keep birds or rats from entering or blocking up the opening. See *IPC* Section 904.6.

Your local code will typically require that a roof-penetrating vent pipe be at least 2 feet from any vertical wall, as shown in Figure 9-37. This ensures that you can properly "dry in" the roof penetration. The *IPC* doesn't cover this type of installation.

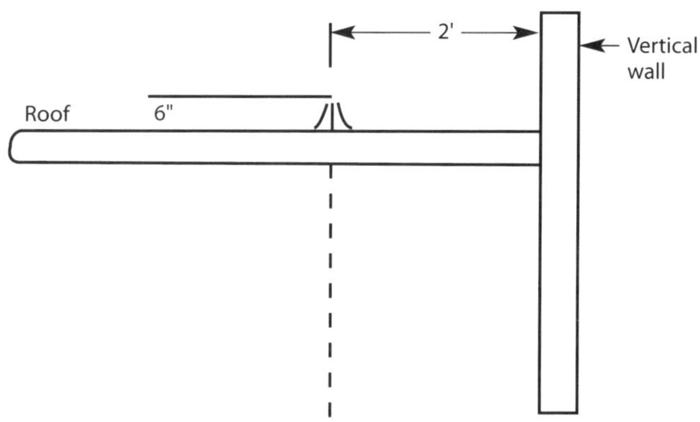


**Figure 9-37**
*Separation of vent pipe terminal from vertical wall*

*IPC* Section 312 tells how to test drainage and vent systems before the piping is covered up. Here are the main requirements:

- The permit holder (most likely the plumbing contractor) must make the arrangements. He gives advance notice to the plumbing inspector that the work is ready for inspection. He furnishes everything (including labor) for the inspection. He makes sure that the full plumbing system is tested by water or air. The plumbing inspector may require you to remove cleanout or test plugs to make sure the pressure reaches all parts of the drainage system. After all plumbing fixtures have been set and the fixture traps filled with water, the entire plumbing system must be submitted to a final test.

- When needed, you'll be smart to test the system in part or in whole with water. The system must be tightly plugged and withstand a 10-foot head of water for at least 15 minutes. In a high-rise building you'll have to test the pipes in sections. Each section must be tested with a 10-foot head of water and must hold for 15 minutes, without leaking.
- When water isn't available, you can arrange to air test a DWV system in part or in whole. Be sure it's tightly plugged and can withstand an air pressure of 5 psi for at least 15 minutes, without leaking.
- The final inspection is visual and the plumbing inspector must be satisfied that it complies with code requirements. If he has any cause for concern, he can request a smoke test. To do this, fill all fixture traps with water and plug all openings. Use a smoke machine to fill the system with pungent thick smoke. While the system is filled with smoke, it must withstand a pressure equal to a 1-inch water column for at least 15 minutes.
- When you test a gravity sewer, plug it up so it'll hold at least a 10-foot head of water for 15 minutes.
- When you test a forced sewer, it must be plugged up and withstand for 15 minutes a pressure that's 5 psi greater than the pumping equipment rating.
- Test a storm drainage system the same way as a DWV system.
- Backflow prevention assemblies and air gaps must have an annual inspection to determine if they are operable.

# *Chapter 9 — Questions and Answers*

**9-1 When installing hubless cast iron pipe and fittings in a drainage system, you may join the pipe and/or fittings with**

(A) a compression gasket joint.

(B) a lead and oakum joint.

(C) mechanical joint couplings.

(D) elastomeric rolling O-rings.

***Answer:* C**

*Code response: Section 705.5.3*
*This book, page 140*

**9-2 When making a joint between asbestos-cement pipe and fittings, the *IPC* requires that you use a**

(A) sleeve coupling of the same material as the pipe.

(B) coextruded-type joint.

(C) mechanical joint coupling.

(D) threaded joint connection.

***Answer:* A**

*Code response: Section 705.3*
*This book, page 141*

**9-3 When wiping lead pipe to a ferrule, the *IPC* requires the ferrule be made of**

(A) a copper and brass alloy.

(B) 50 percent white brass and 50 percent lead.

(C) stainless steel.

(D) red brass.

***Answer:* D**

*Code response: Section 705.18*
*This book, page 145*

**9-4 When working with water or drainage piping, the only *IPC*-approved joints you can use are**

(A) cement or concrete joints.

(B) PVC Schedule 80 threaded joints.

(C) mastic or hot-pour bituminous joints.

(D) saddle-type fittings.

***Answer:* B**

*Code response: Sections 705.2.3 and 707.1*
*This book, pages 141 and 147*

**9-5 The *IPC* prohibits using double sanitary tee pattern fittings in a sanitary drainage system receiving discharge from back-to-back**

(A) lavatories.

(B) showers.

(C) sinks.

(D) water closets.

***Answer: D***

*Code response: Section 706.3*
*This book, page 145*

**9-6 When backfilling a drainage pipe trench, the *IPC* requires that you place fill material in 6-inch layers and tamp the material in place until the crown of the pipe is covered by at least ______ of tamped earth.**

(A) 12 inches

(B) 14 inches

(C) 16 inches

(D) 18 inches

***Answer: A***

*Code response: Section 306.3*
*This book, pages 149 and 150*

**9-7 If there's poor load-bearing soft material in the bottom of a trench for a 4-inch drainage pipe, the *IPC* requires that you over-excavate at least _______ and backfill with fine gravel.**

(A) 5 inches

(B) 6 inches

(C) 7 inches

(D) 8 inches

***Answer: D***

*Code response: Section 306.2.3*
*This book, page 150*

**9-8 When a vent terminal from a drainage system is located directly below an openable window, the *IPC* requires it to extend at least _______ inches above the opening.**

(A) 6

(B) 12

(C) 18

(D) 24

***Answer: D***

*Code response: Section 904.5*
*This book, page 157*

**9-9 The *IPC* does not permit ________ to be used for building sewer piping installed within 5 feet of a water service pipe.**

(A) polyolefin plastic pipe

(B) PVC plastic pipe

(C) ABS plastic pipe

(D) copper-alloy tubing

***Answer:* *A***

*Code response: Section 703.1*
*This book, page 153*

**9-10 According to the *IPC*, you should never connect a dry vent pipe to**

(A) a vent stack.

(B) the side of a horizontal drainage pipe.

(C) the side of a stack vent.

(D) the bottom of a circuit vent.

***Answer:* *B***

*Code response: Section 905.3*
*This book, page 156*

◆ Chapter 10 ◆

# Water Supply and Distribution System

Code authorities always insist that all new buildings connect to a public water main if there is one. But an existing building that has its own private water system (a well) won't have to unless, or until, a building permit is required for an addition or remodeling. When you obtain a permit for new work, most local jurisdictions will require that you connect to the public water system. The water main for a public water distribution system is usually located in the street, alley or dedicated easement beside each privately-owned lot or parcel of land. The public water distribution system is installed, maintained and controlled by the local utility authority.

The water main provides potable water at a pressure that meets required code standards, usually from about 40 to 60 psi. If the pressure will be over 80 psi, install a pressure-reducing valve in the water-service line where it enters the building. See *IPC* Section 604.8.

Sometimes the pressure at the public water main is too low for plumbing fixtures to work properly. This is often because the size of the original water main can't supply enough pressure for new construction in the area. If this happens, the *IPC* says you must install a water pressure booster pump to provide flow pressures at fixture outlets as required in *IPC* Table 604.7, shown in Figure 10-1.

When a permit holder (usually the plumbing contractor) is ready to connect to the public water main, the utility company will tap the water main. They'll install a water meter at the property line to measure water usage. The plumber connects the building water service pipe to the outlet side of the water meter. Figure 10-2 shows a typical water service connection.

## Sizing a Building Water System

National and local codes provide some basics you can follow to figure out the maximum water demand in a small or simple job. But if you run into a large or complicated job, you'll have to count on the expertise of a plumbing engineer. *IPC* Appendix E (Section E101.1.2) says engineers and local authorities can design and size their own water systems without being rigidly tied down to a complicated code process. Here's what the code says:

"Because of the variable conditions encountered in hydraulic design, it is impractical to specify definite and detailed rules for sizing of the water piping system. Accordingly, other sizing or design methods conforming to good engineering practice standards are acceptable alternatives to those presented herein."

**TABLE 604.3**
**WATER DISTRIBUTION SYSTEM DESIGN CRITERIA**
**REQUIRED CAPACITIES AT FIXTURE SUPPLY PIPE OUTLETS**

| FIXTURE SUPPLY OUTLET SERVING | FLOW RATE[a] (gpm) | FLOW PRESSURE (psi) |
|---|---|---|
| Bathtub | 4 | 8 |
| Bidet | 2 | 4 |
| Combination fixture | 4 | 8 |
| Dishwasher, residential | 2.75 | 8 |
| Drinking fountain | 0.75 | 8 |
| Laundry tray | 4 | 8 |
| Lavatory | 2 | 8 |
| Shower | 3 | 8 |
| Shower, temperature controlled | 3 | 20 |
| Sillcock, hose bibb | 5 | 8 |
| Sink, residential | 2.5 | 8 |
| Sink, service | 3 | 8 |
| Urinal, valve | 15 | 15 |
| Water closet, blow out, flushometer valve | 35 | 25 |
| Water closet, flushometer tank | 1.6 | 15 |
| Water closet, siphonic, flushometer valve | 25 | 15 |
| Water closet, tank, close coupled | 3 | 8 |
| Water closet, tank, one piece | 6 | 20 |

For SI: 1 psi = 6.895 kPa, 1 gallon per minute (gpm) = 3.785 L/m.

a. For additional requirements for flow rates and quantities, see Section 604.4.



**Figure 10-1**

*Water distribution system design criteria required capacities at fixture supply pipe outlets*

For jobs that don't require an engineer's design, estimate the maximum demand carefully so there'll be enough water supplied to all fixtures. You want to avoid undersizing, but you also don't want to oversize and add unnecessarily to the cost to the owner.

## *Sizing Those Pipes*

I'll show you a simple way to size a building's water supply system. The method uses the principle of demand load based on the maximum flow rate and pressure for certain fixtures and fittings. Plumbing plans examiners and inspectors use this method to check building water supply pipe sizes. Using this method, you'll be sure to have enough water pressure to supply the highest and farthest fixtures during peak demand. When you size a supply system, make sure you consider the maximum water consumption flow rate and flow pressures required for the fixtures to operate efficiently. If the water pressure at street level is at least 50 psi, you can use this method to size practically all one- and two-family dwellings, most multi-family dwellings not over two stories high, and many small one- or two-story commercial and light industrial buildings.

In these buildings, 50 psi is usually more than enough to overcome the static head and ordinary pipe friction losses. In your local code you'll find sizing tables similar to *IPC* Table 604.3 and Figures 10-4, 10-5 and 10-6.

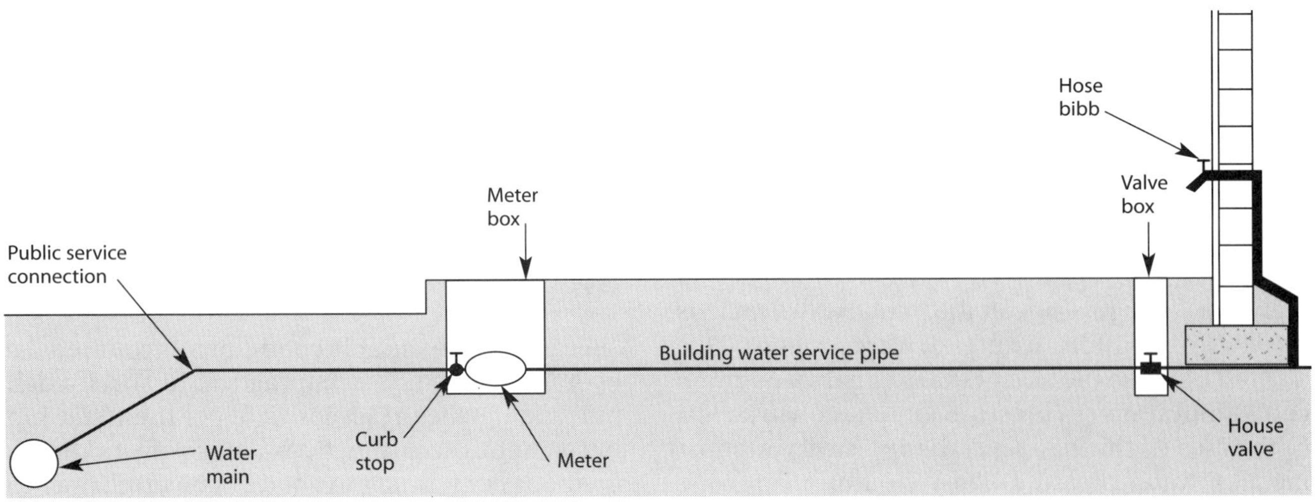


**Figure 10-2**

*Typical water service connection*

**TABLE 604.4**
**MAXIMUM FLOW RATES AND CONSUMPTION FOR PLUMBING FIXTURES AND FIXTURE FITTINGS**

| PLUMBING FIXTURE OR FIXTURE FITTING | MAXIMUM FLOW RATE OR QUALITY[b] |
|---|---|
| Lavatory, private | 2.2 gpm at 60 psi |
| Lavatory, public, (metering) | 0.25 gallon per metering cycle |
| Lavatory, public (other than metering) | 0.5 gpm at 60 psi |
| Shower head[a] | 2.5 gpm at 80 psi |
| Sink faucet | 2.2 gpm at 60 psi |
| Urinal | 1.0 gallon per flushing cycle |
| Water closet | 1.6 gallons per flushing cycle |

For SI: 1 gallon = 3.785 L, 1 gallon per minute = 3.785 L/m, 1 pound per square inch = 6.895 kPa.

a. A hand-held shower spray is a shower head.

b. Consumption tolerances shall be determined from referenced standards.

From the *IPC*® ©2003, with permission of the ICC.

**Figure 10-3**

*Maximum allowable water usage for plumbing fixtures*

First, you need to know the minimum pressure needed at the most remote and highest water outlet. Then find the minimum pressure required for normal fixture use at the highest water outlets during periods of normal flow conditions. For single family homes and one- and two-story residential buildings, most plumbing fixtures require 8 psi. In most commercial buildings, the urinals and water closets will have flush valves and they'll require a minimum working pressure of 15 psi, according to *IPC* Table 604.3 (Figure 10-1). You'll also find the flow rate (gpm) and flow pressure (psi) for other types of water-supplied fixtures in this table.

Even with the new water conservation measures, the code still uses the same minimum pressure for regular fixtures as those shown in the table. But where water conservation is strictly enforced, you must reduce the flow rate shown in Figure 10-1 to the ones in Figure 10-3, *IPC* Table 604.4. But this adjustment probably won't affect the size of water service and water distribution pipes. Check with your local authority on this. If any water supply system you work with uses pipe sizes larger than those we cover in this chapter, leave the sizing to a professional engineer.

## Sizing the Water Service Pipe

For fixtures to operate properly, the water service pipe has to carry enough water, moving fast enough, to the water distribution pipes throughout the building. The water service pipe size continues within the building to the first distribution pipe. Then, whether it's horizontal or vertical, it becomes the building water main. After you connect other distribution branch pipes into the building main, using proper fittings, you can reduce the size of the main. As the main goes through the building, the demand on the line will usually decrease.

*IPC* Section 603.1 says you must use building water service pipe that's at least 3/4-inch. This generally works for single-family residences with up to two bathrooms, or for businesses with no more than two water closets equipped with tanks. Figures 10-4 and 10-5 show recommended water service pipe sizes, meter sizes, and approximate pressure loss, for residential and commercial buildings. These sizes apply only where water main pressure at street level never falls below 50 psi. Figure 10-6 shows the minimum sizes for a fixture supply pipe from the main, or riser, to the wall opening. Here are some limitations on fixtures that you have to consider as you figure pipe size:

- *IPC* Section 604.5 limits the length of a fixture supply tube from a water supply pipe outlet to 30 inches.
- The *IPC* doesn't specify the number of fixtures you can connect to a 1/2-inch water supply pipe. Most local codes allow up to two fixtures to connect to a 1/2-inch water supply pipe.
- Where water main pressure may fall below 50 psi and a building is more than two stories high, it's best to increase pipe sizes to the next larger size.

| Number of bathrooms and kitchens — Tank-type closets | | Diameter of water service pipe | Recommended meter size | Approximate pressure loss, meter and 100' of pipe | Number of bathrooms and kitchens — Flush valve closets | |
|---|---|---|---|---|---|---|
| Copper | Galvanized | (in) | (in) | (psi) | Copper | Galvanized |
| 1-2 | — | $\frac{3}{4}$ | $\frac{5}{8}$ | 27 | — | — |
| — | 1-2 | $\frac{3}{4}$ | $\frac{5}{8}$ | 40 | — | — |
| — | — | 1 | 1 | 30 | 1 | — |
| 3-4 | — | 1 | 1 | 22 | — | — |
| — | 3-4 | 1 | 1 | 24 | — | — |
| — | — | $1\frac{1}{4}$ | 1 | 32 | 2-3 | — |
| — | — | $1\frac{1}{4}$ | 1 | 36 | — | 1-2 |
| 5-9 | — | $1\frac{1}{4}$ | 1 | 28 | — | — |
| — | 5-8 | $1\frac{1}{4}$ | 1 | 32 | — | — |
| — | — | $1\frac{1}{2}$ | $1\frac{1}{2}$ | 29 | 4-10 | — |
| — | — | $1\frac{1}{2}$ | $1\frac{1}{2}$ | 30 | — | 3-7 |
| 10-16 | — | $1\frac{1}{2}$ | $1\frac{1}{2}$ | 17 | — | — |
| — | 9-14 | $1\frac{1}{2}$ | $1\frac{1}{2}$ | 21 | — | — |
| — | — | 2 | $1\frac{1}{2}$ | 26 | 11-18 | — |
| — | — | 2 | $1\frac{1}{2}$ | 32 | — | 8-18 |
| 17-38 | — | 2 | $1\frac{1}{2}$ | 27 | — | — |
| — | 15-38 | 2 | $1\frac{1}{2}$ | 32 | — | — |
| — | — | 2 | 2 | 25 | 19-33 | — |
| — | — | 2 | 2 | 24 | — | 19-24 |
| 39-56 | — | 2 | 2 | 25 | — | — |
| — | 39-45 | 2 | 2 | 24 | — | — |
| — | — | $2\frac{1}{2}$ | 2 | 28 | 34-57 | — |
| — | — | $2\frac{1}{2}$ | 2 | 32 | — | 25-57 |
| 57-78 | — | $2\frac{1}{2}$ | 2 | 28 | — | — |
| — | 46-78 | $2\frac{1}{2}$ | 2 | 32 | — | — |
| — | — | 3 | 3 | 16 | 58-95 | — |
| — | — | 3 | 3 | 19 | — | 58-95 |
| 79-120 | — | 3 | 3 | 16 | — | — |
| — | 79-120 | 3 | 3 | 19 | — | — |

**Figure 10-4**

*Minimum water service pipe size for one- and two-story buildings (hotels, motels, and residential occupancy only)*

Note in Figures 10-4 and 10-5 that you can install more fixtures on a copper system (or plastic system) than on a galvanized iron or steel system of the same size. That's because the interior of the pipe is smoother, so there's less friction. In general, the pressure loss is reduced or stays the same as the size of a water service pipe increases.

## Some Sizing Examples

Example 1 is the two-bath single-family residence piping diagram shown in Figure 10-7. I based the sizing of this piping on *IPC* Table 604.3, and Figures 10-4 and 10-6. Follow along as we size this building. Figure 10-1 gives us the flow rate and flow pressure rate. The maximum flow pressure needed at any one time, and any one fixture, is 8 psi. Looking at Figure 10-4, this two-bath residence requires a minimum $\frac{3}{4}$-inch water service. Figure 10-6 tells us the water supply pipe outlet sizes needed to serve each type fixture. Remember, it's a good idea to limit two fixtures to a $\frac{1}{2}$-inch pipe. Since we have a $\frac{3}{4}$-inch water service, we have to continue this size and connect to the water heater in accordance with the manufacturer's $\frac{3}{4}$-inch opening for the cold water connection. Continue the $\frac{3}{4}$-inch building water main to each bathroom because there are three fixtures in each one. The hose bibbs may be $\frac{1}{2}$-inch. Some local

| Number of fixture units Flush tank water closet | | Service size (in) | Meter size (in) | Approximate pressure loss (psi) | Number of fixture units Flush valve water closet | |
|---|---|---|---|---|---|---|
| Copper | Galvanized iron or steel | | | | Copper | Galvanized iron or steel |
| 18 | — | $\frac{3}{4}$ | $\frac{5}{8}$ | 30 | — | — |
| — | 15 | $\frac{3}{4}$ | $\frac{5}{8}$ | 30 | — | — |
| 19-55 | — | 1 | 1 | 30 | — | — |
| — | 16-36 | 1 | 1 | 30 | — | — |
| — | — | 1 | 1 | 30 | 9 | — |
| 56-84 | — | $1\frac{1}{4}$ | 1 | 30 | — | — |
| — | 37-67 | $1\frac{1}{4}$ | 1 | 30 | — | — |
| — | — | $1\frac{1}{4}$ | 1 | 30 | 10-20 | — |
| — | — | $1\frac{1}{4}$ | 1 | 30 | — | 14 |
| 86-255 | — | $1\frac{1}{2}$ | $1\frac{1}{2}$ | 30 | — | — |
| — | 68-175 | $1\frac{1}{2}$ | $1\frac{1}{2}$ | 30 | — | — |
| — | — | $1\frac{1}{2}$ | $1\frac{1}{2}$ | 30 | 21-77 | — |
| — | — | $1\frac{1}{2}$ | $1\frac{1}{2}$ | 30 | — | 15-52 |
| 226-350 | — | 2 | $1\frac{1}{2}$ | 30 | — | — |
| — | 176-290 | 2 | $1\frac{1}{2}$ | 30 | — | — |
| — | — | 2 | $1\frac{1}{2}$ | 30 | 78-175 | — |
| — | — | 2 | $1\frac{1}{2}$ | 30 | — | 53-122 |
| 351-550 | — | 2 | 2 | 30 | — | — |
| — | 291-450 | 2 | 2 | 30 | — | — |
| — | — | 2 | 2 | 30 | 176-315 | — |
| — | — | 2 | 2 | 30 | — | 123-227 |
| 551-640 | — | $2\frac{1}{2}$ | 2 | 30 | — | — |
| — | 450-580 | $2\frac{1}{2}$ | 2 | 30 | — | — |
| — | — | $2\frac{1}{2}$ | 2 | 30 | 316-392 | — |
| — | — | $2\frac{1}{2}$ | 2 | 30 | — | 228-343 |
| 641-1340 | — | 3 | 3 | 22 | — | — |
| — | 581-1125 | 3 | 3 | 22 | — | — |
| — | — | 3 | 3 | 22 | 393-940 | — |
| — | — | 3 | 3 | 22 | — | 344-895 |

**Figure 10-5**

*Minimum water service pipe size for one- and two-story buildings (commercial use)*

| Type of fixture or device | Pipe size (in) | Type of fixture or device | Pipe size (in) |
|---|---|---|---|
| Bathtub | $\frac{1}{2}$ | Laundry tray - up to three compartments | $\frac{1}{2}$ |
| Combination sink and tray | $\frac{1}{2}$ | | |
| Drinking fountain | $\frac{3}{8}$ | Shower (single head) | $\frac{1}{2}$ |
| Dishwasher (domestic) | $\frac{1}{2}$ | Sink (service, mop) | $\frac{1}{2}$ |
| Hose bibbs | $\frac{1}{2}$ or $\frac{3}{4}$ | Sink (flushing rim) | $\frac{3}{4}$ |
| Hot water heater | $\frac{3}{4}$ min. | Urinal (flush tank) | $\frac{1}{2}$ |
| Kitchen sink, residential | $\frac{1}{2}$ | Urinal (direct flush valve) | $\frac{3}{4}$ |
| Kitchen sink, commercial, one or more compartments | $\frac{3}{4}$ | Water closet (tank type) | $\frac{3}{8}$ |
| Lavatory | $\frac{3}{8}$ | Water closet (flush valve type) | 1 |

**Figure 10-6**

*Size of fixture supply pipe*

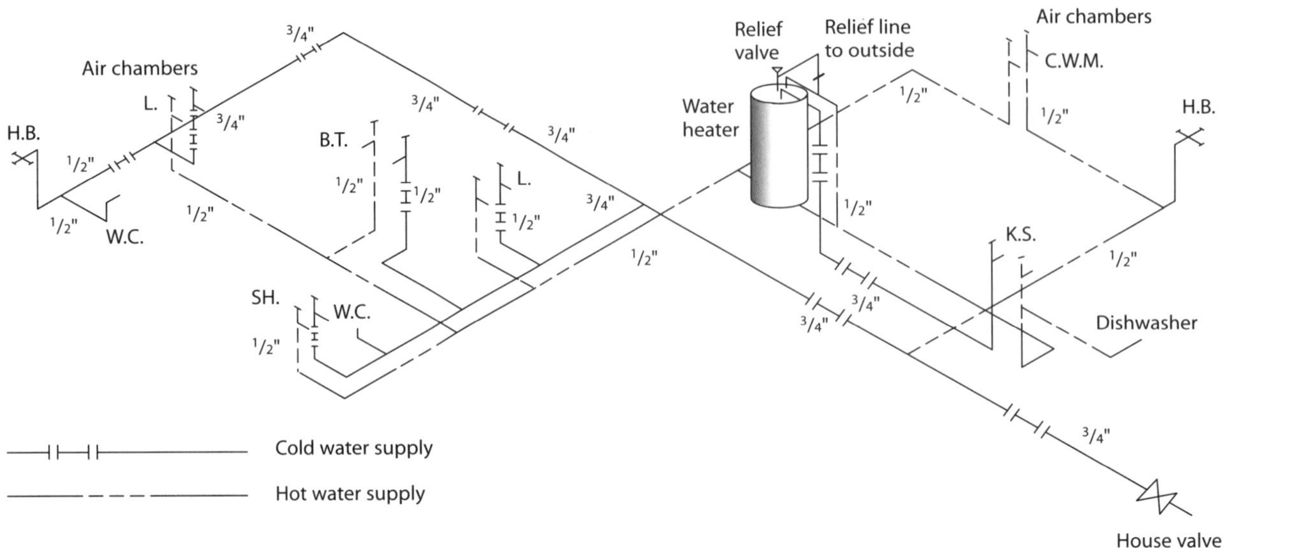


**Figure 10-7**

*Example 1 — Water distribution piping diagram*

codes require $^3/_4$-inch hose bibbs. If you aren't sure, check with your inspector.

A $^1/_2$-inch hot water pipe will probably work for this job. Be sure you show the difference between hot and cold water pipes in piping diagrams. A solid line with one dash stands for cold water and a solid line with two dashes stands for hot water, as shown in Figure 10-7.

Example 2 is the two-bath residence shown in Figure 10-8. Let's figure on a street level pressure of 50 psi. We'll choose copper as our piping material. There are 18 individual sections of pipe to size, called A through R. Use Figure 10-4 to size the water service pipe and the building water main. Use Figure 10-6 to size the fixture water supply pipes. Start at the remote outlet (A) and work back toward the building water service pipe (R).

Done? Now look at Figure 10-9 for the answers.

Example 3 is the apartment building shown in Figure 10-10. We'll use Figure 10-4 as our guide. The single water supply riser serves eight kitchens and eight baths with tank-type water closets (two on each floor). We'll choose copper as our water piping material. The street level water main pressure is 55 psi. Without looking ahead to find the answers, figure the pipe sizes for Sections A through D. Start at the most remote outlet (A) and work back toward the building main (E), adding up the demand for the apartment units on each floor. In the first column in Figure 10-4, you'll see that you should use the next larger pipe size because the building is over two stories high. After you've figured your answer, look at Figure 10-11 to see if you're correct.

Example 4 is the apartment building shown in Figure 10-12. Again, we'll use Figure 10-4 as our guide. Let's say you're sizing the cold water service piping in this apartment building. You don't have to consider the hot water piping to size the water service and distribution lines. The horizontal and vertical piping in this example serves 18 kitchens and 18 baths. We'll assume the street level water pressure is 55 psi. This time, let's use galvanized water piping. Try sizing the 20 sections of pipe lettered A through T in Figure 10-12 without looking ahead to find the right answer. Be sure to start at the most remote outlet (A) and work toward the water meter (T). Add up the total demand for the apartment units. After you've worked it out on your own, look at my answers in Figure 10-13, on page 171.

It's very important that you know how to use *IPC* Table 604.3, and Figures 10-4 through 10-6 and other relevant code sections that apply to sizing the piping for water supply systems — especially those buildings that aren't designed by an engineer.

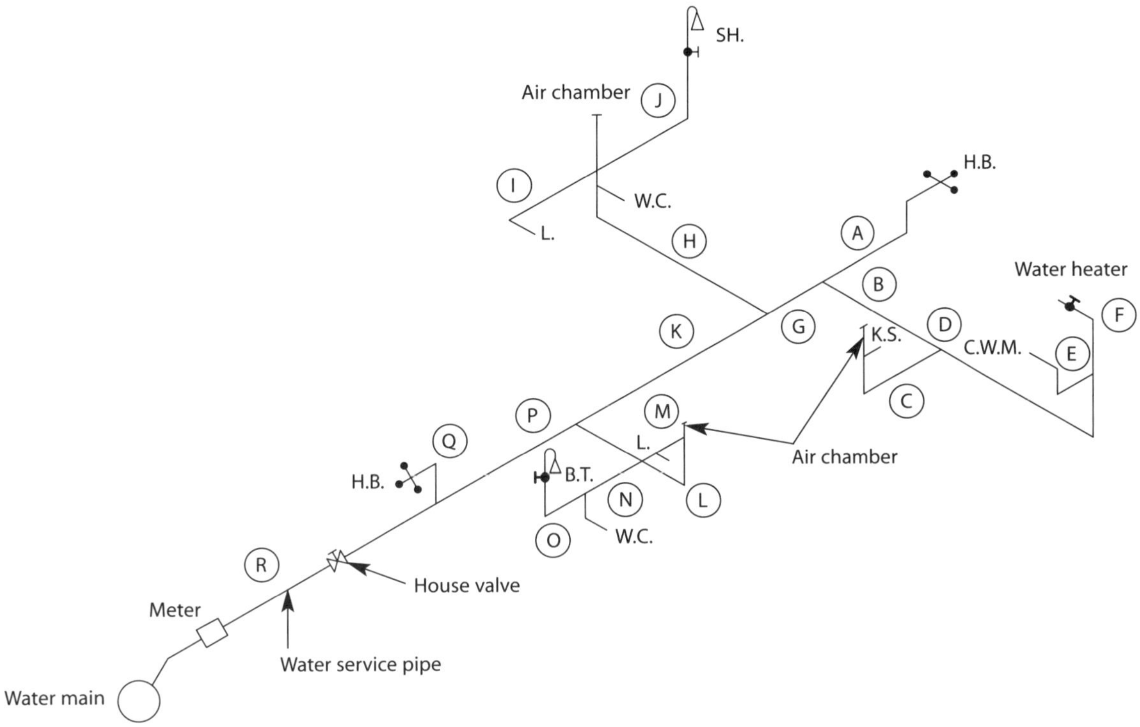


**Figure 10-8**

*Example 2 — Two-bath residence piping diagram*

| | | | | | |
|---|---|---|---|---|---|
| A | hose bibb | ½" or ¾" | J | one fixture | ½" |
| B | branch line | ¾" | K | building water main | ¾" |
| C | one fixture | ½" | L | bathroom supply | ¾" |
| D | branch line | ¾" | M | over 2 fixtures | ¾" |
| E | one fixture | ½" | N | two fixtures | ½" |
| F | water heater | ¾" | O | one fixture | ½" |
| G | building water main | ¾" | P | building water main | ¾" |
| H | bathroom supply | ¾" | Q | hose bibb | ½" or ¾" |
| I | one fixture | ½" | R | water service pipe | ¾" |

**Figure 10-9**

*Pipe sizes for the building in Figure 10-8*

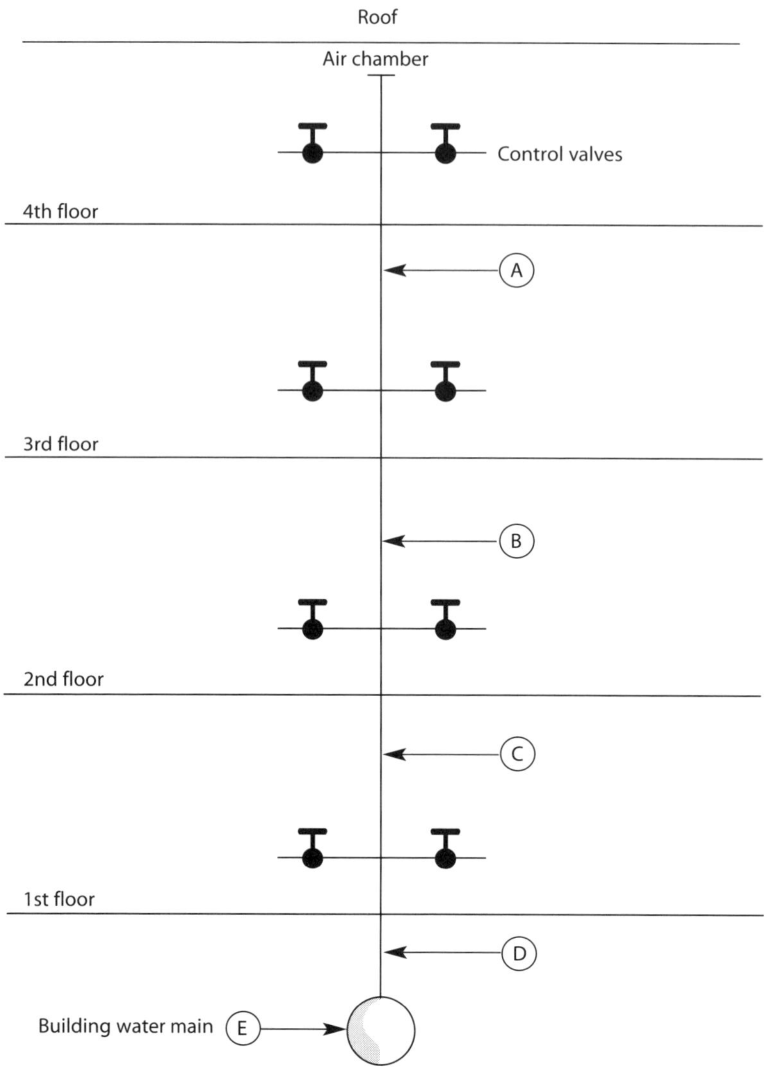


**Figure 10-10**
*Example 3 — Apartment building piping diagram*

| | | |
|---|---|---|
| A | 2 one-bath units | 1" |
| B | 4 one-bath units | 1¼" |
| C | 6 one-bath units | 1½" |
| D | 8 one-bath units | 1½" |

**Figure 10-11**
*Pipe sizes for the building in Figure 10-10*

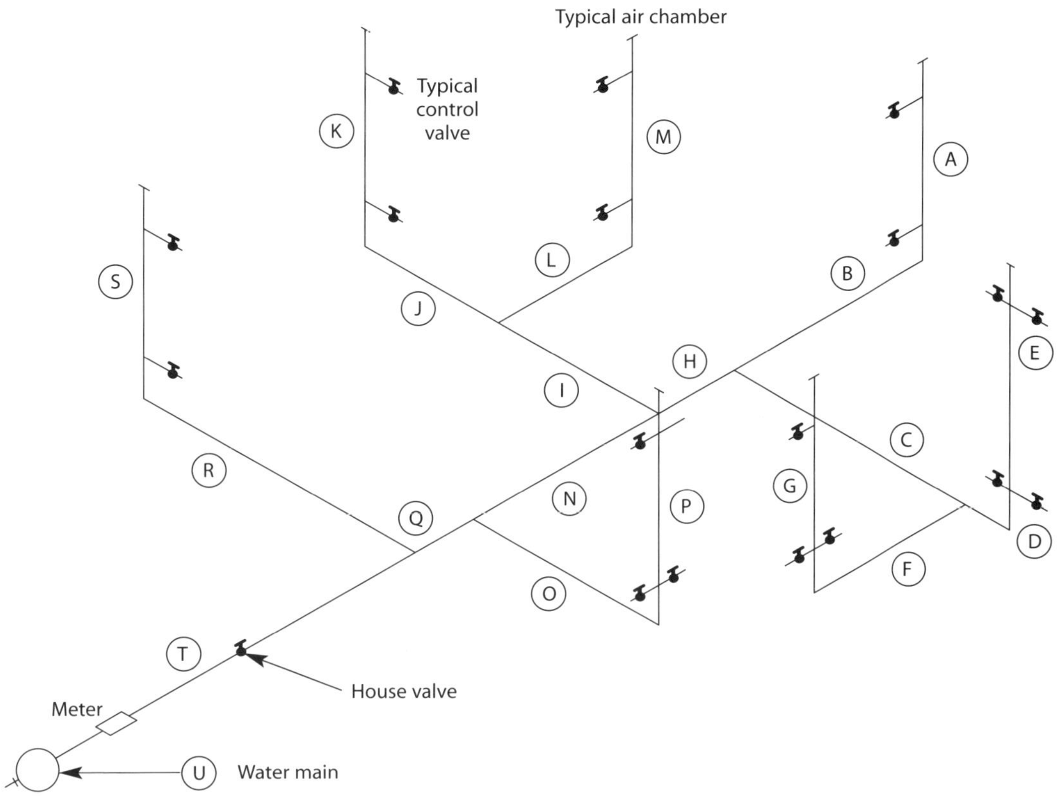


**Figure 10-12**

*Example 4 — Apartment building cold water piping diagram*

| | | | | | |
|---|---|---|---|---|---|
| A | one-bath unit | 3/4" | K | one-bath unit | 3/4" |
| B | two-bath units | 3/4" | L | two-bath units | 3/4" |
| C | seven-bath units | 1 1/4" | M | one-bath unit | 3/4" |
| D | four-bath units | 1" | N | thirteen units | 1 1/2" |
| E | two-bath units | 3/4" | O | three-bath units | 1" |
| F | three-bath units | 1" | P | one-bath unit | 3/4" |
| G | one-bath unit | 3/4" | Q | sixteen units | 2" |
| H | nine-bath units | 1 1/2" | R | two-bath units | 3/4" |
| I | four-bath units | 1" | S | one-bath unit | 3/4" |
| J | two-bath units | 3/4" | T | eighteen units | 2" |

**Figure 10-13**

*Pipe sizes for the building in Figure 10-12*

# *Chapter 10 — Questions and Answers*

**10-1 The *IPC* requires that you install a pressure-reducing valve in the public water service pipe if the pressure is more than**

(A) 50 psi.

(B) 60 psi.

(C) 80 psi.

(D) 100 psi.

***Answer: C***

*Code response: Section 604.8*
*This book, page 163*

**10-2 If the pressure at the public water main is too low for fixtures to function properly, the *IPC* requires that you**

(A) substitute special low-pressure fixtures provided they are code-approved.

(B) install a water pressure booster pump.

(C) reduce the number of fixtures connected to a single main.

(D) install a centrifugal pump.

***Answer: B***

*Code response: Section 604.7*
*This book, page 163*

**10-3 Most residential plumbing fixtures will function at a minimum ______ pressure.**

(A) 8 psi

(B) 10 psi

(C) 12 psi

(D) 15 psi

***Answer: A***

*Code response: Table 604.3*
*This book, pages 164 and 165*

**10-4 According to the *IPC*, the minimum working pressure for a water closet equipped with a flushometer tank is**

(A) 12 psi.

(B) 15 psi.

(C) 18 psi.

(D) 20 psi.

***Answer: B***

*Code response: Table 604.3*
*This book, page 164*

**10-5 For a building with one bathroom, a kitchen sink, and no washing machine, the water service pipe must be at least**

(A) $^{1}/_{2}$ inch.

(B) $^{5}/_{8}$ inch.

(C) $^{3}/_{4}$ inch.

(D) 1 inch.

***Answer:*** ***C***

*Code response:* *Section 603.1*
*This book, page 166*

**10-6 When installing water pipe to a fixture, the *IPC* specifies a maximum distance of _______ from the water supply outlet to the fixture.**

(A) 24 inches

(B) 26 inches

(C) 28 inches

(D) 30 inches

***Answer:*** ***D***

*Code response:* *Section 604.5*
*This book, page 165*

**10-7 When sizing water piping for a large building, the *IPC* says the design must follow or meet**

(A) the master plumber's design.

(B) the plans examiner's approval.

(C) the code official's approval.

(D) good engineering practice.

***Answer:*** ***D***

*Code response:* *Appendix E (Section E101.1.2)*
*This book, page 163*

**10-8 With the new water conservation measures now enforced by the *IPC*, the maximum flow rate for a residential lavatory is**

(A) 0.5 gpm at 60 psi.

(B) 1.0 gpm at 65 psi.

(C) 2.2 gpm at 60 psi.

(D) 2.5 gpm at 80 psi.

***Answer:*** ***C***

*Code response:* *Table 604.4*
*This book, page 165*

**10-9 According to the *IPC*, a residential sink requires a _______ minimum flow rate.**

(A) 1.6 gpm

(B) 2.5 gpm

(C) 3 gpm

(D) 3.5 gpm

***Answer: B***

*Code response: Table 604.3*
*This book, page 164*

**10-10 Since 8 psi is the flow pressure for a water closet tank, close coupled, you'll have to use at least a _______ flow rate.**

(A) 1.6 gpm

(B) 2.5 gpm

(C) 3 gpm

(D) 3.5 gpm

***Answer: C***

*Code response: Table 604.3*
*This book, page 164*

◆ *Chapter 11* ◆

# Hot Water Supply Systems

*IPC* Section 607.1 says you must install a hot water system in any occupied building so that all plumbing fixtures (including bathing, washing, culinary, cleansing or building maintenance equipment) are served. However, in nonresidential occupancies, you must supply hot or tempered water for bathing and washing purposes. The *IPC* specifically requires that you use certain installation methods and safety devices, but it's not specific about sizing or designing hot water distribution systems. Local authorities and professional engineers will take on this task. *IPC* Section 604.1 says that any design of a water distribution system (which includes the hot water distribution system) must "conform to accepted engineering practice." Also, the sizing of hot water piping must be approved by the powers that be.

In the case of a large commercial building, a professional engineer will design and size the hot water supply system. Your only job will be to install it according to the plans. But if it's a residential or small commercial building, it's highly probable that you (the installing plumber) will design and size the hot water system. It's my hope that this chapter will help you design safe and functional hot water systems. For starters, the two principal objectives in designing and sizing a hot water system are satisfying hot water demand for the occupants, and including appropriate safety features. You'll find several sample piping diagrams at the end of this chapter to help you accomplish these objectives.

Another thing to remember is that even if a professional engineer designed the system, he can make mistakes. If you install a system that turns out not to be adequate, or that doesn't meet code, you can point out that it was the engineer's mistake all you like — you're not going to come out of it unscathed. There's still the expectation that you, as a licensed plumber, should have some idea what's right and what's wrong.

## Water Heaters

Before you select and install a water heater, you must make sure it's properly listed and approved. *IPC* Section 501.5 requires all water heaters to be third-party certified. *IPC* Section 501.7 spells out certain requirements for water heaters and hot water storage tanks. The maximum allowable working pressure for any heater or storage tank must be indelibly stamped in the metal or on the plate welded to the product. The *IPC* also requires these markings be on the outside of the tank, easily visible for any inspection.

Some installation instructions are pretty clearly set forth in the code. But you also need to be aware of any not mentioned in the code. I'll include

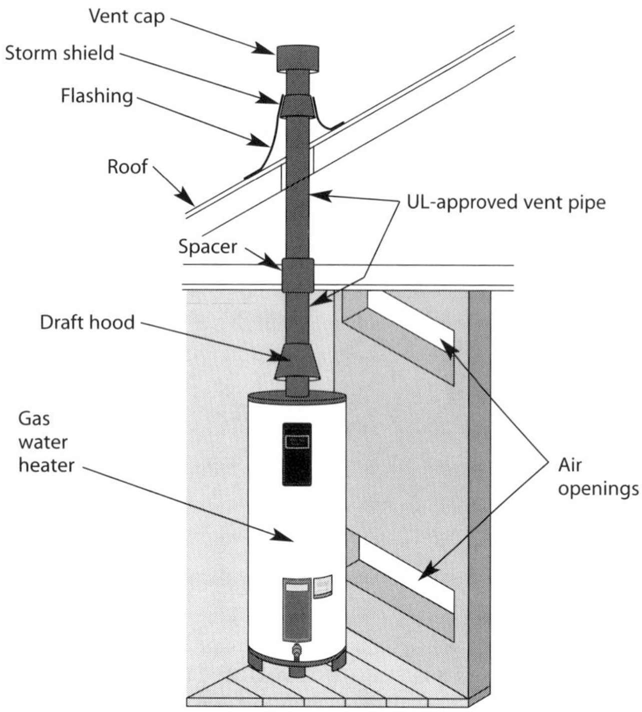


**Figure 11-1**
*Installing a fuel-burning water heater*

acknowledgements of *IPC* requirements when they apply. We'll also discuss the standard requirements by local authorities and good engineering practice (not included in the code) because this will be an important part of your work as a plumber.

The normal design temperature for a domestic water heater is preset at the factory to about 150 degrees F. That's hot enough for most purposes, but cool enough to prevent scalding of the skin. Most direct hot water heating units you install will use gas, oil, or electricity. All residential and commercial hot water storage tanks approved for today's use must have the required insulation to prevent excessive heat loss. The manufacturer will attach this information to the heater jacket. According to *IPC* Section 505.1, the heat loss can't be more than 15 Btu per hour per square foot of external tank surface.

Electric water heaters are clean and attractive enough to install nearly anywhere in a building. Gas or oil-fired water heaters must be put in a well-ventilated area with flues to carry away combustion gases. Any confined space must have two permanent openings, one starting within 12 inches from the top of the heater and one 12 inches from the bottom, as shown in Figure 11-1. The openings must provide enough air for combustion, ventilation and dilution of flue gases. We'll discuss in more detail the installation and venting of gas appliances in Chapter 15.

If an architect's design uses a hot water heating unit to supply hot water to a space heating system, your plumbing work will be a bit more difficult. You'll have to install hot water piping from the hot water heating unit to each hot water space heater. Then you need a circulating system with a circulating pump to move the water, as it cools, back to the hot water heating unit. The reheated water makes the rounds again to each hot water space heater.

You must insulate the hot water circulating line to prevent heat loss. *IPC* Section 501.2 says you have to have a tempering valve wherever space heater temperatures are greater than 140 degrees F. It's never safe to allow potable hot water temperatures to get any hotter than that.

When you design and install a hot water system for a one- or two-family dwelling, try to put the hot water heater as close to the center of the dwelling as possible. The water in the pipes will cool when it's not being used, but if you've put the heater in the center of the dwelling, it won't take as long for hot water to reach the fixtures when needed. If the residence is very large and spread out, consider installing more than one hot water heater to satisfy the owner's needs. This will also help you maintain the energy efficiency requirements of *IPC* Section 607.2.

Where you have a central hot water generating system supplying several areas, you must have a hot water feed and a hot water return line equipped with a circulating pump. The pump must have an automatic shut off, or a manual shut off for when the hot water system isn't being used. See *IPC* Section 607.2.2. Insulate the hot water return circulating system according to local code requirements.

Hot water rises to its highest level and expands as it's heated. This won't have an effect on the average small dwelling unit with short runs. But when you design a hot water system for a larger job, you have to allow for pipe expansion and contraction. In large units with hot water riser supply pipes and circulating lines, expansion and contraction can pose a serious problem. The type of piping materials is a key factor. A 100 linear-foot-run at 100 degrees F may expand as little as $^3/_4$ inch, or as much as 2 inches. An expansion variance of $1^1/_2$ inches for each 100 feet of piping materials will probably handle this. If you use a common four-elbow job-constructed loop, as shown in Figure 11-2, you'll have enough protection to prevent excessive expansion in a hot water piping system.

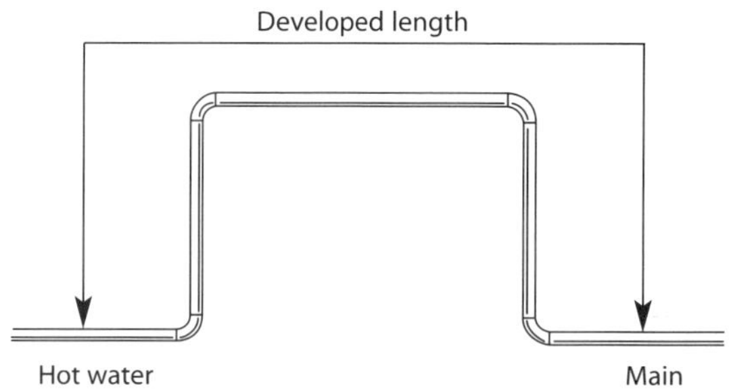


**Figure 11-2**
*Horizontal U bend (four-elbow loop)*

*IPC* Section 607.3 says you have to provide thermal expansion control for large units that have a central hot water generating system. *IPC* Sections 607.3.1 and 607.3.2 say you should install a pressure-reducing valve downstream so the pressure in the hot water line can't be more than the main water supply pressure. Without this protection, hot water can be forced into the cold water supply pipes, damaging fixtures or the water meter. As another reminder of "life in plumber's lane," I remember going to a house on a service call many years ago. We didn't have much to go on in the way of a plumbing code in those days. The hot water heater in this home overheated, and hot water backed up into the water closet tanks (which were steaming!). In fact, the water was steaming all the way to the city water meter. My heart skipped a few beats when I saw the existing danger. The relief valve hadn't worked to relieve the pressure. It was a miracle the heater hadn't exploded!

## Three Common Hot Water Circulating Systems

You need to know how to design and install simple hot water circulating systems because an engineer isn't always available. In fact, he's usually not required to design small systems. Here are the three most common ones you're likely to run into. They're called the inverted upfeed system, the looped system, and the downfeed system.

The inverted upfeed system shown in Figure 11-3 carries hot water from the hot water generator (located on the roof or top floor) to the lowest part of the system. A circulating pump returns the cooling water through the upfeed risers to the hot water generator to be heated again. This provides hot water for the users at all times.

The downfeed system shown in Figure 11-4 works just the opposite way. The hot water generator is in the basement or ground floor. The upfeed riser carries hot water across the building top floor or ceiling to the downfeed risers. A circulating pump returns the cooling water to the hot water generator to be heated again. This maintains a supply of hot water for each user.

The hot water looped system shown in Figure 11-5 is simple and economical. It's used most often in small apartment buildings that are two or three stories high. You should install the hot water feed and return directly below the bathrooms of each apartment. The risers are short and have small pipes, so very little hot water is lost. A circulating pump returns the cooling water to a high-recovery hot water heater, usually fuel burning type.

Hot water piping materials and installation methods are the same as those described for water distribution in Chapter 12, *Installing Water Supply and Distribution Systems*. The only difference is that hot water piping supports must allow for anticipated expansion.

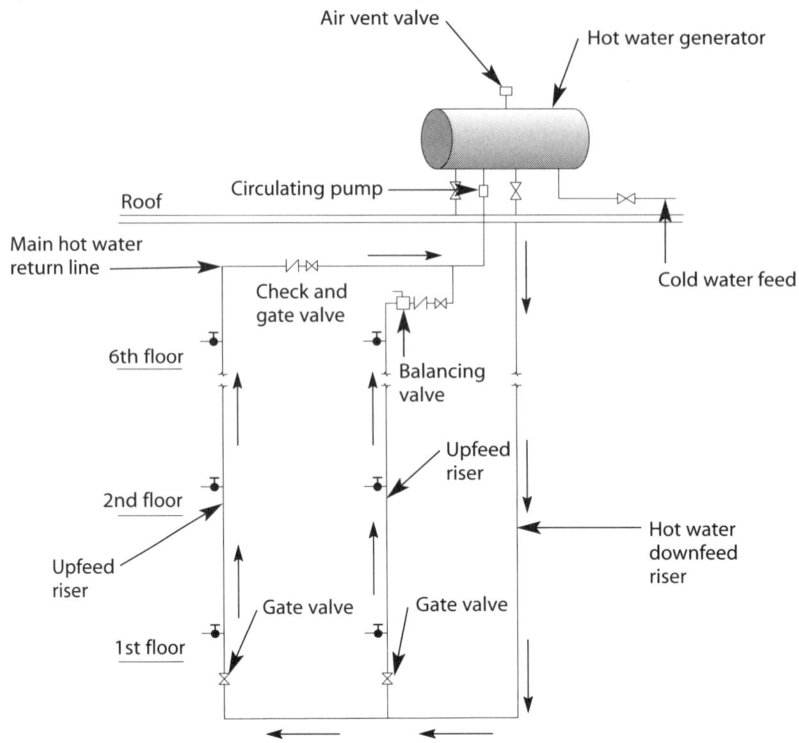


**Note:** The inverted upfeed system carries hot water to the lowest part of the system. The circulating pump returns the water through the upfeed risers to the hot water generator.

**Figure 11-3**

*Inverted upfeed system*

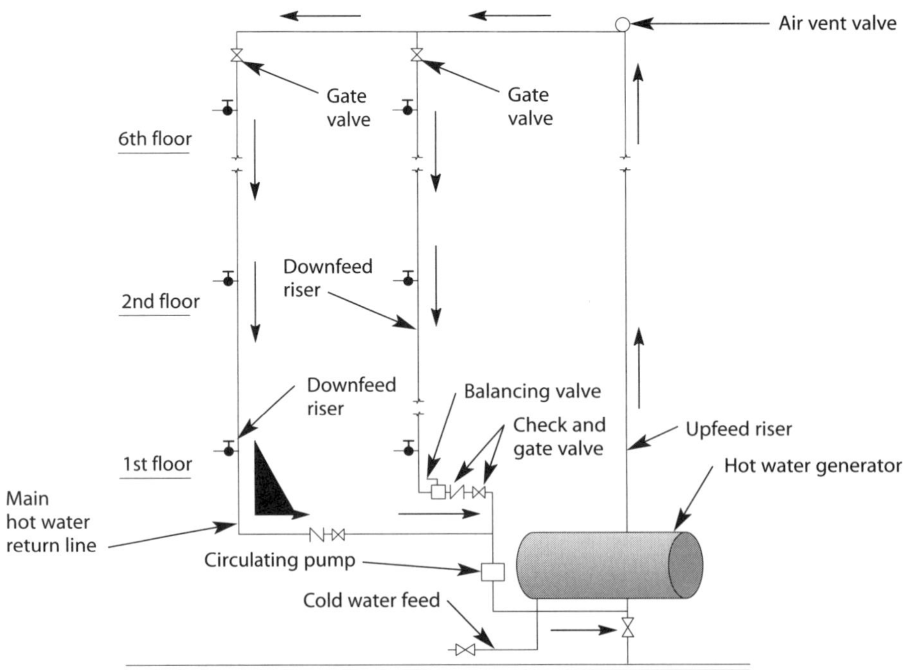


**Note:** The upfeed riser from the hot water source conveys hot water to the downfeed risers. The circulating pump returns this water to the hot water generator.

**Figure 11-4**

*Downfeed system*

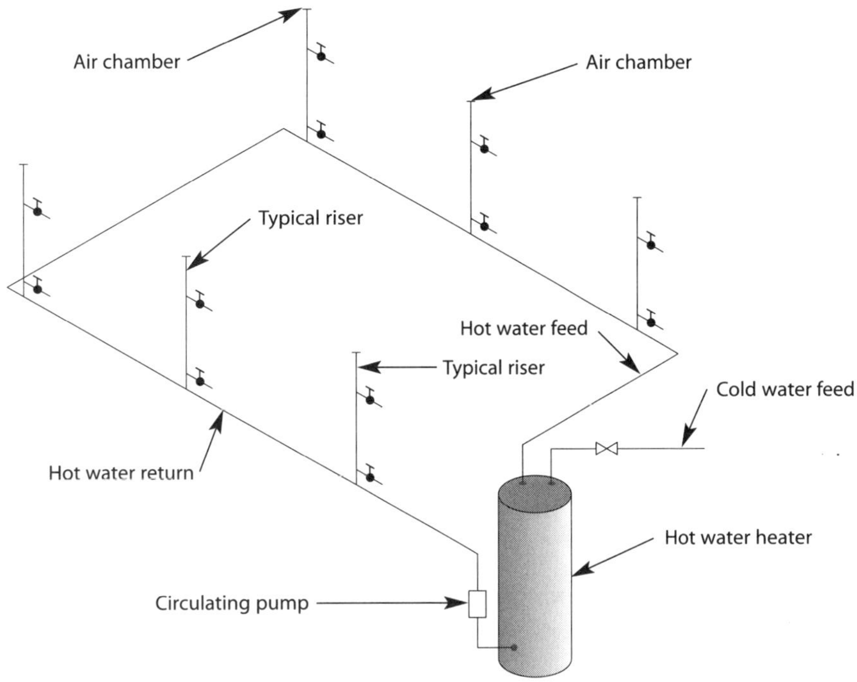


**Figure 11-5**
*Simple hot water looped system*

## Water Heater Capacity

The *IPC* doesn't recommend any particular water heater capacity. This is such an important part of your work that we'll run through the sizing methods that most local authorities will approve. Most modern electric and fuel-burning water heaters have small storage tanks. That's all they need because they have enough heating capacity to keep water at acceptable temperatures (140 to 150 degrees F) during peak draw periods. The peak draw period is usually assumed to be one hour, though in many homes it's as little as 20 minutes. When you size for the proper hot water tank capacity, consider peak draw periods. Here are some peak draw period examples:

- *One-hour peak draw period:* 75 percent of the tank's hot water supply is available during the peak draw period of one hour. For example, a 40-gallon water heater under normal working conditions should provide at least 30 gallons of hot water at 140 to 150 degrees F.
- *Two-hour peak draw period:* 37.5 percent of the tank's capacity is available per hour. That means a 40-gallon water heater should provide 15 gallons per hour.
- *Three-hour peak draw period:* 25 percent of the tank's capacity is available per hour (75 percent over the entire three-hour period). A 40-gallon water heater should provide 10 gallons per hour.

For a hot water heater storage tank, the heating capacity per hour determines the amount of hot water available during peak demand periods. The recommended tank storage capacities shown in Figure 11-6 are economical and satisfactory for an average dwelling unit as long as the peak draw period isn't more than one hour. If there are special demands you know about, consider moving up to the next larger size.

Use the number of bedrooms in a dwelling, not bathrooms, to figure out the hot water heater size you'll need. See Figure 11-6. For each additional

| Fuel type | Number of bedrooms | Storage (gal) |
|---|---|---|
| Gas | | 20 |
| Electric | 1 | 30 |
| Oil | | 30 |
| Gas | | 30 |
| Electric | 2 | 40 |
| Oil | | 30 |
| Gas | | 30 |
| Electric | 3 | 50 |
| Oil | | 30 |
| Gas | | 40 |
| Electric | 4 | 66 |
| Oil | | 30 |

**Figure 11-6**
*Recommended hot water storage tank capacity*

bedroom over four, use the next larger size hot water heater. If the runs are too long, say over 100 feet, consider installing two water heaters.

In my plumbing business, we used a 75-gallon storage tank packaged with a high recovery rate heater for up to 12 multi-family units and a 100-gallon storage tank similarly-equipped for up to 18 multi-family units. These capacities worked fine. For any dwelling units larger than these, get an engineer to design the hot water system.

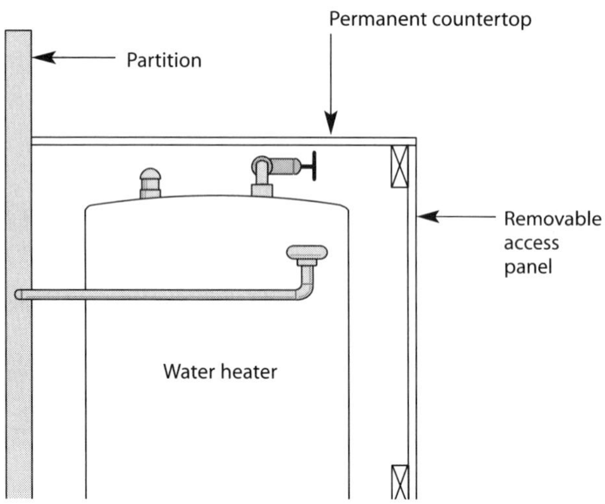


**Figure 11-7**
*Installation of an undercounter hot water heater*

## Safety Devices

According to *IPC* Sections 504.4, 504.4.1 and 504.5 there has to be a combination pressure and temperature relief valve on every hot water heater storage tank. The pressure portion of the relief valve will be 75, 100, 125 or 150 psi. Always install a water heater so the plate with the important data (like maximum working water pressure) is visible. Locate the heater where it can be serviced or replaced without removing any permanent part of the building. As an example, Figure 11-7 shows an undercounter water heater that's accessible through a removable panel.

The discharge rate for a pressure relief valve must limit the pressure rise for any given heat input to within 10 percent of the pressure that opens the valve. Most standard hot water storage tanks are manufactured to withstand working pressures up to 150 psi. The pressure relief valve should open at 25 psi above the maximum working pressure. Under no circumstances should the valve opening pressure exceed 175 psi.

Now let's look at the temperature requirements of relief valves. The temperature portion of a valve releases at 210 degrees F. This valve must be the reseating type and rated by its Btu capacity. The Btu rating of a temperature relief valve must always be greater than the Btu rating of the appliance it serves. This keeps the valve from opening prematurely. In some cases you'll know the wattage of an electric water heater but not its Btu rating. To figure out the Btu rating of an electric water heater when you know its wattage, multiply its wattage by 3.4 (since 1 watt equals 3.4 Btu).

A domestic hot water heater and temperature relief valve doesn't need a separate pressure and temperature relief valve. A combination pressure and relief valve is preferable and will satisfy the code. It must be a self-closing lever type and conform to the ANSI Z21.22 standard. But don't use a relief valve to control thermal expansion. See *IPC* Section 504.4.

*IPC* Section 504.4.1 requires that you install a relief valve so the temperature-sensing element is immersed in the top 6 inches of the tank. See Figure 11-8. Since heated water rises, that's where the hottest water is. The code here is very clear that you can't install a check valve or shutoff valve between the relief valve and the water heater or storage tank it serves. Why is this a concern? One possibility is that the check valve fails to operate (it does happen sometimes). Another possibility is that someone closes the shut-off valve. In either case, the relief valve won't work. It's a bit like having a live bomb that could explode at any time. I was at a workshop where they had a water heater explode as a demonstration. The destruction was awesome. I bet there's not a plumber who was at that workshop who doesn't now check relief valves with extra care.

## Other Safety Features

*IPC* Section 501.3 says all hot water heaters and hot water tanks must have a drain cock (valve). Drain cocks must conform to ASSE 1005 Standards. The drain valve is for flushing sediment from the tank and emptying the tank for repairs or replacement.

The cold water supply pipe to a hot water heater must have at least a 3/4-inch shut-off valve in an accessible location. The shut-off valve must have a cross-sectional area equal to 80 percent of the nominal size of the pipe it's installed in. The minimum size cold water supply pipe for a hot water heater, regardless of its capacity, is 3/4-inch diameter.

*IPC* Section 504.1 calls for installing a cold water "dip" tube with a hole in its top or a vacuum relief valve in the cold water supply line above the top of a water heater or tank. This protects the public water system from siphoning the contents of the hot water heater/tank into its own system in case of a drop in pressure. When you work with a bottom-fed water heater/tank, *IPC* Section 504.2 says you have to install a vacuum relief valve that meets ANSI Z21.22 standards.

The electrical contractor must provide an approved method to disconnect a hot water heater from its energy source in case of emergency or a need to replace it. When working with a fuel-burning water heater, you must also install an accessible separate valve to shut off the fuel supply for replacement purposes or in case of emergency. See *IPC* Section 504.3.

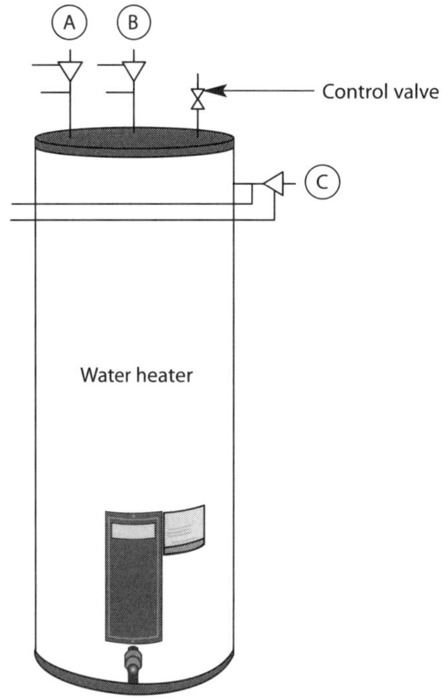


**Note:** Positions shown at A, B, C, are acceptable.

**Figure 11-8**
*Locate relief valves in the top 6 inches of a water heater*

*IPC* Section 607.4 says all hot water outlets to fixtures must be installed on the *left* (from facing the fixture). But you'd be surprised how many hot water outlets are erroneously installed on the right side of the fixture.

## Relief Valve Discharge Lines

Never connect relief valve discharge lines (*popoff* lines) directly to a building drainage system. See *IPC* Section 504.6. It's a long shot, but if a water heater is improperly installed, it could contaminate the potable water system. If a relief valve pops off and it goes unnoticed, a great deal of good water could, literally, go down the drain. Never terminate a relief valve above a water closet, urinal, bidet, bathtub or shower stall. Someone could be scalded. See *IPC* Section 504.6.1. Popoff lines for most buildings should terminate outside the building, at an observable point within 6 inches of ground level. Don't thread relief valve discharge lines. Don't make

| Btu rating | Inside diameter pipe size (in) |
|---|---|
| Up to 69,000 Btu | 3/8 |
| Over 69,000 to 127,000 Btu | 1/2 |
| Over 127,000 to 340,000 Btu | 3/4 |
| Over 340,000 to 600,000 Btu | 1 |

**Figure 11-9**
*Relief valve discharge lines*

it easy for some well-meaning "plumbing illiterate" to connect something to this pipe. However, when a building covers an entire lot, you may terminate the popoff line above a floor drain or other suitable fixture approved by your local authority. Don't install any trapped sections in the line. Terminate the line above the floor drain and install an air gap where it's clearly visible. See *IPC* Section 504.6.1.

All relief valve discharge piping lines are sized by their inside diameters. The discharge piping should never be smaller than the relief valve discharge opening. For a single water heater installed with a single discharge line, the discharge line should be at least the same size as the relief valve discharge opening. When you're installing heaters in a multi-story building using a manifold line, size the line according to the combined Btu rating of the appliances. If you know the total Btu rating, you can use Figure 11-9 to figure out the relief discharge lines for both single-family residences and multi-story buildings. The approved materials for relief valve discharge lines are the same as those used in the water distribution system. You'll find more information on these materials in Chapter 12.

## Hot Water Heater Installation

The *IPC* lists a number of code requirements for locating and installing hot water heaters. Let's take a closer look at them and some illustrations.

If you put a hot water heater in a private or public garage, be sure the heater ignition source is at least 18 inches above the garage floor. This applies to both fuel-fired and electric water heaters. You should never install a fuel-fired hot water heater in a room containing air-handling equipment if that room is used as a plenum. See *IPC* Section 502.2. Never install a fuel-fired water heater in a bedroom, bathroom, or a closet. See *IFGC* Section 303.3.

However, the *IFGC* allows some exceptions for direct vent water heaters. Gas appliances rated at 20 Btu or less per hour per cubic foot of room space need no vent. Let's look at an example. Suppose you want to install a hot water heater with an input rating less than 5,000 Btu in a room that's normally not closed. The room is 7 feet long by 5 feet wide with an 8-foot ceiling. Is a vent required? No. And here's why:

*7′ × 5′ × 8′ = 280 CF*
*5,000 Btu divided by 168 CF*
*= 17.86 Btu per hour*

Since the water heater Btu rating is less than 20 Btu, no vent is needed. The same principle works for small, freestanding gas space heaters and gas logs for fireplaces. They don't need a vent either. However, a hot water heater with less than 5,000 Btu is rarely used, even if you can find one.

If you install a hot water heater in an earthquake zone, install special seismic straps to keep it in place in the event of seismic activity. Specific requirements for this fall under the jurisdiction of the *International Building Code*. The job architect should have the support requirements on the building plans. If not, talk to your local authority for details.

If an owner wants his hot water heater tucked away in the attic, you should advise him that it's usually not a good idea. The owner (and the architect if there's one working on the job) should understand that the ceiling joists must support the water heater. Since water weighs 8 1/3 pounds per gallon, you're looking at 330 pounds *plus* the weight of the average 40-gallon hot water heater. That's about the same weight as a good football lineman. If you do put a water heater in an attic, *IPC* Section 502.3 has specific requirements for access. There must be an access door that's large enough to install and remove a hot water heater. It has to be at least 20 inches by 30 inches according to the dimensions in the 2003 *IPC*, Section 502.3. The passageway from the access opening to the water heater can't be longer than 20 feet. The passageway must have a continuous solid flooring at least 24 inches wide and an open area at least 30 inches high and 22 inches wide. Then there must be a 30-inch-deep × 30-inch-wide service space in front of the hot water heater.

## Hot Water Heater Drain Pans

*IPC* Section 504.7 says you have to install a drain pan with any hot water heater or storage tank where leakage can cause damage. See Figure 11-10.

There are additional drain pan guidelines that most local jurisdictions will accept. A drain pan must be:

- made of high-impact plastic at least $^{1}/_{16}$-inch thick. Galvanized sheet steel or other corrosive-resistant metals at least 24 gauge are also approved.
- at least $1^{1}/_{2}$ inches deep.
- constructed with a minimum $^{3}/_{4}$-inch drain outlet located $^{1}/_{2}$ inch above the bottom of the pan.
- large enough to provide a 2-inch clearance between the drain pan sides and the heater base.

When water heaters are installed directly over one another in a multi-story building, the local authority may let you use one riser for both the water

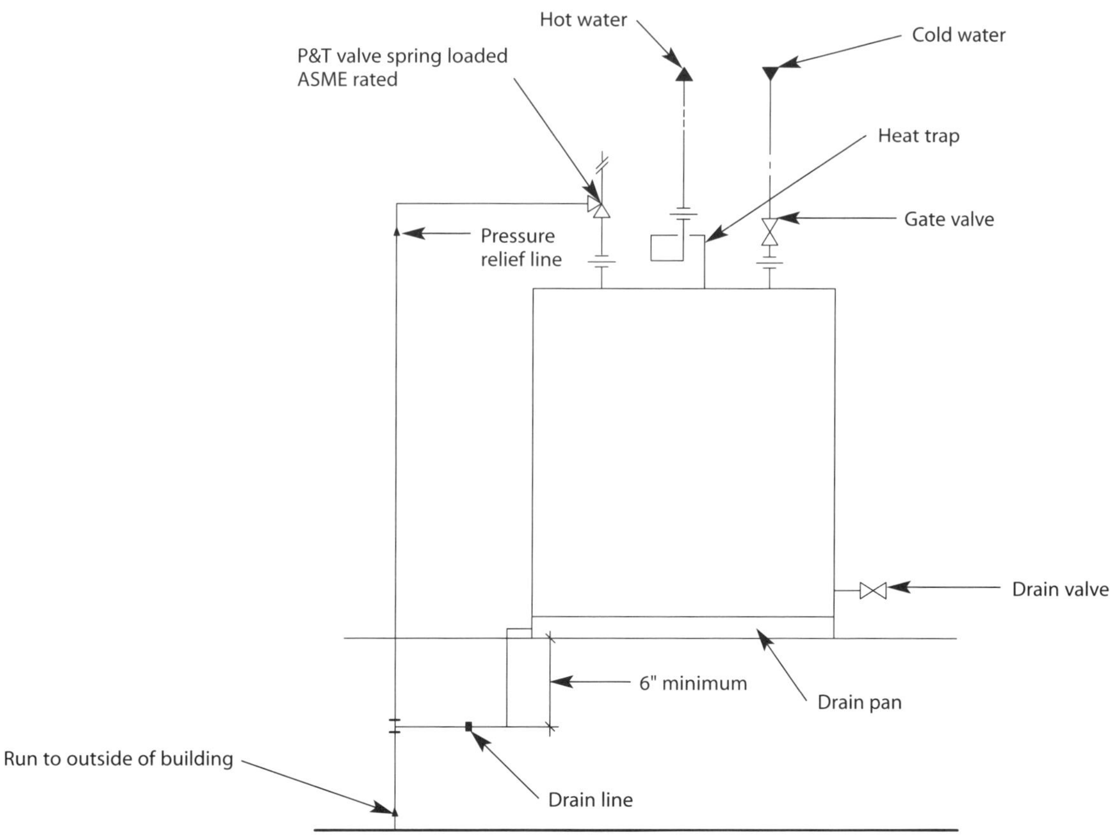


**Figure 11-10**
*Electric water heater detail and drain pan*

heater pressure relief line and the drain pan. In this case, make sure the drain from the drain pan runs downward at least 6 inches before you connect into the main riser. See Figure 11-10.

You can select your materials for the drainpipes leading away from drain pans from the approved list for water distribution piping in Chapter 12. In most cases, your design and installation will be much like that of a plastic system.

## Sizing Drain Pan Risers

The code doesn't give much help on sizing drains for water heater drain pans. Let me share with you what most local authorities will accept. Check with your local plumbing chief before you take the information as "gospel." The number of water heaters you need depends on the size of the riser. You may connect a:

- 1-inch riser to a maximum of three drain pans.
- 1¼-inch riser to a maximum of ten drain pans.
- 1½-inch riser to 11 or more drain pans.

When you install a drain pan waste pipe under a concrete floor slab, it's a good idea to increase the pipe by one size. Your local authority may permit the common vertical riser pipe to serve both the hot water heater popoff line and the drain pan. Whether individual or combined, drain pan lines must terminate over a suitable waste receptor or extend outside the building in an observable location. The termination point must be at least 6 inches, but no more than 24 inches above ground level. See *IPC* Section 504.7.2.

## Energy Conservation

The topic of saving energy is a familiar theme. A hot water heater is one of the worst offenders because it uses a great deal of energy. In fact it uses about 25 percent of the total energy used in the average household. Here are energy-saving requirements typically found in local codes today:

- Hot water has a tendency to rise several feet into a riser or travel horizontally into a hot water line when not in use. Make a simple loop in the hot water line leading from the hot water heater to prevent "hot water creep." This saves about 2 percent of the cost of generating hot water. Manufacturers now make what is called a heat trap in a one-piece unit. Most local authorities now require a heat trap be installed on all hot water heaters or storage tanks. See Figure 11-10.

- Local authorities require that hot water heaters be equipped with an energy shut-off device. This device automatically cuts off the supply of heat energy (electric or fuel burning) to the water tank to prevent temperatures rising to more than 210 degrees F.

- Today's hot water heaters are required to have efficient heating devices and controls. There are also stringent standards for hot water heater insulation to prevent excessive heat loss.

Just these few requirements on one item alone — a water heater — have added up to considerable energy savings.

# *Chapter 11 — Questions and Answers*

**11-1 The *IPC* requires that you install hot water service to**

(A) all residential plumbing fixtures.

(B) plumbing fixtures used for culinary purposes.

(C) all commercial plumbing fixtures.

(D) all fixtures in any occupied structure.

***Answer: B***

*Code response: Section 607.1*
*This book, page 175*

**11-2 The *IPC* says that any water heater must be**

(A) sized properly.

(B) approved by the code official.

(C) third-party certified.

(D) installed by a licensed plumber.

***Answer: C***

*Code response: Section 501.5*
*This book, page 175*

**11-3 The *IPC* requires that a water heater's maximum working pressure be ______ on the heater itself or on a plate welded to the heater.**

(A) indelibly stamped

(B) color coded

(C) embossed

(D) imprinted

***Answer: A***

*Code response: Section 501.7*
*This book, page 175*

**11-4 When installing a water heater that supplies hot water to space heaters, you must install a tempering valve to limit the water temperature to less than _____ degrees F.**

(A) 120

(B) 130

(C) 140

(D) 150

***Answer: C***

*Code response: Section 501.2*
*This book, page 176*

**11-5 Hot water heater drain pans must be**

(A) equipped with minimum 1/2-inch drains.

(B) 20-gauge galvanized sheet steel.

(C) drained by a direct waste pipe.

(D) 1.5 inches deep.

***Answer: D***

*Code response: Section 504.7.1*
*This book, page 183*

**11-6 The *IPC* requires you to install a temperature relief valve in the top _____ inches of any hot water heater tank.**

(A) 4

(B) 6

(C) 8

(D) 10

***Answer: B***

*Code response: Section 504.4.1*
*This book, page 180*

**11-7 The *IPC* says any hot water heater you install must have**

(A) a registered engineer's seal.

(B) a built-in tempering valve.

(C) two heating elements.

(D) a drain valve.

***Answer: D***

*Code response: Section 501.3*
*This book, page 181*

**11-8 The *IPC* says that a plumbing fixture's hot water outlet must be**

(A) located on the left.

(B) no less than 18 inches from finished floor.

(C) of material conforming to ASTM80.

(D) accessible.

***Answer: A***

*Code response: Section 607.4*
*This book, page 181*

**11-9 According to the *IPC*, a relief valve discharge line shall never**

(A) be directly connected to a drainage system.

(B) discharge indirectly into a floor drain.

(C) discharge on a building's floor.

(D) discharge to the outside of a building.

***Answer: A***

*Code response: Section 504.6*
*This book, page 181*

**11-10 The *IPC* states that a drain pan beneath a hot water heater must have a drain outlet no smaller than**

(A) $^{1}/_{4}$ inch.

(B) $^{1}/_{2}$ inch.

(C) $^{3}/_{4}$ inch.

(D) 1 inch.

***Answer: C***

*Code response: Section 504.7.1*
*This book, page 183*

◆ Chapter 12 ◆

# Installing Water Supply and Distribution Systems

The *IPC* regulates the materials, size and installation methods for water piping. The *IPC* also establishes safeguards to prevent contaminating the potable water supply. We'll discuss them in this chapter, as well as other requirements, limitations and restrictions.

All building water service and distribution system pipes must resist corrosive/degrading action from the potable water they carry. See *IPC* Section 605.1. Piping materials for any water system must comply with the standards listed in Figure 12-1. It's always a good idea to consider the type of water supply available in your area before you choose a material. Does your community's water supply corrode or leave deposits on the interior walls of certain piping materials? Will the soil or fill corrode the exterior of certain piping materials? *IPC* Section 605.1 says you can't install building water service pipe in soils that contain solvents, fuels, organic compounds or other detrimental materials. These substances can allow fluids to get into the interior of the piping, or they can corrode the exterior of certain types of pipes so the system fails. If these conditions exist on your job, be sure to get a chemical analysis of the soil and ground water to find out what type of material will be acceptable.

Depending on any particular harmful conditions, your local authority may specify a certain type of material for you to use, or require a different routing. There are many water piping materials that meet *IPC* requirements. Figure 12-2 shows common water system materials and whether they can be used aboveground or belowground. However, there are two kinds of pipe you *can't* use in a potable water system:

- pipes that can leave toxic substances in potable water
- pipe and fittings that have been used for purposes other than a potable water supply system. For example, you can't reuse pipe and fittings previously used in a gas system.

You also need to consider lead that might enter a potable water system. *IPC* Section 605.2 limits the lead content level in pipe, fittings, valves and faucets to 8 percent. But the only way to enforce this is by regulating manufacturers. So any items you purchase from your supplier should meet code requirements.

## Water Service Supply Pipe Outside a Building

Figure 12-1 lists materials approved for water service supply piping for buildings. These materials should conform to *IPC* standards. Plastic water service pipe and fittings must carry permanent identification markings from the ASTM standards (American Society for Testing Materials) or another

| Materials | Standards |
|---|---|
| **Ferrous pipe:**<br>Ductile iron water pipe | AWWA C151; AWWA C115 |
| **Ferrous fittings:**<br>Cast iron | ASME B16.4; ASME B16.12 |
| Gray iron ductile iron | AWWA C110; AWWA C153 |
| **Nonferrous pipe:**<br>Brass pipe<br>Copper or copper-alloy pipe<br>Copper or copper-alloy tubing (Type K, WK, L, WL, M or WM) | ASTM B43<br>ASTM B42; ASTM B302<br>ASTM B75; ASTM B88; ASTM B251; ASTM B447 |
| **Nonferrous fittings:**<br>Copper or copper-alloy | ASME B16.15; ASME B16.18; ASME B16.22, ASME B16.23; ASME B16.26; ASME B16.29, ASME B16.32 |
| **Nonmetallic pipe:**<br>Asbestos-cement pipe | ASTM C296 |
| **Plastic pipe:**<br>ABS plastic pipe<br>Polybutylene (PB) plastic pipe and tubing<br>Polyethylene (PE) plastic pipe<br>Polyethylene (PE) plastic tubing<br>Cross-linked polyethylene (PEX) plastic tubing<br>Cross-linked polyethylene/aluminum/cross-linked polyethylene (PEX-AL-PEX) pipe<br>Polyethylene/aluminum/polyethylene (PE-AL-PE) pipe<br>Polyvinyl chloride (PVC) pipe | ASTM D1527; ASTM D2282<br>ASTM D2662; ASTM D2666; ASTM D3309; CSA B137.8<br>ASTM D2239; CSA CAN/CSA-B137.1<br>ASTM D2737; CSA B137.1<br>ASTM F876; ASTM F877; CSA CAN/CSA-B137.5<br>ASTM F1281; CSA CAN/CSA B137.10<br>ASTM F1282; CSA CAN/CSA-B137.9<br>ASTM D1785; ASTM D2241; ASTM D2672; CSA CAN/CSA-B137.3 |
| **Plastic fittings:**<br>Acrylonitrile butadiene styrene (ABS) plastic<br>Chlorinated polyvinyl chloride (CPVC) plastic<br>Polyethylene (PE) plastic<br>Polyvinyl chloride (PVC) plastic | ASTM D2468<br>ASTM F437; ASTM F438; ASTM F439<br>ASTM D2609<br>ASTM D2464; ASTM D2466; ASTM D2467; CSA CAN/CSA-B137.2 |
| **Galvanized steel pipe** | ASTM A53 |
| **Galvanized steel pipe fittings:**<br>Malleable iron<br>Galvanized steel | ASME B16.3<br>ASME B16.9; ASME B16.11; ASME B16.28 |

**Note:** Not every code book will contain these exact standards for pipe, tube, and fittings. Always check your local code standard tables for the complete listings in your locale. See Figure 9-1 for abbreviations for plumbing-related organizations.

**Figure 12-1**

*Plumbing material standards for potable water systems*

| Materials | Water service pipe | Underground within buildings | Aboveground within buildings | Cold water piping | Hot water piping |
|---|---|---|---|---|---|
| ABS pressure rated plastic pipe | X | X | | X | |
| Asbestos-cement pressure pipe | X | | | | |
| Brass pipe | X | X | X | X | X |
| Cast iron water pipe | X | X | | | |
| Copper water tube Type K,L,M | X | X | X | X | X |
| CPVC plastic water pipe | X | X | X | X | X |
| Ductile iron pipe cement lined | X | | | | |
| Galvanized steel pipe | X | X | X | X | X |
| PB plastic water distribution system | X | X | X | X | X |
| PE plastic pipe | X | | | | |
| PVC plastic pipe | X | | | | |
| Cross-linked (PEX) plastic tubing | X | X | X | X | X |
| Cross-linked (PEX-AL-PEX) pipe | X | X | X | X | X |

**Note**: Potable water piping requires at least 160 psi at 73.4° F.
Hot water piping must be within the limits of its listed standard *and* the manufacturer's recommendations.
Some local codes may list materials not included in this chart.
Some local codes do not permit installation of M copper underground within buildings.
Always check local code requirements.

**Figure 12-2**

*Materials and locations for common potable water systems*

national standard of acceptance. For most of your jobs you'll use one these piping materials:

- galvanized (inside and outside) steel threaded pipe and fittings
- copper pipe or tubing, Type K, L or M
- pressure-rated plastic pipe, tubing and fittings, including ABS, PVC, CPVC, PE, PEX and PB.

All water service piping installed underground outside a building must be able to withstand at least 160 psi. If the working water pressure on a job could be more than 160 psi, be sure to use piping materials that can handle more pressure.

Usually you'll use ductile iron water piping for the main water supply pipes for large projects. Typically, the code specifies that this material be cement mortar lined according to AWWA (American Water Works Association) standards. Other more familiar materials will do fine for the branch piping.

## Water Supply Pipe

Here are some important points to remember when you install water supply piping:

- Always place the water service supply pipe on a solid shelf excavated at one side of the common trench and at least 12 inches above the sanitary sewer line, as shown in Figure 12-3. Note the exceptions in *IPC* Section 603.2.
- Install the water service supply pipe with a minimum of joints. Keep joints as far as possible from the building sewer. Use full-length 21-foot pipe and 50-foot to 100-foot coils when possible.

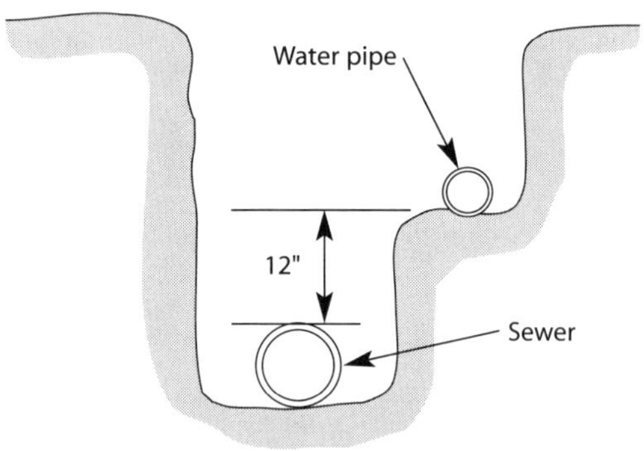


**Figure 12-3**

*Water supply pipe and sewer pipe in a common trench*

- The *IPC* requires at least 5 feet of undisturbed soil between the water service supply pipe and building sewer pipe that are installed in separate trenches. See *IPC* Section 603.2, and Figure 12-4.
- Never install a building water supply service in, under or above cesspools, septic tanks, septic tank drainfields or seepage pits. See *IPC* Section 603.2.1.

If you have to install metallic water piping on filled corrosive soil or where hydrogen sulfide gas is known to be present, be sure to protect the pipe. You can paint it with one or two coats of asphaltum paint or other approved coating. Also protect the fittings by painting or wrapping with an approved material.

Always install water service supply piping in open trenches, laid on a firm bed of earth for the entire length of the piping. Make sure it's securely supported to prevent sagging, misalignment and breaking. If you must install water service supply or distribution piping by tunneling, jacking or a combination of both, protect the pipe from damage during installation and later uneven loading. You must provide enough supporting structures to prevent future settling or caving. See *IPC* Section 306.4. Check with your local authority about details for this type of installation.

Where buried pipe can freeze, make the trench with the water service supply pipe at least 6 inches below the frost line. Insulate pipe that enters a building above ground or in areas not protected from the cold. See *IPC* Section 305.6. Where freezing isn't a problem, there are still several precautions you need to take. Dig the trench deep enough to protect the pipe from being damaged by edgers or other sharp tools. Be sure it's buried deep enough so the water it carries isn't heated by the sun on warm days. If the water service supply pipe

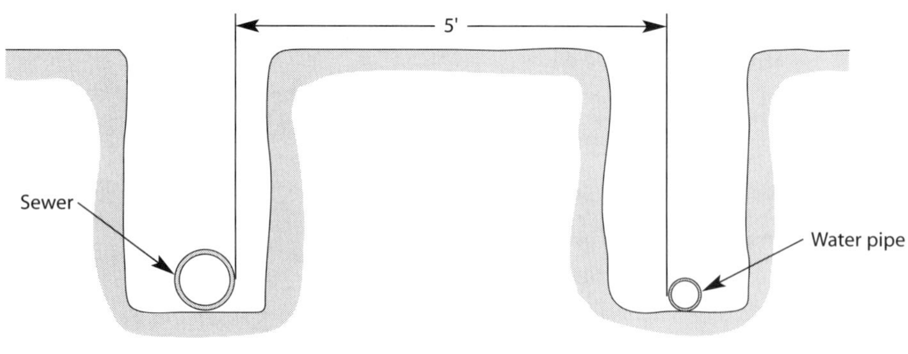


**Figure 12-4**

*Sewer and water pipe in separate trenches. The IPC Section 603.2 says there must be at least 5 feet of undisturbed or compacted soil between the water service supply pipe and the building sewer pipe installed in separate trenches*

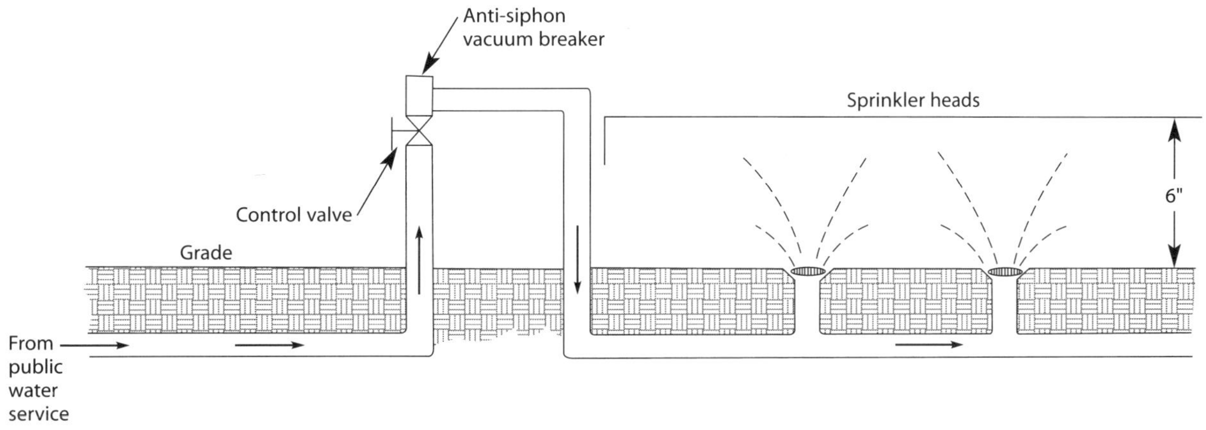


**Figure 12-5**

*Installing an anti-siphon vacuum breaker*

is constructed of plastic, the code requires at least a 12-inch cover.

Remember that a building will settle. If water piping passes under a building foundation, make sure there's a 2-inch clearance from the bottom of the footing to the top of the pipe. If the piping passes through a slab or foundation wall or other cast-in-place concrete, you'll need to sleeve it two pipe sizes larger than the pipe itself. See *IPC* Section 305.5. This prevents damage or breakage due to the building settling, or normal expansion and contraction of the pipe. It also protects the pipe from the corrosive effects of concrete. Be sure to caulk the sleeve opening with an approved flexible sealing compound to prevent crawling insects from entering the building. See *IPC* Sections 305.3 and 305.4.

If you have to connect a lawn sprinkler system to the potable water supply, you must take preventive action against backflow. You can install an atmospheric-type vacuum breaker, a pressure-type vacuum breaker or a reduced pressure principle backflow preventer. Any one of these should eliminate the possibility of a cross connection. See Figures 12-5 and 12-6. It's good plumbing to install the backflow preventer at least 6 inches above the highest sprinkler head, and not less than 6 inches above the surrounding ground. Also, you can't install a valve downstream from an atmospheric vacuum breaker. If there's any possibility that chemicals will get into the sprinkler system, install a reduced pressure principle backflow preventer. It's very important that you always protect the potable water system. See *IPC* Section 608.16.5.

Water service supply piping must be electrically isolated from all other pipe, conduit, soil pipe, building steel and steel reinforcing. If contact with the above is unavoidable, wrap the piping with

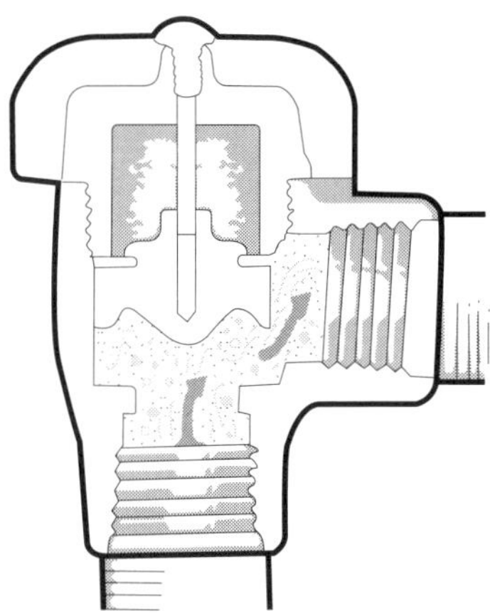

**Figure 12-6**

*Atmospheric type anti-siphon vacuum breaker*

some approved material. When replacing a metallic water service supply pipe that's used as an electrical ground for a building, be sure to use another metallic piping material. Never replace it with a non-metallic material unless other grounding means have been approved. See *IPC* Section 601.3.

Never connect a private water supply, such as a domestic well, with the water service to a building that has public water. See *IPC* Section 608.6.1.

## Water Distribution Pipe Inside a Building

You can use brass, copper Type K, L or M, wrought iron, or steel for water piping above first-floor slabs. All wrought iron and steel pipe and fittings must be galvanized inside and out. Use only approved fittings for the type of pipe you install. When you install water piping that's permanently inaccessible (such as under floor slabs), you must use:

- cast iron water pipe, brass pipe, copper tubing Type K, L, or M.
- wrought iron and steel pipe and fittings (galvanized inside and out; used with appropriate approved fittings).
- chlorinated polyvinyl chloride (CPVC).
- polybutylene (PB).
- ABS pressure-rated plastic pipe.
- cross-linked polyethylene (PEX) plastic tubing.
- cross-linked polyethylene/aluminum/cross-linked polyethylene (PEX-AL-PEX) plastic pipe (always use approved fittings or bends with plastic pipe or tubing).

### *Protecting Piping Inside a Building*

Always make certain you support underground water distribution piping inside a building for the entire length of the piping on well-compacted fill. This prevents sagging and keeps the piping aligned. It's extremely important to avoid sags, or you'll have traps or depressions that collect sediment or mineral deposits. Some mineral deposits solidify over a period of time and reduce the flow of water. This can make a premature failure of a system and an unwelcome expense for you. Always use fine material for backfill in the trench so compaction doesn't damage the pipe. If at all possible, install an entire system so it will drain dry. This is very important if you have to repair the piping or replace lost air in air chambers.

If you install water pipe under a concrete slab, always protect it from corrosion. The most likely cause for this type of corrosion is that the building is constructed on filled ground containing corrosive material such as lime and acid from concrete dumping and cinders. Another source is hydrogen sulfide gas. If you work in a coastal area, you're more likely to run into this. If you should encounter corrosive substances, you can paint the pipe and fittings with two coats of asphaltum paint or other approved coatings. Or wrap the pipe with an approved material. See *IPC* Section 305.1. If you're aware of corrosive danger in advance, it's an even better idea to run the piping overhead. In fact, some local authorities require it.

If you work in a climate zone where water pipes might freeze, don't install water piping in crawl spaces or other unheated areas unless you protect it with insulation. See *IPC* Section 305.6. What about restaurants, packing houses and other commercial buildings with walk-in cold storage facilities? Don't even think about installing water piping through a cold storage room.

Always make certain that your water supply pipe doesn't come in contact with other ferrous pipes or materials. These include electrical conduits, metal partition studs and reinforcing steel. They're all good conductors of electricity. Where physical contact is absolutely unavoidable, wrap the pipe with heavy tarpaper or other approved material to provide the necessary insulation.

It's bad news when metal hot and cold water pipes come in contact with each other. Be careful to keep these pipes as separate as possible when you install them underground or within partitions. Contact, or even close proximity, can transfer the heat and cold from one line to the other. Make certain you prevent any flow between the hot and cold systems at the point of interconnection between hot and cold water supply piping to fixtures, appliances or devices. See *IPC* Section 604.2.

*IPC* Section 301.6 says you must never install water supply piping under elevators or in elevator equipment rooms.

## Control Valves

*IPC* Section 606.2 says you must install an independent control valve for each separate apartment or store in a building. Otherwise, you must install independent control valves for *each fixture* in that apartment or store. Also, hotels, motels, office buildings, hospitals, clinics, places of public assembly and manufacturing plants require either separate fixture control valves, or a single accessible control valve for each group of fixtures in a single room.

For each water closet and urinal supply, you'll need to install an independent control valve above the floor. These fixtures require more-than-average maintenance. Using separate control valves, via the code, you can repair or replace them without disrupting the whole building.

Residential buildings with two or more units and commercial buildings and a single water service must have a separate control valve for each hose bibb (sill cock). This allows for repairs without interrupting the water supply to the individual units. This is usually not required by most local authorities for single-family residences.

Refrigerators with automatic ice makers and dishwashers must also be valved. Clothes washing machines don't need separate shutoff valves. See *IPC* Section 606.2.

Some local authorities require a control valve with a drip valve near the point the water supply pipe enters a building. A drip valve is essential in cold climates as it drains off water that might freeze in the pipes causing them to burst. It's especially useful for vacation homes.

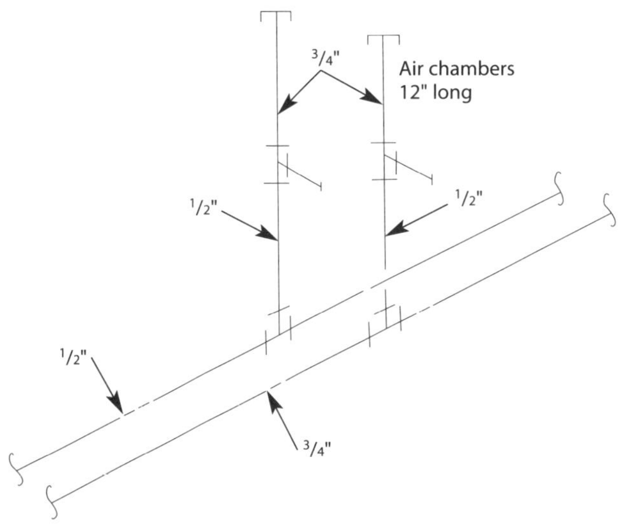


**Figure 12-7**
*Proper location of air chambers*

You need to protect all water pipe installations from water hammer with properly-located water-hammer arrestors, known in the trade as air chambers. See *IPC* Section 604.9. Always install the arrestors on the highest hot and cold water pipes serving a plumbing fixture in each bathroom, preferably the lavatory pipes. See Figure 12-7. You should also put air chambers on water pipes serving individual fixtures such as kitchen sinks and clothes washing machines. If you can't get factory-constructed water-hammer arrestors, you can make your own on the job. They're acceptable, not nearly as expensive and work very well. Make an air chamber from a pipe-like material that's 12 inches long. Be sure it's one pipe size larger than the pipe it serves and capped on one end. Install it in the locations as described.

## Water Pipe and Fittings

In Chapter 9 we talked about piping, fittings and joints for drainage, waste and vent systems. Let's now explore the same for water systems. Use only approved fittings for the type of pipes you install. The fittings must conform to the respective pipe fitting standards listed in Figure 12-1. Don't use any fitting that has ledges, shoulders or reductions that could slow down the flow of water. *IPC* Section 605.5 specifies that fittings for ductile and gray iron water systems be cement mortar lined.

## Plastic Pipe

Plastic products (being less expensive and easier to install) have made plumbing installations more affordable. They're also long lasting because they resist corrosion, scale, sediment buildup, and aren't affected by soil conditions or electrolysis. You can use several types of approved plastic pipe and fittings aboveground within buildings. CPVC was the first plastic piping material to gain favorable regard from plumbing experts. It's a high-temperature vinyl plastic designed for both hot and cold water systems. It comes only in rigid lengths, can withstand pressures up to 100 psi and temperatures up to 180 degrees F.

The *IPC* accepts, with no limitations, polybutylene pipe, tubing and fittings. You can get PB plastic pipe and fitting resins (manufactured from 85 percent butene-1) in both rigid piping and flexible coils. This material, like CPVC, has a number of benefits. It can withstand temperatures up to 180 degrees F and pressures up to 100 psi. Polybutylene has a unique advantage over metallic piping and fittings and some approved plastic piping and fittings because it's not likely to expand. This means you don't have to make any stress/expansion adjustments, as you would for rigid CPVC plastic or metal pipes. Also, when you're joining it to other PB pipe and fittings, there are several approved methods (compression fittings, flare connections, crimp-type fittings and heat fusion) you can use. PB pipe is unharmed by freezing temperatures. You can also use cross-linked polyethylene (PEX) plastic tubing and cross-linked polyethylene/aluminum/cross-linked polyethylene (PEX-AL-PEX) underground and aboveground within buildings. If you install a plastic water service, *IPC* Section 605.3 says it must end within 5 feet of the building's entry point.

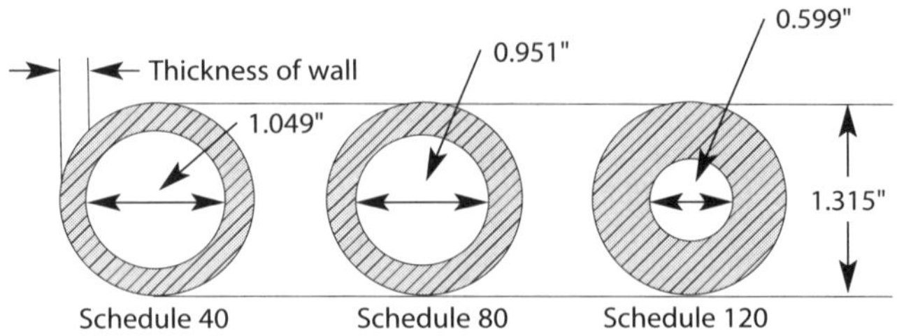


**Figure 12-8**

*Three weights of 1" nominal size pipe*

## Galvanized Steel Pipe

This piping is still one of the most common materials used for water piping systems. It's coated inside and out with zinc to protect it from corrosion. It comes in 21-foot lengths with threads on each end and nominal inside diameter sizes as small as 1/8 inch. Most sizes come in three weights or strengths — Schedule 40 standard, Schedule 80 extra strong and Schedule 120 double extra strong. See Figure 12-8. You'll probably use standard weight pipe and fittings the most, as they usually work for the typical job. Heavier weight pipe has the same outside diameter as standard weight pipe but its walls are thicker so you can use standard threading dies on all three weights. Inside diameters of heavier pipe are smaller, so they cut down the flow of liquids. Galvanized steel pipe usually has a pipe coupling screwed to one end of each length of pipe. This is a free fitting. Use it to join two pieces of pipe in long or short runs unless a fitting is required.

Galvanized pipe has plenty of advantages. Because of its strength, durability and resistance to trench loads, it's very good to use for both hot and cold water. It's especially good for outside water service installations. It's so strong that you seldom have to worry about the depth of placement, except to protect it against freezing.

However, galvanized steel pipe has some serious disadvantages. It's more subject to corrosion than copper or plastic pipe. Water with a high acid content will rust the inside of galvanized pipe. Hard water can leave scales or deposits clinging to the inside of galvanized pipe. These sediments harden in galvanized steel pipe more readily than incopper or plastic piping. Also if you're in a low water-pressure area, the rougher inside surface of galvanized steel pipe will make more friction on water moving in the pipe than the smoother interior of copper or plastic. But if a job doesn't have these conditions, galvanized steel pipe and fittings will provide a long service life.

Galvanized malleable iron threaded fittings are usually used in water supply systems to connect lengths of pipe, change the direction of flow, or reduce pipe diameter. They are pressure rated. The name of a fitting usually describes its shape and use. A plumber apprentice will generally learn the names of these fittings in the first few weeks of work.

*IPC* Section 605.12.3 says that all standard pipe and fitting threads must conform to ASME B1.20.1 standards. The standard pipe thread has a taper of $3/4$ inch per foot. Figure 12-9 shows information on various size pipes up to 4 inches. Use the proper size wrench when joining pipe and fittings on threaded pipe. A wrench too small for the job will require unnecessary effort for your hands, arms and back. A wrench too large will force the fitting too far onto the threaded pipe. This can result in a bad joint or a cracked fitting. Use Figure 12-10 to pick the proper wrench for the pipe sizes shown. Always apply pipe-joint compound to male threads only. And don't apply it to interior pipe threads because the excess will go into the pipe and eventually clog up any faucet aerators and water closet valves. Galvanized steel pipe joints may be of the mechanical type. The joints must be made with approved elastomeric seals. Naturally, they must be installed according to the manufacturer's recommendation. See *IPC* Section 605.18.2.

Manufactured galvanized steel pipe nipples come in lengths ranging from close (an all-thread nipple) to 6 inches. They must conform to ASTM A 733 standards. See *IPC* Section 605.8.

| Pipe size (in) | Threads per inch | Approximate length of threads (in) | Approximate thread engagement (in) |
|---|---|---|---|
| 1/4 | 18 | 5/8 | 3/8 |
| 3/8 | 18 | 5/8 | 3/8 |
| 1/2 | 14 | 13/16 | 1/2 |
| 3/4 | 14 | 13/16 | 1/2 |
| 1 | 11 1/2 | 1 | 9/16 |
| 1 1/4 | 11 1/2 | 1 | 5/8 |
| 1 1/2 | 11 1/2 | 1 | 5/8 |
| 2 | 11 1/2 | 1 1/16 | 11/16 |
| 2 1/2 | 8 | 1 9/16 | 15/16 |
| 3 | 8 | 1 5/8 | 1 |
| 4 | 8 | 1 3/4 | 11/16 |

**Figure 12-9**
*Standard pipe threads*

| Pipe size (in) | Pipe wrench size (in) |
|---|---|
| 1/8, 1/4, 3/8 | 6 or 8 |
| 1/2, 3/4 | 10 |
| 1 | 14 |
| 1 1/4, 1 1/2 | 18 |
| 2 | 24 |
| 2 1/2, 3 | 36 |
| 4 | 48 |
| * Larger sizes usually use chain tongs or special equipment. | |

**Figure 12-10**
*Pipe wrench sizes*

### Brass Pipe and Fittings

Brass piping is approved for water pipe distribution but it costs so much that most people can't afford it. You probably won't have many chances to work with these materials except on very special applications. You can use the same pipe-threading compound as for galvanized steel. Nipples come in lengths from close to 6 inches. Nipples must conform to ASTM B 687 standards. See *IPC* Section 605.8.

## Joints

### Brass and Copper Joints

*Brazed joints* — When you make a brazed joint, make certain that all surfaces are as clean as a new silver dollar. Then apply approved flux to the specific area you're going to braze. The filler material (the stuff that joins the pipe and fittings together) must conform to AWS A5.8. See *IPC* Section 605.12.1. According to *IPC* Sections 605.14.1 and 15.1, you can also use brazed joints for both copper pipe and copper tubing. Use the same procedure for cleaning and fluxing as described for brass pipe.

*Mechanical joints* — *IPC* Sections 605.12.2, 605.14.2 and 605.15.3 say you can use mechanical joints with brass pipe, copper pipe and copper tubing. But you do have to follow the manufacturer's instructions. It would be smart to check out the specifics with your local authority.

*Welded joints* — You can use welded joints for brass pipe. Make sure all areas you're going to weld are clean. Use only approved filler metal. See *IPC* Section 605.12.4

*Soldered joints* — You can use soldered joints for copper pipe and copper tubing. Guidelines are in ASTM B 828. Be sure the cut ends are square and reamed to the full inside diameter of the pipe or tubing. Make certain that all surfaces are clean. Use only flux conforming to ASTM B 813 standards and solder that complies with ASTM B 32. Use solder that has less than 0.2 percent lead.

*Flare joints* — Copper tubing is soft and flexible so it can take a flare easily when you use a tool designed for that purpose. See *IPC* Section 605.15.2.

### ABS Plastic Joints

*Mechanical joints* — ABS plastic pipe and fittings may be linked by mechanical means. However, note two cautions: (1) the mechanical joint can be used only on underground water systems and (2) the joints must use elastomeric seals conforming to ASTM D 3139. You must install these joints according to manufacturer's instructions. See *IPC* Section 605.10.1.

*Solvent cement joints* — All surfaces of the pipe and interior of the fittings must be clean and dry. Use solvent cement that conforms to ASTM D 2235. You have only a few moments to join the pipe and fitting before the cement hardens. You'll have to work like greased lightning, making any necessary adjustments while the cement is still wet. You can't daydream when working with plastic joints. Solvent cement joints are acceptable for above or below ground installations. See *IPC* Section 605.10.2.

*Threaded joints* — You can thread ABS plastic pipe if it's Schedule 80 or heavier. However, the dies aren't the same as those used for galvanized steel pipe. They're specifically designed for threading plastic pipe. It's very important that you seal the threads with approved lubricant or tape (not pipe joint compound) to the male threads only. See *IPC* Section 605.10.3.

### PVC Plastic Joints

*Mechanical joints* — Install joints for PVC plastic pipe and fittings exactly as for ABS plastic pipe and fittings. However, you can install mechanical joints in PVC plastic pipe above ground, if you get the approval from your local authorities. See *IPC* Section 605.21.1.

*Solvent cement joints* — Make certain the surface of the pipe and the interior of the fittings are clean. Use a purple primer that conforms to ASTM F 656. Then use solvent cement (not purple) that conforms to

ASTM D 2564 or CSA CAN/CSA B 137.3. Apply it to all the joint surfaces. Time is critical, so you need to make any joint as quickly as possible while the cement is wet. You can use solvent-cement joints above or below ground. See *IPC* Section 605.21.2.

*Threaded joints* — You can thread PVC plastic pipe that's Schedule 80 or heavier. Use dies especially designed for threading plastic pipe in compliance with ASME B1.20.1. The pressure rating for PVC threaded plastic pipe is reduced by 50 percent. Apply lubricant or tape to only male threads. See *IPC* Section 605.21.3.

## CPVC Plastic Joints

It created quite a stir in plumbing circles when CPVC plastic pipe and fittings first gained approval for use within a building. And even more unbelievable, it could be used for hot as well as cold water systems. You can join CPVC pipes much the same as other types of plastic, with a few exceptions.

*Mechanical joints* — Make mechanical joints on CPVC plastic pipe and fittings following the manufacturer's instructions. See *IPC* Section 605.16.1. It's always best to get details from your local plumbing official or code.

*Solvent cement joints* — Make sure the surface of the pipe and interior of fittings are clean and dry. Use solvent cement (orange) that conforms to ASTM F 493 standards. Apply it to all surfaces you're joining. In accordance with ASTM D 2849 standards, make any fitting adjustments while the cement is still wet. You can install CPVC pipe and fittings above or below ground. You may skip the primer if the following conditions are met:

- Use yellow, third-party certified cement that conforms to ASTM F 493.
- Use solvent cement to join only CPVC pipe and fittings that are between 1/2 inch and 2 inches in diameter, according to *IPC* Section 605.16.2.
- Use CPVC pipe and fittings manufactured according to ASTM D 2846 standards.

*Threaded joints* — You can thread Schedule 80 or heavier CPVC. Use dies that are specifically for threading plastic pipe. Threading CPVC plastic pipe cuts its pressure rating 50 percent. Use approved thread lubricant or tape-on male fittings only. See *IPC* Section 605.16.3.

## Polyethylene Plastic Joints

Use a flared joint if the manufacturer recommends it. Use a special flaring tool to accomplish this. See *IPC* Section 605.20.1. For heat-fusion joints, be sure all surfaces are clean and dry. Then heat the joint surfaces to melting temperature. After you make the joint, let it cool undisturbed so as not to weaken it. Follow ASTM D 2657 standards. See *IPC* Section 605.20.2. It's OK to use mechanical joints with polyethylene pipe and tubing if you follow the manufacturer's instructions. See *IPC* Section 605.20.3.

The use of asbestos-cement pipe for water mains or water service piping is limited. If you do use it, you must seal the joint between pipe and fittings with an elastomeric ring that conforms to ASTM D 1869. See *IPC* Section 605.11.

If you use gray iron and ductile iron joints, follow the manufacturer's instructions. You may also have to get details from your code official. These joints must comply with AWWA C111 standards. See *IPC* Section 605.13.

## Other Plastic Joints

You can use flared or mechanical joints for cross-linked polyethylene plastic tubing and fittings. For a flared joint, you must use a specially-designed tool. See *IPC* Section 605.17.1. For a mechanical joint, you must use metallic lock rings and insert fittings as described in ASTM F 1807. It's also important that you follow the manufacturer's installation instructions. See *IPC* Section 605.17.2.

You can use flared joints for polybutylene plastic. Cut pipe ends square and flare using a special tool designed for this purpose. See *IPC* Section 605.19.1. For heat-fusion polybutylene plastic joints, make sure joint surfaces are clean and dry. You must make socket-fusion or butt-fusion joints. It's important to heat joint surfaces to melting temperature and then leave them undisturbed until they cool. Make joints in accordance with ASTM D 2657, ASTM D 3309 or CSA CAN3-B137.8 standards. See *IPC* Section 605.19.2. For mechanical joints, use metallic lock rings with insert fittings for mechanical joints on polybutylene plastic. Follow the manufacturer's installation recommendations and ASTM D 3309 or CSA CAN3-B137.8 standards. See *IPC* Section 605.19.3.

### *Prohibited Joints and Connections*

*IPC* Section 605.9 prohibits these joints and connections in a water supply system:

- cement or concrete joints
- joints made with fittings not approved for the specific installation
- solvent-cement joints to connect ABS and PVC
- saddle-type fittings.

## Shutoff Valves

In each building, you must install a separate full-open gate valve that's independent of the water meter valve. The control valve must be accessible and located at, or near, the foundation line outside the building, either above ground or in a separate approved box with a cover. Look back in Chapter 10 at Figure 10-2. *IPC* Section 606.1 outlines other locations for accessible control valves. Your local code may or may not require these valves:

1. near the curb
2. on the discharge side of every water meter
3. on the base of each water pipe riser in buildings that are three or more stories high
4. on the top of every water supply down-feed pipe in buildings other than one- and two-family residential units
5. on the water supply to any gravity or pressurized water tank
6. on the water supply to all water heaters
7. on the water distribution pipe at the entrance of a structure
8. on the water supply pipe to a dwelling unit, except for a single fixture equipped with individual stops.

## Avoiding Cross-Connections

*IPC* Section 608.1 prohibits any arrangement of piping that could possibly promote connection of a potable water supply system to a contaminated liquid source. Some of the more common situations that might create cross-connections are:

- a garden hose attached to an unprotected hose bibb (without backflow protection) with the end of the hose lying in contaminated water.
- a garden hose attached to an unprotected laundry sink faucet (without backflow protection) with the end of the hose submerged in a tub of dirty water, as shown in Figure 12-11.
- a private water supply, such as a well, interconnected with the public water supply.
- failure to keep the correct air gap between fixtures and faucets.
- a leaking water supply pipe in the ground near a leaking sewer line or septic tank.

You may wonder how dirty water from a submerged hose can be pulled into a water system when the latter is under pressure. Well, it isn't very likely to happen, but it can. And if it does, the whole public water system can be contaminated. Here are a couple of possible scenarios that could trigger a complete drop in water pressure and cause contamination:

1. a break in a water line or a water line that's under repair, and

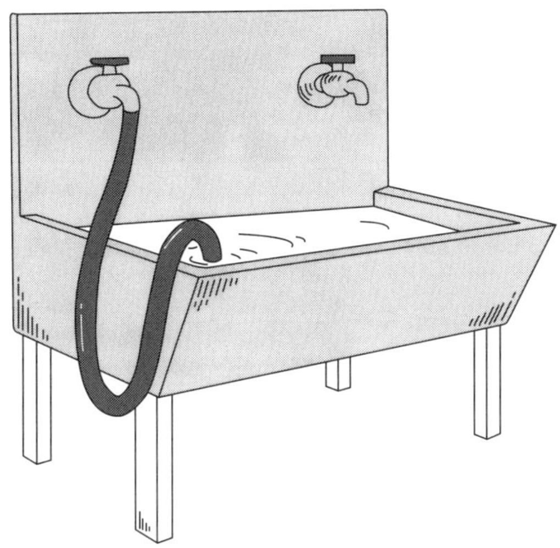

**Figure 12-11**

*A situation that creates a cross-connection*

Hose bibb vacuum breaker
for frost-proof hydrants

Hose bibb vacuum breaker

**Figure 12-12**

*Use vacuum breakers to avoid contamination of potable water*

2. a fire situation in which a pumper truck places heavy demand on the system. When water pressure fails, it pulls the water out of existing piping in the direction opposite to normal flow.

Now about the common garden hose. *IPC* Section 608.15.4.2 says you must install a hose bibb vacuum breaker on every hose bibb to isolate garden hose applications and protect the potable water supply from contamination. See Figure 12-12.

*IPC* Section 608.15.1 says you must terminate the potable water supply outlet (faucet) by using an air gap between the opening and the fixture flood level rim. Most authorities require at least a 1-inch air gap. This keeps any fixture contents from siphoning back into the potable water system. For water closets with a flush valve, install your vacuum breaker at least 6 inches above the overflow rim of the fixture. You can get a pretty clear picture of this in Figure 12-13. Also see *IPC* Section 608.15.4.

Below-rim water supplied fixtures are prohibited. See Figure 12-14. A stoppage in the waste line could cause a cross-connection. If a negative pressure develops in the supply piping, the liquid in the fixture bowl can rise to the flood-level rim of the fixture and be siphoned into the water-supply system through the submerged water outlets. However, some special fixtures can't function properly without a below-rim water supply. Some examples of this are water-cooled compressors, degreasers, portable cleaning equipment, or any tank, vat, pump or other equipment subject to back pressure. Potable water connections must be protected by a reduced-pressure-principle backflow preventer. These are described in various points in *IPC* Section 608.16.

You probably won't ever have to install a barometric loop (see Glossary). In the event that you do, extend it to a vertical height of 35 feet from the point of connection. You can use it only as an atmospheric-type or pressure-type vacuum breaker. See *IPC* Section 608.13.4. Also see the illustration in the Glossary.

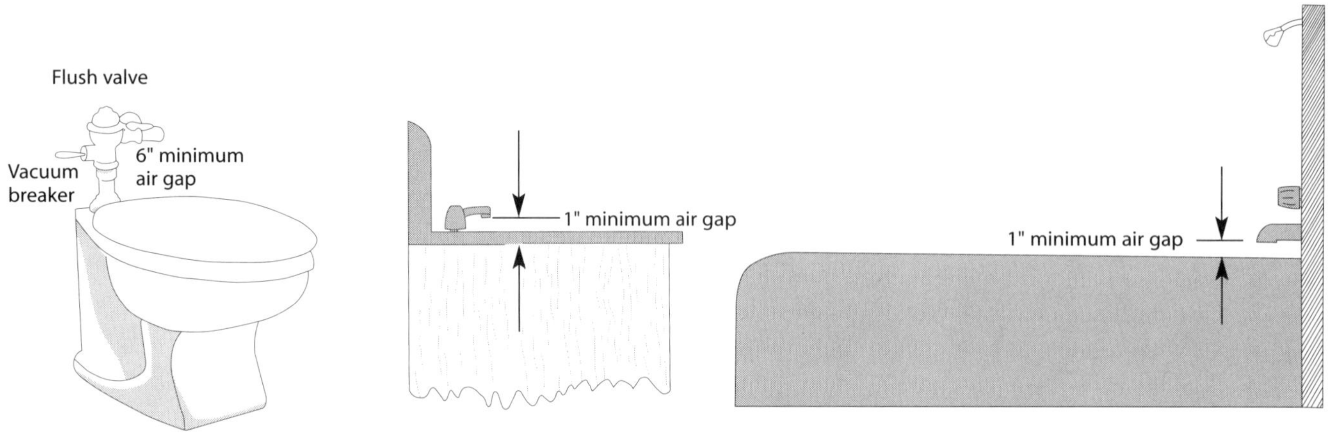


**Figure 12-13**

*Three plumbing fixtures that provide the minimum air gap permitted by code*

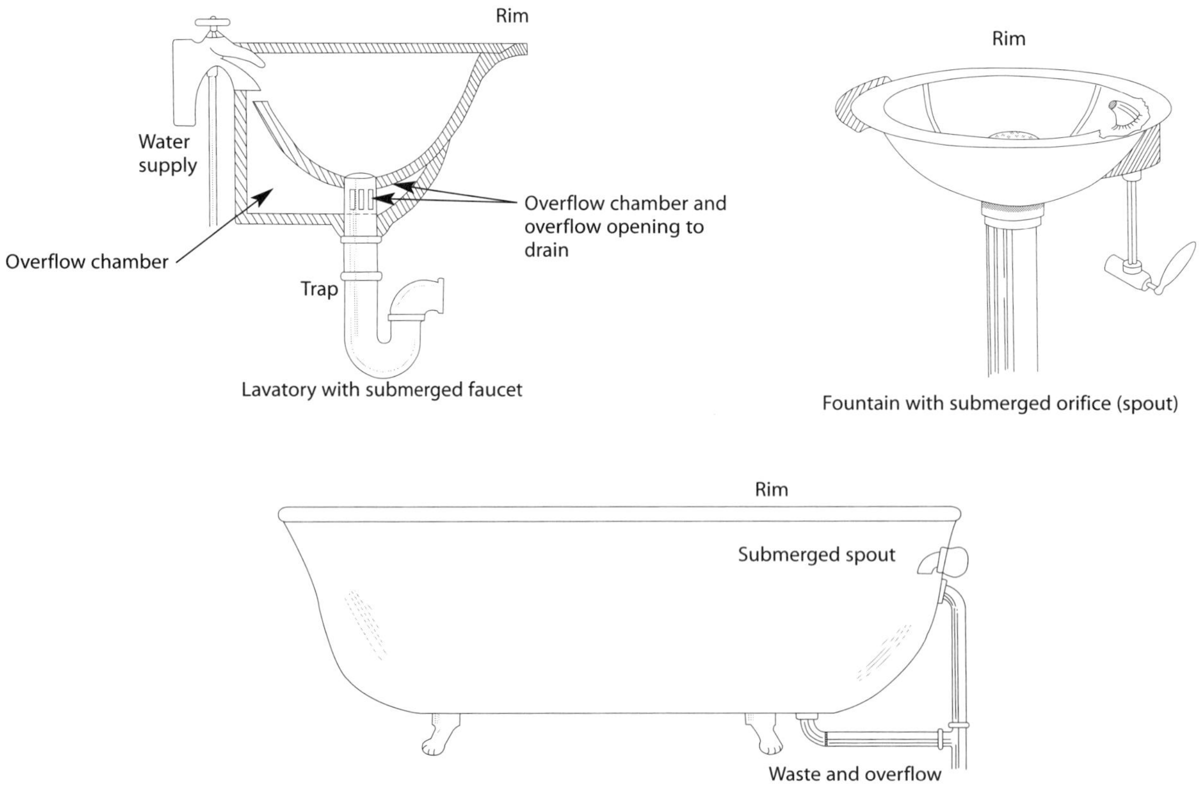


**Figure 12-14**

*Fixtures with supply outlets below the overflow rim are prohibited*

| Maximum distances between supports for horizontal water supply piping installed above ground | |
|---|---|
| **Type of pipe** | **Approximately** |
| ABS and PVC pipe | every 4 feet |
| aluminum tubing | every 10 feet |
| brass pipe | every 10 feet |
| cast-iron pipe | every 5 feet[a] |
| copper or copper-alloy pipe | every 12 feet |
| copper or copper-alloy tubing 1¼-inch dia. and smaller | every 6 feet |
| copper or copper-alloy tubing 1½-inch dia. and larger | every 10 feet |
| cross-linked polyethylene plastic pipe (PEX and PEX-AL-PEX) | every 2.67 feet |
| CPVC plastic pipe or tubing 1-inch dia. and smaller | every 3 feet |
| CPVC plastic pipe or tubing 1¼-inch dia. and larger | every 4 feet |
| steel pipe | every 12 feet |
| lead pipe | continuously |
| PB plastic pipe or tubing | every 2.67 feet |
| polyethylene plastic pipe (PE-AL-PE) | every 2.67 feet |

a. Means the maximum horizontal spacing of cast-iron pipe hangers must be increased to 10 feet when installing 10-foot lengths of pipe. See *IPC* Table 308.5.

**Figure 12-15**

*Maximum distances for horizontal pipe supports*

| Maximum distances between supports for vertical water supply piping installed above ground | |
|---|---|
| **Type of pipe** | **Approximately** |
| ABS and PVC pipe | every 10 feet[b] |
| aluminum tubing | every 15 feet |
| brass pipe | every 10 feet |
| cast-iron pipe | every 15 feet |
| copper or copper-alloy pipe | every 10 feet |
| copper or copper-alloy tubing 1¼-inch dia. and smaller. | every 10 feet |
| copper or copper-alloy tubing 1½-inch dia. and larger | every 10 feet |
| cross-linked polyethylene plastic pipe (PEX) | every 10 feet[b] |
| cross-linked polyethylene plastic pipe (PEX-AL-PEX) | every 4 feet |
| CPVC plastic pipe or tubing 1-inch dia. and smaller | every 10 feet[b] |
| CPVC plastic pipe or tubing 1¼-inch dia. and larger | every 10 feet[b] |
| steel pipe | every 15 feet |
| lead pipe | every 4 feet |
| PB plastic pipe or tubing | every 4 feet |
| polyethylene plastic pipe (PE-AL-PE) | every 4 feet[b] |

b. Means you have to provide mid-story guides for sizes 2 inches and smaller. This is to keep the pipe or tubing in alignment. See *IPC* Table 308.5.

**Figure 12-16**

*Maximum distances for vertical pipe supports*

## Pipe Supports

Use pipe straps or another approved method to secure water piping inside walls and partitions. This will prevent pipes from moving as they contract or expand. Any water supply pipes you install above ground or in a vertical position must be securely supported by the building structure. Take a look at code requirements for horizontal pipe supports for different sizes and types of material. The *maximum* distances between supports for horizontal water supply piping installed above ground are given in Figure 12-15.

Let's look now at code requirements for different types of water piping materials and pipe sizes installed vertically. Figure 12-16 gives the *maximum* distances between supports for vertical water piping installed above ground.

### *Retaining Structural Integrity*

Any time you install or repair any part of a plumbing system, it's important to maintain the building's structural integrity. The finished floors, walls, ceilings, tile work or any part of a building that's changed or replaced must be left in a safe structural condition. See *IPC* Section 307. The *IPC* provides some guidelines to make sure you're in compliance with the *International Building Code* requirements. Here are code requirements for cutting, notching or boring holes in floor joists and studs:

- *Boring a floor joist* — You can only bore a floor joist. Notching is prohibited. The joist must have at least a 2-inch undisturbed area on the top and on the bottom. For example, if a joist is 6 inches deep, the maximum-size hole you can bore in its center is 2 inches, and it will have to be *exactly* in the center in order to meet the 2-inches-on-the-top-and-bottom requirement.
- *Boring nonbearing studs* — The largest-size hole you can bore is 60 percent of the stud's width. There must be at least 5/8-inch of undisturbed wood on each side of hole. The width of the stud will determine the largest-size pipe you can use.
- *Boring bearing studs* — The largest-size hole you can bore is 40 percent of the stud's width. There must be at least 5/8-inch of undisturbed wood on each side of hole. The width of the stud will determine the largest-size pipe you can install.
- *Notching nonbearing studs* — The largest notch depth that you can make for a 2 × 4 stud is 1 7/16 inches. For a 2 × 6 stud, the largest notch depth is 2 1/4 inches.
- *Notching bearing studs* — The largest notch depth that you can make for a 2 × 4 stud is 7/8 inch. For a 2 × 6 stud, the largest notch depth is 1 3/8 inches.

You must protect copper pipe or tubing and plastic pipe or tubing you install in notched partitions from later damage by lathers or drywall installers. It's easy to drive a lath nail right through one of these pipes. Always protect pipe by covering notched openings with a metal stud guard. Look back to Figure 9-28 in Chapter 9.

## How to Measure Pipe

I always like to include this little tidbit in my books. Even though it's quite simple, you'd be amazed how many plumbers don't seem to have a clue. Before you attempt to cut and join a pipe, you'd better know the correct way to measure it. There are two basic types of pipe measurement — straight run and offset. Also, there are several ways to measure individual pieces of pipe. It's easier to accomplish an exact fit if you know the fitting dimensions and the length of thread engagement. See Figure 12-17.

Let's look at Figure 12-18, which shows several ways to measure threaded pipe. For an end-to-end measure you measure the full length of the pipe, including both threads. For an end-to-center measure you measure from the center of the fitting plus the length of thread engagement. This measurement is used for pipe that has a fitting screwed on one end only. For a center-to-center measure, pipe length is equal to the measurement minus the sum of the end-to-center dimensions of the fittings, plus twice the length of the thread engagement. This measurement is used for a length of pipe which has fittings screwed on both ends. For a face-to-face measurement, pipe length is equal to the measurement plus twice the length of the thread engagement.

| Pipe size (in) | Threads per inch | Approximate length of threads (in) | Approximate thread engagement (in) |
|---|---|---|---|
| 1/4 | 18 | 5/8 | 3/8 |
| 3/8 | 18 | 5/8 | 3/8 |
| 1/2 | 14 | 13/16 | 1/2 |
| 3/4 | 14 | 13/16 | 1/2 |
| 1 | 11 1/2 | 1 | 9/16 |
| 1 1/4 | 11 1/2 | 1 | 5/8 |
| 1 1/2 | 11 1/2 | 1 | 5/8 |
| 2 | 11 1/2 | 1 1/16 | 11/16 |
| 2 1/2 | 8 | 1 9/16 | 15/16 |
| 3 | 8 | 1 5/8 | 1 |
| 4 | 8 | 1 3/4 | 11/16 |

**Figure 12-17**
*Standard pipe threads*

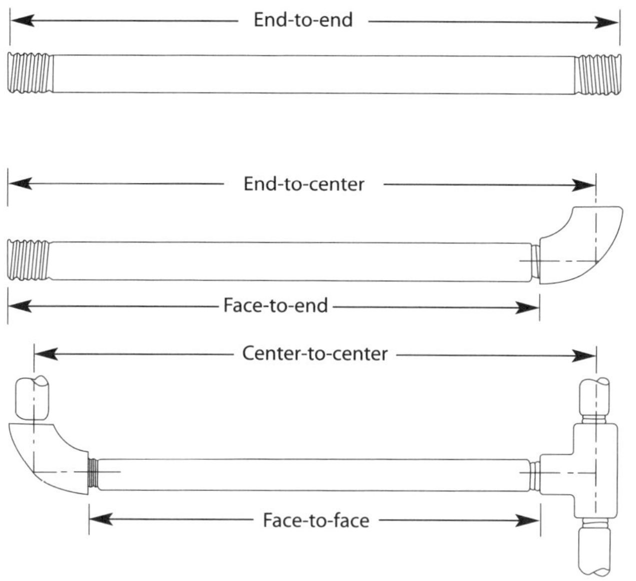


**Figure 12-18**

*Threaded pipe measurements*

| Degree of offset | When C = 1, D = | When D = 1, C = | When C = 1, E = |
|---|---|---|---|
| 60 | 0.5773 | 1.7320 | 1.1547 |
| 45 | 1.0000 | 1.0000 | 1.4142 |
| 30 | 1.7320 | 0.5773 | 2.0000 |
| 22½ | 2.4140 | 0.4142 | 2.6131 |
| 11¼ | 5.0270 | 0.1989 | 5.1258 |
| 5⅝ | 10.1680 | 0.0983 | 10.2170 |

**Figure 12-19**

*Finding the length of pipe to connect two parallel runs*

## *Measuring Offsets*

Take it from me, you need to know about offsets. Simple straight runs or 90-degree turns won't strain the brain, but offsets can. We'll use Figure 12-19 to figure out the exact distance between centers of fittings for various type offsets. And as an example, we'll use Figure 12-20, which shows a common pipe offset between parallel runs of pipe. The letters in the headings of Figure 12-19 refer to the piping segments in Figure 12-20.

Let's suppose the center-to-center offset distance C is 10 inches and that pipes A and B are connected with 45-degree elbows so the degree of offset is 45 degrees. To find the length of pipe E in Figure 12-20, look in Figure 12-19 along the 45 degrees of offset row for the amount we need to multiply C by to get the length of pipe E. In the C = 1, E = column, we see that the multiplier is 1.4142. Since C is 10 inches, E must be 10

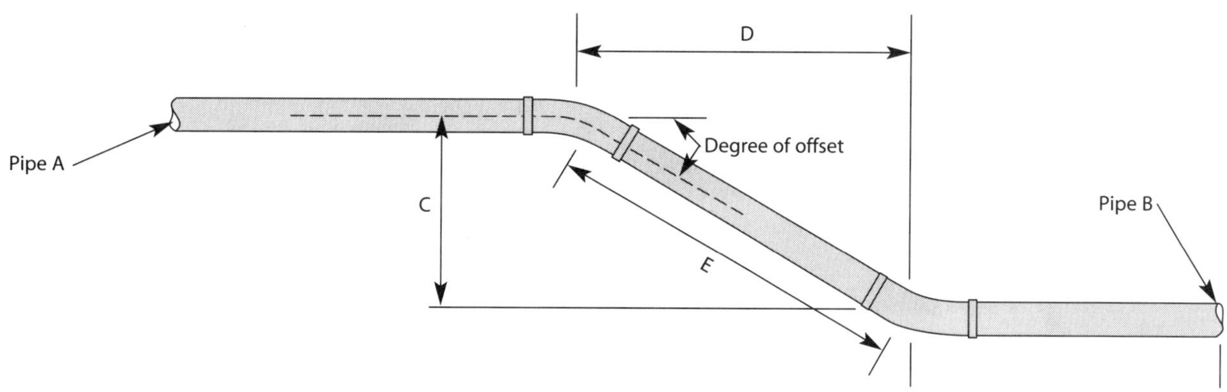


**Figure 12-20**

*Common pipe offset between parallel runs of pipe*

times 1.4142, or 14.14 inches. So pipe E must be cut and threaded end-to-end at 14.14 inches.

Now for a different example, let's suppose C is 10 inches and pipes A and B are connected with 22½-degree elbows and we need to know the length of pipe E. In this case, the multiplier is 2.6131. So pipe E must be cut and threaded end-to-end at 26.13 inches.

## Testing the Water System

You may test a water system in full or in part. It must be tight under maximum working pressure. Non-plastic piping systems may be tested with 55 psi of pressure. If testing with water, it must be from a potable source. See *IPC* Section 312.5.

# *Chapter 12 — Questions and Answers*

**12-1 According to the *IPC*, to make CPVC joints, use only ____________ solvent cement.**

(A) clear

(B) purple

(C) orange

(D) yellow

***Answer:* *C***

*Code response: Sections 605.16 and 605.16.2*
*This book, page 199*

**12-2 If you find contaminated ground water when installing water service piping, the *IPC* says you must**

(A) use galvanized piping.

(B) use glass-lined piping.

(C) install Schedule 80 plastic piping.

(D) get a chemical analysis of the soil.

***Answer:* *D***

*Code response: Section 605.1*
*This book, page 189*

**12-3 The *IPC* specifies a maximum _______ -percent lead content for all materials used in a water supply system (pipe, fittings, valves and faucets).**

(A) 4

(B) 6

(C) 8

(D) 10

***Answer:* *C***

*Code response: Section 605.2*
*This book, page 189*

**12-4 The *IPC* states that all water service piping you install must withstand a working pressure of at least _____ psi.**

(A) 150

(B) 160

(C) 170

(D) 180

***Answer:* *B***

*Code response: Section 605.3*
*This book, page 191*

**12-5 The *IPC* allows you to install water service piping in the same trench as a building sewer if you install the water piping**

(A) at least 12 inches above highest point of the sewer line.

(B) without joints.

(C) at least 10 inches above highest point of the sewer line.

(D) with the code official's approval.

***Answer: A***

*Code response: Section 603.2 (Exception 1)*
*This book, page 191*

**12-6 According to the *IPC*, if you're installing a plastic water service pipe in an area that can freeze, you must install it at least _______ inches below the frost line.**

(A) 2

(B) 4

(C) 6

(D) 8

***Answer: C***

*Code response: Section 305.6*
*This book, page 192*

**12-7 When installing a 2-inch water pipe through a building foundation, you must install a sleeve that has at least a _____-inch diameter.**

(A) $2^1/_2$

(B) 3

(C) $3^1/_2$

(D) 4

***Answer: D***

*Code response: Section 305.5*
*This book, page 193*

**12-8 If you can't install an irrigation well, the *IPC* will permit you to connect an irrigation system to the public water system using any of the following to prevent backflow *except***

(A) a double-swing check valve.

(B) a pressure-type vacuum breaker.

(C) an atmospheric-type vacuum breaker.

(D) a reduced pressure principle backflow preventer.

***Answer: A***

*Code response: Section 608.16.5*
*This book, page 193*

**12-9 When you install PVC Schedule 40 water service piping, the *IPC* says it must terminate no more than _____ feet from where it entered the building.**

(A) 4

(B) 5

(C) 6

(D) 10

***Answer: B***

*Code response: Section 605.3*
*This book, page 196*

**12-10 If you must install a barometric loop, the *IPC* says it must go vertically _____ feet from the point of connection.**

(A) 20

(B) 25

(C) 30

(D) 35

***Answer: D***

*Code response: Section 608.13.4*
*This book, page 201*

◆ *Chapter 13* ◆

# Private Water Wells

More than you may know, private water supply wells make up a significant portion of the water usage in homes in America. You may not be familiar with this water source and ask, "What difference can it possibly make to me as a plumber?" It may make a whole lot of difference. For one thing, your plumber's license qualifies you to do this type of work, and millions of people still depend upon well water for domestic use. I personally didn't have access to public water until I entered the Navy in World War II. After I left the service, I was fortunate enough to live in city areas where public water was available, until I moved to my present location in Florida. Guess what? I found that most of the homes in our county were on domestic wells. My family was obliged to use well water for the next six years. Finally, public water was brought to our area and we could tap in to the public water supply. We wasted no time in doing so!

A few years ago, the Water Quality Association estimated that nearly one home in five wasn't connected to a public water supply. That figure may have changed by now, but there are still many suburban and rural homes that get their water from private wells. The rapid growth of urban communities has often outstripped the capacity of public water distribution systems to handle their needs. On the other hand, many smaller towns and cities can't afford a public water system. And then there are all those fashionable people who've taken up country-style living the past few years. They too must depend entirely on well water.

Your local health department, and sometimes the Department of Environmental Resource Management (DERM), regulate private wells. They set the rules, and the local plumbing inspector enforces them. Irrigation wells may or may not be covered by the same rules. There are guidelines on the depth and separation distance between potable water supply wells (and in some cases irrigation wells) and any source of contamination. These guidelines are incorporated into your local codes. They're pretty standard across the nation. So you'll find this chapter to be very useful, regardless of where you do your plumbing work.

Most well water is hard water. That's because it's high in mineral content. Raw well water may have a distinctive taste (and sometimes a smell) about it. Well water users have to get used to the sometimes moss-like, woody taste because well water has none of the chlorine or other added chemicals public water system users are accustomed to. Some folks are bothered that well water doesn't make a good lather and they never feel squeaky clean after using well water.

### Irrigation Wells

Irrigation well water has a high iron content. It can leave a dark reddish brown stain on exterior building walls and sidewalks. Untreated well water used for domestic purposes will also stain the surface of plumbing fixtures. Today, we have chemicals that will remove most of these stains. Ideally, with proper measures, you can avoid most exterior stains. Place sprinkler heads where they won't sprinkle walls and walks. Also, a good water softener will prevent some or all of these unwanted minerals.

### Hot Water Tanks

Dissolved minerals in well water can cause a scale buildup in water heater tanks and water distribution pipes. This can make a water heater fail prematurely. It's good business to install a good water softener to cut down on mineral buildup in well water. It will also improve the taste of the water.

### Water Softeners

There are many different types and sizes of water softeners available. I installed a twin tank softener in my own home, as it provides soft water 24 hours a day, without interruption. A single-tank system can't do this.

The hardness in water is removed when water containing the hard ions (calcium and magnesium) flows through a bed of resin beads that exchanges the hard ions for soft sodium ions from salt. As the resin bed gradually loads up, it's regenerated with brine, rinsed and backwashed to put it back in service. In a one-tank system, the water conditioner usually shuts down during this process and the building is out of water until regeneration is complete. With a two-tank system, when the tank being used goes into regeneration, it automatically switches to the other one. That's pretty neat, because you're never out of water to use. The local authority should approve all backwash disposal methods.

## Well Installation

You must get a permit to drill, drive, bore or dig any well, whether the well is for domestic or irrigation purposes. Each well will be inspected to make sure the owner has complied with applicable regulations. A professional well driller is generally called in to install the well, suction line, water pressure pump and water softener. These workmen are certified and licensed as specialty contractors, permitted to do only this type of work. A plumbing contractor is also licensed to do this work, but is rarely called upon to do it. Even if your plumbing contract certifies you to do a complete well system, you'll most likely subcontract that part of the job to a professional well driller. Still, it's important for you to be familiar with well systems. First, many of the people you deal with will assume you know something about wells. Second (and maybe more important if you're seeking a master's license), plumber's examinations usually include questions on well systems.

As you'd expect, local authorities will determine the depth of potable water supply wells based on the depth of the water table. Even when the underground water table is within a few feet of the ground surface (as is true in many areas of Florida), your code authority may require a 30-foot minimum depth. The *IPC* sets separation distance from pollution sources and other requirements for well casings.

As a plumber, you're expected to know the installation methods and local code requirements for well installations, even if you never put in a well. Here are a few universally-accepted code requirements:

- Don't locate a well within any building or projection of any building or structure unless it's specifically approved by the local authority.

- A well casing must be continuous, made of new steel pipe, and terminate in a suitable aquifer. Well casing pipe that's 6 inches or smaller in diameter is generally required to be galvanized steel.

- To divert surface water away from a well, some local authorities require a concrete collar at least 4 inches thick and 36 inches wide poured around the top of the well casing. Slope the concrete collar away from the well casing so surface water pollutants can't get into the well casing and the water reservoir. See Figure 13-1.
- Install a tee the same size as the well casing at the top of a well. This is for access to the well casing for inspection, adding disinfecting agents, measuring well depth and testing the static water level. The pipe passing through the collar must be protected from corrosion and be watertight.

## Proper Well Location and Depth

*IPC* Section 608.17.1 sets these requirements for well locations and pump suction lines to prevent possible contamination:

- 100 feet from any barnyard
- 25 feet from a farm silo
- 100 feet from a pasture
- 2 feet from a pumphouse if there's a drain inside which drains to the outside
- 50 feet from any seepage pit
- 25 feet from a septic tank
- 10 feet from a building sewer
- 50 feet from a septic tank drainfield
- 50 feet from any sub-surface pit

The *IPC* gives two conditions which require you to increase the distance beyond those shown above: (1) in areas where creviced rock or limestone exists and (2) in areas where the ground water is moving from a contamination source toward the planned well location. The *IPC* doesn't spell out specifically how much to increase distances, so you need to check with your local authority to get it right. If the underlying rock structure is limestone or fragmented shale, consult your local or state health department regarding well location.

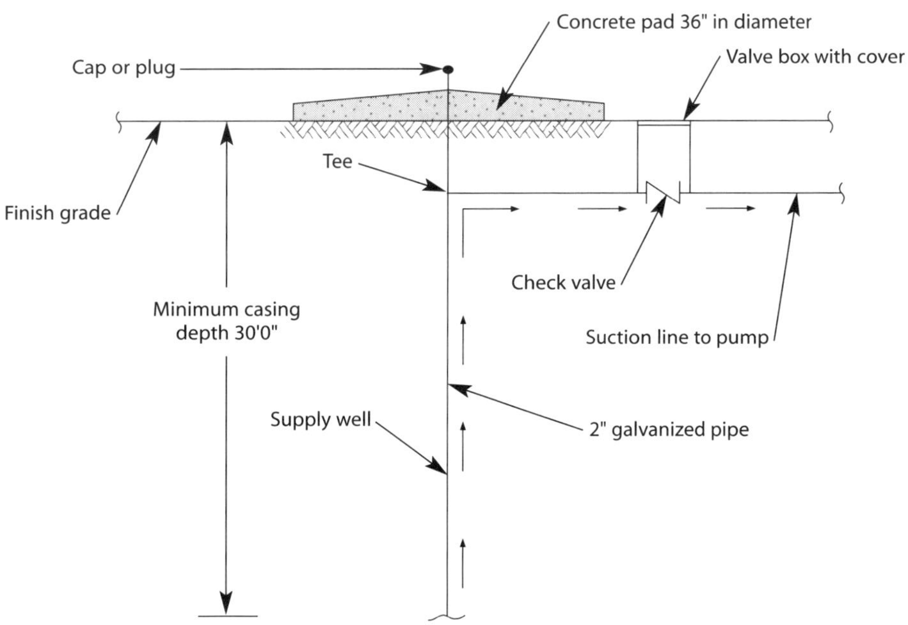


**Figure 13-1**
*Isometric supply well detail (potable water)*

Section 608.17.2 says that well sites must be positively drained (no standing water in the area) and must be at a higher elevation than any potential contamination source. Potable water-supply wells must draw water from below the first 10 feet under the ground surface. See *IPC* Section 608.17.3. That means that a well casing must extend at least below that point.

## Driven and Drilled Wells

Wells that are driven are classified as shallow wells. You'll usually find them where the water table is within 22 feet of the ground surface. The well must penetrate deep enough below the existing water table to ensure a dependable supply of water, even in very dry seasons.

There are two types of driven wells. One type has an open-end casing that extends into the rock formation. The other type has a casing equipped with a well point that ends in sand or shale rock. The casing must be steel or other approved material, and must be large enough to permit the installation of a separate drop pipe if one is needed. Extend the casing at least 6 inches above the well platform. Fill the spacing between the outside of the well casing and the earth with at least 10 feet of cement grout. See *IPC* Sections 608.17.4 and 608.17.5.

Drilled wells are classified as deep wells and may penetrate hundreds of feet into the earth. Water that comes from these wells may be a bit safer, as there's less chance of contamination. This type of well is deep enough that its water level is hardly affected by seasonal rainfall or dry seasons. When you put in a drilled well, you use materials and installation methods very much like the ones for a driven well. Figure 13-2 shows a drilled well extending considerably below the dry weather water table.

## Dug or Bored Wells

Sometimes an owner wants a dug or bored well. You'll find the *IPC* requirements in Section 608.17.6. The well casing for this must be watertight and made of concrete, tile, or galvanized or corrugated metal pipe to at least 10 feet below the ground surface. If the water table is more than 10 feet from the ground surface, extend the casing

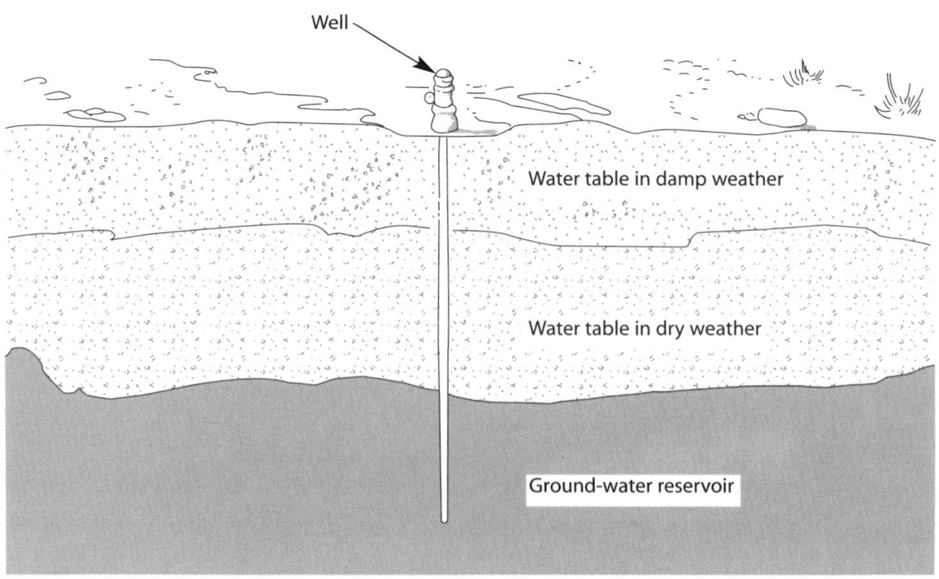


**Figure 13-2**

*Drilled well extending beyond dry weather water table*

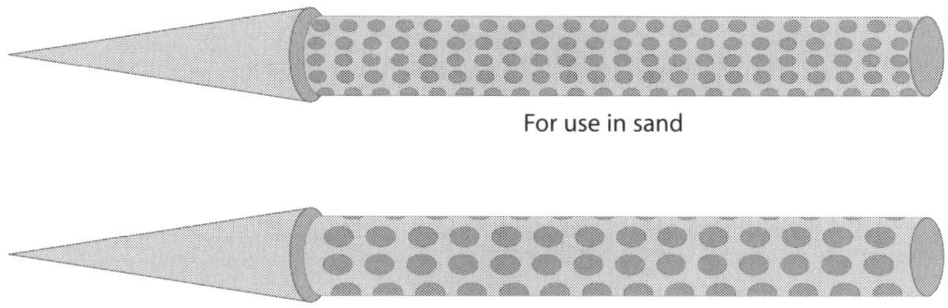


**Figure 13-3**
*Well points*

below the water surface. The *IPC* gives no specifics on distance. The code also requires that the hole be 6 inches greater than the well casing for this particular type of installation. The 6 inches between the outside of the casing and the ground must be filled with grout to a 10-foot depth.

### Well Points

If your well casing ends in loose shale or sand, you'll need to use a well point. There are two types of well points. In sand, use one that has a screen or fine perforations. In gravel, use one that has larger openings. See Figure 13-3. Attach the appropriate well point to the separate drop pipe (usually made of plastic) and lower it down the well casing to the proper depth — usually about three feet below the water surface.

### Well Covers

To prevent contaminated water from seeping down a potable well casing and getting drawn up by the pump and into the distribution lines, *IPC* Section 608.17.7 requires an overlapping cover at the top of the well casing. You must extend the cover downward, over the outside of the casing, at least 2 inches. If the cover is for a dug well, be sure you provide a sleeve that can be used to withdraw the pump suction pipe, cylinder or jet body so the well cover isn't disturbed. If the pump suction pipe or discharge pipe enters or leaves the side of the well casing, be sure to make the opening watertight.

## Suction Lines

This is an area that the *IPC* leaves to the discretion of local authorities. The rest of this chapter will provide some of the guidelines that your local jurisdiction will most likely accept. When in doubt, check with your local authority before installing a system.

When you size a suction line, be sure it's large enough to provide the water volume and pressure required to operate the plumbing fixtures in the building. Don't install a suction line or water service pipe from a well to a pump that's smaller than 1 inch. Install the suction line with a pitch toward the well. If a building requires a suction line longer than 40 feet, increase the size of the suction pipe to the next pipe size. Let's do a simple example for a residence that has 25 dfu and a suction line 55 feet long using Figure 13-4. For 25 dfu, you need a suction pipe that's $1^1/_4$ inches. However, since the suction line is longer than 40 feet, you need the next larger size, and that's $1^1/_2$ inches. Figure 13-4 only works for residences or small commercial buildings equipped with flush tank type water closets. Any larger job that depends on well water will have to be sized by an engineer.

As a rule, local codes require that you install a soft seat check valve rated at 200 pounds water test as close as possible to a well. The code does permit the use of either a spring-loaded or flapper-type brass constructed check valve up to 2 inches in diameter. You must install a union or slip coupling close to the pump for repair or replacement purposes.

| Drainage fixture units (dfu) | Supply required (GPH) | Diameter of suction pipe | Diameter pressure pipe | Diameter service pipe | Size of tank | Horsepower | Well size |
|---|---|---|---|---|---|---|---|
| 23 | 720 | 1 | 3/4 | 3/4 | 42 | 1/2 | 1 1/2 |
| 30 | 900 | 1 1/4 | 1 | 1 | 82 | 3/4 | 2 |
| 40 | 1200 | 1 1/2 | 1 | 1 | 120 | 3/4 | 2 |

**Figure 13-4**
*Tank and pump size requirements*

Materials and installation methods for suction pipe are usually the same as for water service piping described in Chapter 12.

## Pressure Tanks

There are two types of pressure tanks that are commonly used today. The older type (usually visible from the road as you travel through the country) is the hydropneumatic tank. The other type is the diaphragm pressure tank, typically used for urban residences.

### The Hydropneumatic Tank

The hydropneumatic tank has a good track record, so long as it's working properly. This tank uses compressed air to provide enough water pressure for a number of plumbing fixtures to work properly. The compressed air acts like a pressure plate pushing down on the surface of the water, forcing water out of the tank to the fixtures. The pressure switch is usually set so the pump starts when pressure drops to about 20 psi. The pump is preset to stop usually when pressure reaches about 40 psi. Air chargers and air volume control devices maintain the compressed air automatically, as long as they're serviced regularly. Figure 13-5 shows a hydropneumatic tank.

As water is drawn from the tank, the air pressure gradually lessens. Eventually there won't be enough air available to pressurize the tank. The lost air must be automatically replaced. Since water itself is almost incompressible, if a tank loses air, the water will be pressurized only when the pump is running. So each time a faucet is opened, the pump cycles on and off. If this is allowed to go on too long, the pump will burn out.

If the automatic controls don't keep a tank charged with air, you can manually drain the water from the tank. As you put water back into the empty tank, it will refill with air at atmospheric pressure. The compressed air will again supply the proper operating pressure for the water system. Be sure to check the automatic air devices to find out if they need to be repaired or replaced.

When you size a hydropneumatic tank, be sure it's large enough so the pump won't cycle too often. A tank should supply 6 gallons of water and maintain an operating pressure between 20 to 40 psi. Choose a pump that's large enough to quickly refill the tank. Do your sizing based on the dfu in the building. You may wish to look at Figure 3-10, back in Chapter 3. For other tank sizes and specific requirements, see Figure 13-4. For a single-family residence, make sure the hydropneumatic tank is at least 42 gallons.

### The Diaphragm Pressure Tank

The diaphragm pressure tank, also known as a breather tank, came into use in the 1950s. The tank shell is made of lightweight drawn steel, with an epoxy finish and a bonded plastic lining for extra corrosion resistance. It has a replaceable air charge

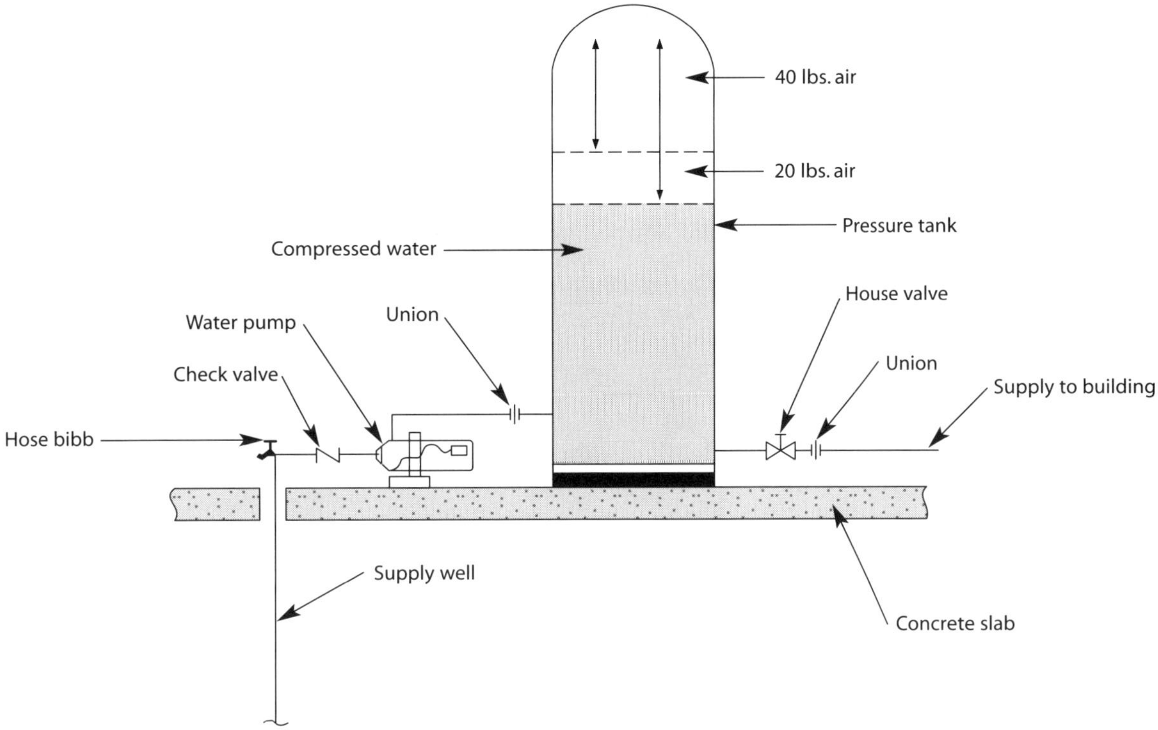


**Figure 13-5**
*Typical pressure tank (hydropneumatic) installation*

valve at the top of the tank for safe and easy operation, and a strong, flexible butyl diaphragm with a locking retainer ring to separate air and water. As a rule, local officials will give the nod to your using a diaphragm pressure tank. See Figure 13-6 for the working details of a diaphragm pressure tank.

A diaphragm tank has several advantages. It comes from the factory pre-charged with air, and it's more attractive and occupies less space than a hydropneumatic tank. It's suitable for residences and small businesses. You can usually count on it to be trouble-free because it doesn't require air to be injected into the tank. It can be as small as 14 gallons or as large as 119 gallons. Base the size of a diaphragm tank on the dfu of the building. It's a smart move to allow for a little extra tank capacity. It won't add much to the cost of a system and can prevent excessive pump starts.

## Piping a Pressure Tank

No matter what type of tank you install, here are some guidelines to follow:

- Make certain the piping from pump to tank is no smaller than the discharge outlet of the pump.
- If the tank capacity is more than 42 gallons, install a gate valve, with the handle removed to prevent someone shutting it off, between the pump and the tank.
- Install a 3/4-inch gate valve on the discharge side of the tank. This serves as the house valve to control water in the building.
- Always set the tank level on a concrete pad that's large enough to support the pump and tank as a single unit. Place both so they can be serviced when necessary.

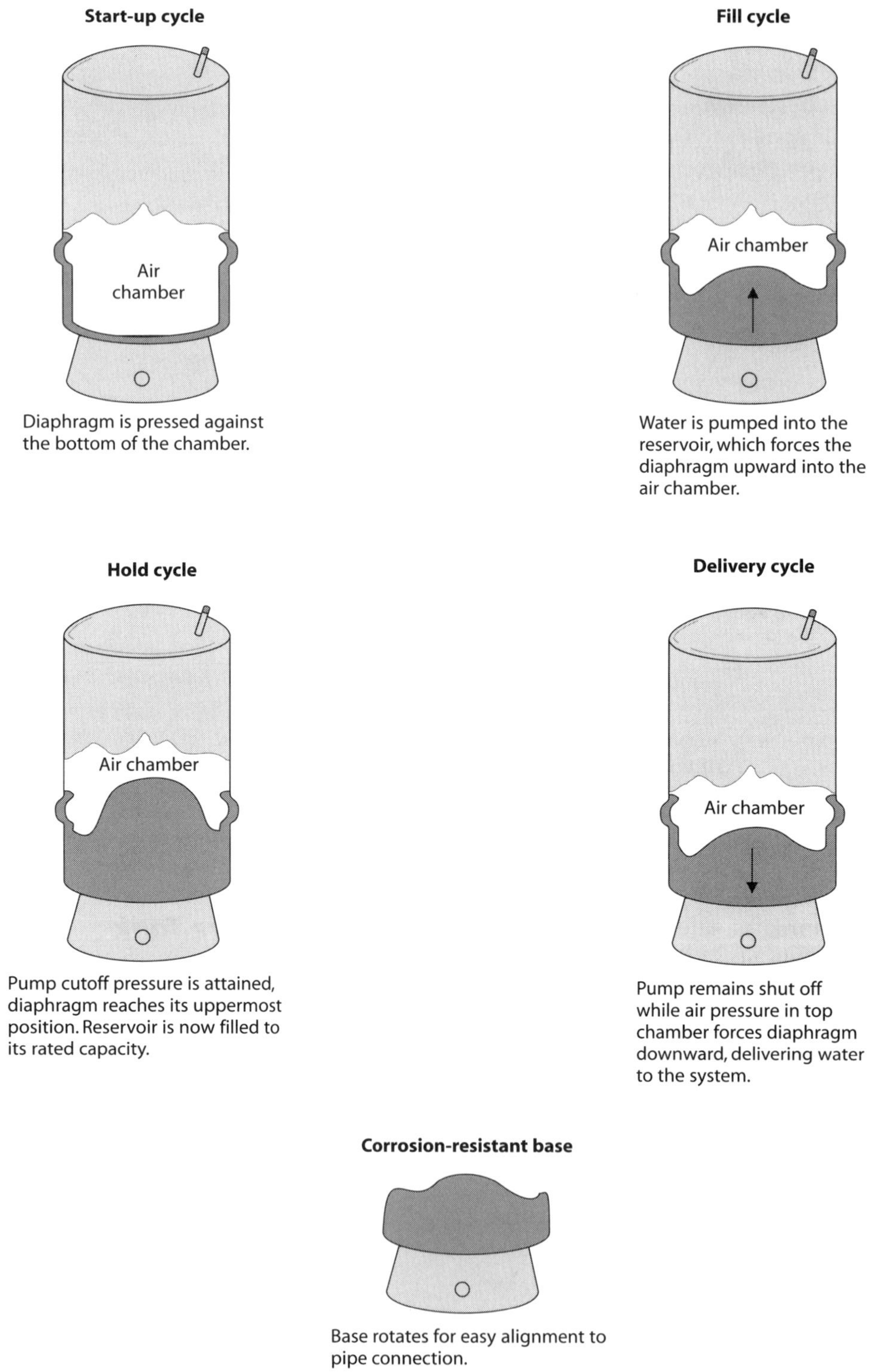


**Figure 13-6**

*How a diaphragm tank works*

# *Chapter 13 — Questions and Answers*

**13-1 If you install a well in limestone, the *IPC* says you must**

(A) seek advice from a certified well driller.

(B) drill the well an extra 30 feet deep.

(C) consult the code official.

(D) consult the local or state health department.

***Answer: D***

*Code response: Section 608.17.1*
*This book, page 213*

**13-2 According to the *IPC*, you can install a domestic well no closer than _____ feet from a seepage pit or subsurface disposal field.**

(A) 35

(B) 50

(C) 75

(D) 100

***Answer: B***

*Code response: Table 608.17.1*
*This book, page 213*

**13-3 The *IPC* states that domestic well sites must be**

(A) located on a level site.

(B) positively drained.

(C) at least 100 feet from any building.

(D) protected from damage.

***Answer: B***

*Code response: Section 608.17.2*
*This book, page 214*

**13-4 The *IPC* specifies that potable water from a domestic well must come from a water table that's at least _____ feet below the ground surface.**

(A) 10

(B) 12

(C) 15

(D) 30

***Answer: A***

*Code response: Section 608.17.3*
*This book, page 214*

**13-5 The *IPC* says that a domestic well casing must extend at least _____ inches above the well platform.**

(A) 2

(B) 4

(C) 5

(D) 6

***Answer:* *D***

*Code response: Section 608.17.4*
*This book, page 214*

**13-6 One material the *IPC* says you may *not* use for the well casings of a dug or bored domestic well is**

(A) tile.

(B) watertight concrete.

(C) Schedule 80 plastic pipe.

(D) corrugated metal pipe.

***Answer:* *C***

*Code response: Section 608.17.6*
*This book, page 214*

**13-7 To meet *IPC* requirements for a well with a 2-inch casing, you must drill a hole _____ inches in diameter.**

(A) 6

(B) 10

(C) 12

(D) 14

***Answer:* *D***

*Code response: Section 608.17.6*
*This book, page 215*

**13-8 The *IPC* says you must locate any domestic-use well at least _____ feet from any farm pasture.**

(A) 75

(B) 100

(C) 150

(D) 200

***Answer:* *B***

*Code response: Table 608.17.1*
*This book, page 213*

**13-9 One *IPC*-approved material you can use when you drill a domestic well to receive a 2-inch well casing is ______**

(A) steel.

(B) cast iron.

(C) extra-heavy brass conforming to ASTM 40.

(D) wrought iron.

***Answer: A***

*Code response: Section 608.17.5*
*This book, page 214*

**13-10 The *IPC* says you must install a/an ______ to keep contaminated water from seeping down a domestic well casing.**

(A) French drain

(B) anti-siphon device

(C) stainless steel collar around the casing

(D) overlapping watertight cover

***Answer: D***

*Code response: Section 608.17.7*
*This book, page 215*

◆ *Chapter 14* ◆

# Plumbing Fixtures

In this chapter we'll take you through code requirements for the materials, design, and installation of plumbing fixtures and plumbing fixture trim. We'll discuss the type of occupancy, and the minimum number of fixtures for each type of occupancy. It's your responsibility to install only high-quality plumbing fixtures that follow the standards in Figures 14-1A and 14-1B. But always check your local code standards too. Never use substandard fixtures (even if you can get away with it) that will detract from the quality of work you do. Even if it's the fixture that fails, you're the one who put it in. Don't think that the customer won't feel you're at least partly to blame.

A typical single-family residence has approximately 300 feet of piping concealed under the floor and stashed away in partitions. This part of a plumbing system is known as the *rough plumbing*. It includes the drainage and waste piping, the vents and the hot and cold water lines. Installing rough plumbing is possibly the single most important step where plumbers display their skill, or lack thereof. Proper roughing-in of the waste and water outlets requires both knowledge and good workmanship. The roughing-in will determine when and where to install fixtures.

*IPC* Section 402 specifies that plumbing fixtures be made of high-quality materials. There can be no defects or concealed fouling surfaces. Fixture surfaces must be smooth and non-absorbent, and designed so that all surfaces are accessible for cleaning. Porcelain enameled fixture surfaces must be acid resistant. The code says specialty fixture materials must be stainless steel, soapstone, chemical stoneware or plastic. Or, they must be lined with lead, copper-base alloy, nickel-copper alloy, corrosion-resistant steel or other code-approved material especially suited to the fixture's use. According to *IPC* Section 402.3, if you use sheet copper for general plumbing, it must weigh at least 12 ounces per square foot. Section 402.4 says that sheet lead for pans must weigh at least 4 pounds per square foot and coated with asphaltum paint or other approved material.

Be sure to pay attention to plumbing fixture flow rates and flush volume standards (including fixture valves and faucets). Follow the strict water conservation specifications built into *IPC* Table 604.4, shown in Figure 10-3 back in Chapter 10.

## Water Closets

There are four types of code-approved water closets. Although all of these water closets are installed in the same way, they differ in flushing action. The first three water closets in Figure 14-2 are two-piece (better known as closet combinations). The fourth water closet is one-piece (tank and bowl manufactured in one unit).

| | |
|---|---|
| **Bathtubs** | ANSI Z124.1, ASME A112.19.1, ASME A112.19.4, ASME A112.19.9, CSA B45.2, CSA B45.3, CSA B45.5 |
| **Bidets** | ASME A112.19.2, ASME A112.19.9 or CSA B45.1 |
| **Dishwashers (domestic)** | ASSE 1006 |
| **Dishwashers (commercial)** | ASSE 1004 and ANSI/NSF 3 |
| **Drinking fountains** | ASME A112.19.1, ASME A112.19.2 or ASME A112.19.9 and ARI 1010 |
| **Floor drains** | ASME A112.21.1, ASME A112.3.1 or CSA CAN-B79 |
| **Food grinders (domestic)** | ASSE 1008.73 |
| **Laundry trays** | ANSI Z124.6, ASME A112.19.1, ASME A112.19.3, ASME A112.19.9, CSA B45.2 or SCA B45.4 |
| **Lavatories** | ANSI Z124.3, ASME A112.19.1, ASME A112.19.2, ASME A112.19.3, ASME A112.19.4, ASME A112.19.9, CSA B45.1 CSA B45.2, CSA B45.3 or CSA B45.4 |
| **Showers (prefabricated)** | ANSI Z124.2, ASME A112.19.9 or CSA B45.5 |
| **Sinks** | ANSI Z124.6, ASME A112.19.1, ASME A112.19.2, ASME A112.19.3, ASME A112.19.4, ASME A112.19.9, CSA B45.1, CSA B45.2, CSA B45.3 or CSA B45.4 |
| **Urinals** | ASME A112.19.2, CSA B45.1 or CSA B45.5 |
| **Water closets** | ANSI Z124.4, ASME A112.19.2, CSA B45.1, CSA B45.4 or CSA B45.5 |
| **Whirlpool bathtubs** | ASME A112.19.7 or CSA B45.5 and CSA CAN/CSA-B45 (Supplement 1) |

**Figure 14-1A**

*Plumbing fixtures and authorized approvals*

| | |
|---|---|
| **Faucets and fixture fittings** | ASME A112.18.1, CSA B125 and ANSI/NSF 61 |
| **Waste fittings** | ASTM F409 |
| **Hose spray** | ASSE 1025 |
| **Hand shower** | ASSE 1014 or CSA CAN/CSA-B125 |
| **Shower valve** | ASSE 1016 or CSA CAN/CSA-B125 |
| **Hose-connected outlets** | ASME A112.18.3 |
| **Flushometer valves and tanks** | ASSE 1037, ASSE 1001 or CSA CAN/CSA-B64.1.1 |
| **Ball cocks** | ASSE 1002 or CSA B125 |
| **Sheet copper** | ASTM B152 |
| **Flush pipes and fittings (nonferrous)** | ASME A 112.19.5 or CSA CAN/CSA-B125 |
| **Manual food and beverage dispensing equipment** | ANSI/NSF 18 |

**Figure 14-1B**

*Plumbing fixture trim and standards*

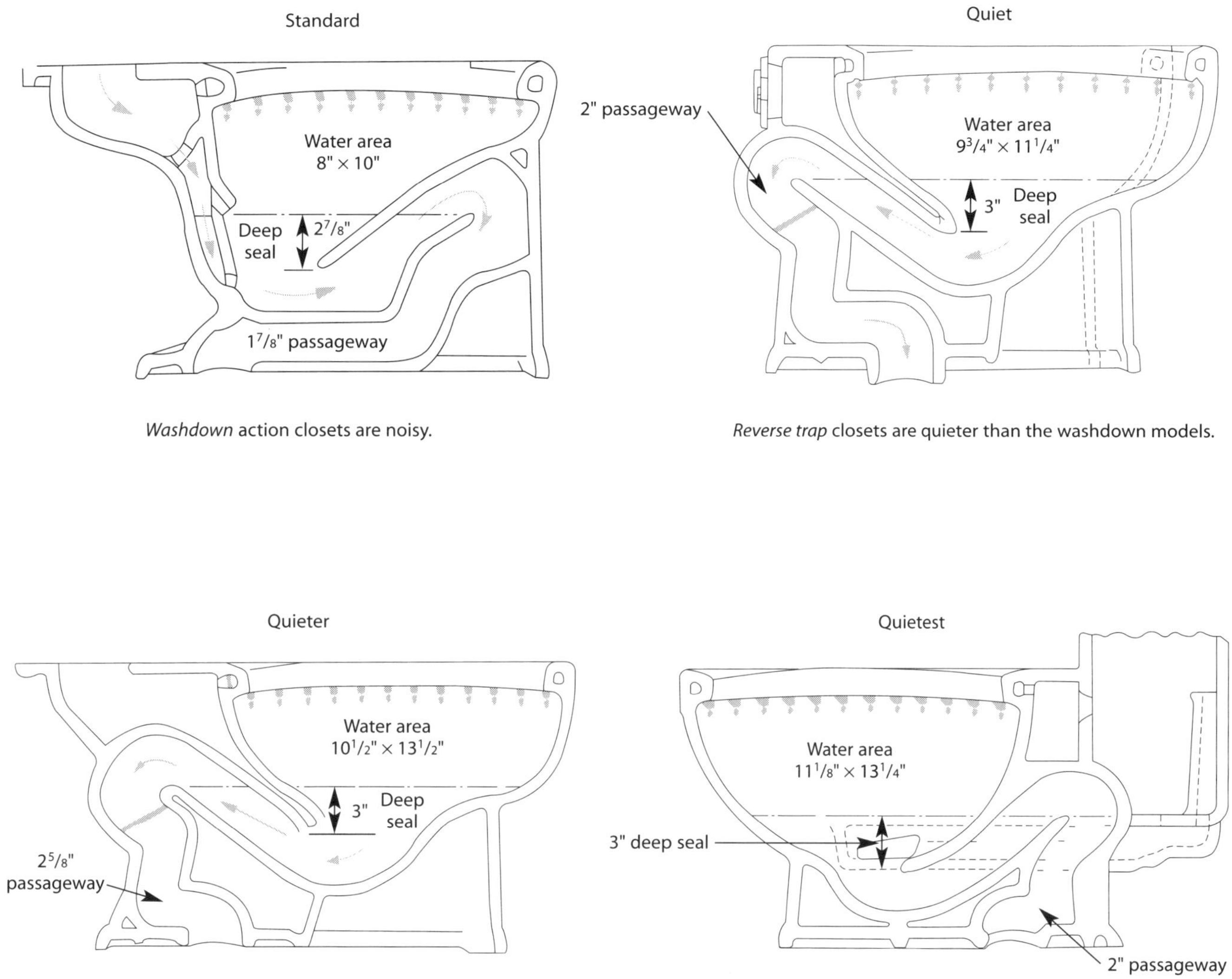


*Washdown* action closets are noisy.

*Reverse trap* closets are quieter than the washdown models.

*Siphon jet* closets are the next in quietness.

*Siphon action* closets are the quietest toilets.

**Figure 14-2**
*Types of water closets*

The washdown bowl is the simplest and least expensive type of all water closets. But it's not accepted by most modern codes. The trap is located in the front of the bowl, which is flushed by small streams of water running down from its flush rim (not a very efficient method). You may find washdown water closets in older installations, but you'll want to check with your local authority first if someone asks you to install one.

The reverse trap bowl is similar to the washdown bowl except that the trap is in the rear of the bowl. This makes for a larger water surface, which provides more cleaning with each flush. It also operates more quietly.

It's difficult to tell a siphon jet bowl from a reverse trap bowl just by looking at it. What makes it unique is the small hole in the bottom of the bowl that delivers a jet of water. At the beginning of each flush, this water combines with water delivered from the bowl's rim, creating a siphoning action and a cleaner bowl.

A siphon action bowl is the most efficient, quietest and most expensive bowl. It looks like the siphon jet bowl. However, a siphon action bowl holds more water, so more of the bowl surface is submerged when the water's at rest in the bowl. This results in a cleaner bowl.

Manufacturers make water closets that have both round and elongated fronts. Typically, you can use a round-front bowl only for residential units. The code permits only the elongated type for public or employee use. See *IPC* Section 420.2. You must install water closet bowls that have hinged type seats, made of smooth, non-absorbent material. The seat must properly fit the water closet bowl. That means you can't install a seat with a round front on an elongated bowl or vice versa. Seats for public use must have an open front. If a manufacturer designs any water closet with an integral seat, the seat must be of like material. See *IPC* Section 420.3. Use only water closets equipped with anti-siphon ball cocks and flushometers.

## Installing Water Closets

Be sure to grout all water closets, wall-hung or floor-mounted, with white cement or other approved sealants. You must make a watertight seal where fixtures come in contact with walls or floors. This stops odor-causing materials from accumulating and keeps insects from taking up residence at these ideal locations. See *IPC* Section 405.5.

According to *IPC* Section 420.4, you can use a 4-inch by 3-inch closet bend. If you choose a straight 3-inch 1/4 bend for a water closet connection, use a 4-inch by 3-inch water closet flange to receive the fixture horn. Check this type installation with local authorities first.

Brass water closet floor flanges must be at least 0.125 inches thick. Plastic floor flanges must be at least 0.25 inches thick. Cast-iron or galvanized malleable iron floor flanges must also be at least 0.25 inches thick, with at least 2 inches of caulking. Lead

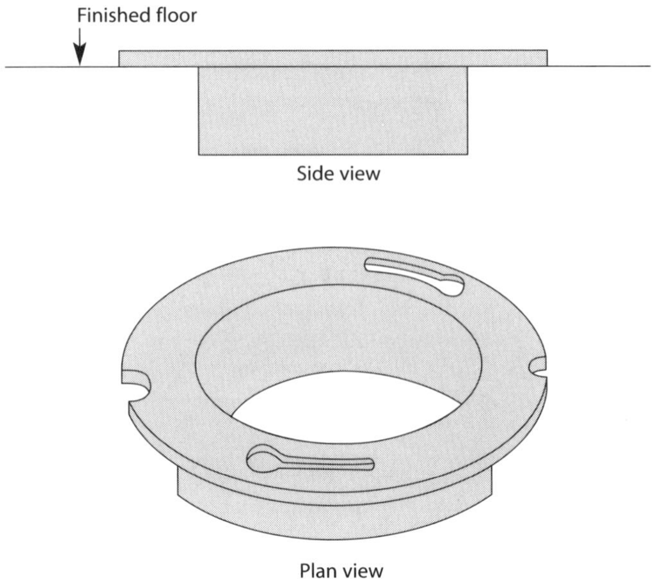


**Figure 14-3**

*Standard design for a cast iron or plastic closet floor flange*

floor flanges must weigh at least 1 pound 9 ounces and be made of lead alloy, with a minimum 7.75 percent antimony weight. See *IPC* Section 405.4.1 and Figure 14-3. It's important to use corrosion-resistant closet screws and bolts (brass) when you secure any floor-mounted fixture to the floor or floor flange. See *IPC* Section 405.4.2.

Support wall-hung water closets to a concealed metal carrier with brass bolts. Make sure the carrier is attached to the building structure, so there's no strain on the water closet connector or other parts of the plumbing system. Make your connection between a wall-hung water closet and the drain with an approved extension nipple or horn adapter. Seal the joint with an approved elastomeric gasket or approved setting compound. See *IPC* Sections 405.4 and 405.4.3.

A water closet with a ballcock-designed tank must refill after each flushing and then close tight when the tank is full. After the flush valve is activated manually, the flushing operation is automatic. The flushing device and connection between tank and bowl should have enough flow capacity for the water to flush all surfaces of the bowl. The ballcock's refill tube turns downward into the tank's overflow tube, automatically restoring the bowl's water seal as the tank refills. Make sure the ballcock has an antisiphon valve built into the unit at least 1 inch above the overflow pipe. This will prevent any possible cross-connection. See *IPC* Sections 425.3 and 425.3.1.

A water closet tank must be equipped with an overflow pipe that's sized to prevent tank overflow and remove excess water at the rate it enters the tank. This is to keep the bathroom from flooding in case the flush ball is so securely in place on the flush valve seat that the ballcock hangs up in an open position. The flush valve seat must be at least 1 inch above the rim of the water closet bowl. See *IPC* Section 425.3.2.

When a water closet uses a flushometer valve, be sure the valve is equipped with a vacuum breaker. Most local codes require that the vacuum breaker be located at least 6 inches above the rim of the closet bowl. It must complete the normal flushing cycle automatically after being manually activated. It must deliver enough water to flush all surfaces of the bowl. The valve should open fully and close tight under the normal water working pressure. Each flushometer valve can serve only one fixture and must be accessible for servicing. The valve must have some way to regulate water flow and automatically refill the water closet trap seal after each flushing cycle. See *IPC* Section 425.2. The code requires all water closets, tanks or flushometers to have a maximum average consumption rate of 1.6 gallons of water per flush. Look back to *IPC* Table 604.4, shown in Chapter 10, Figure 10-3.

Public and employee toilet rooms must have water closets installed in separate compartments. They must have a door enclosing the fixture for privacy. When there's a single-occupant toilet room with a lockable door, the water closet doesn't have to be installed in a separate compartment. See *IPC* Section 310.4.

The general contractor on a job must provide toilet facilities for all construction workers. The company providing portable toilets must keep them in a sanitary condition with regular pump outs. The most common construction toilets are the non-sewer type. See *IPC* Section 311.1.

*IPC* Section 410.2 prohibits installation of a drinking fountain in a public restroom.

## Urinals

Local authorities usually have the final say as to type, number and installation methods for urinals. The code doesn't allow trough urinals in new installations. You must rigidly support wall-hung urinals with concealed metal carriers or other approved backing. This is to keep any strain off the waste pipe connection. Make a watertight seal where a urinal touches the wall by grouting the seam with white cement or other approved materials.

If a urinal has a flushometer valve, make sure it works like one for a water closet. Once it's manually activated, the normal flushing cycle must be automatic and deliver enough water to flush all surfaces of the urinal. The valve must open fully, and close tightly at normal water pressure. The valve must regulate the water flow. See *IPC* Section 425.2. Section 425.1.1 says a flush valve can't serve more than one fixture. Urinal flush valves must have an average maximum flow rate of 1 gallon of water per flush. See *IPC* Table 604.4.

In a public toilet room, you may substitute urinals for up to 67 percent of the required number of water closets. See *IPC* Section 419.2. Check with your local authority as to whether these requirements hold for ladies' public toilet rooms. Yes, there are urinals designed for women. But time will tell how this new luxury item will be accepted and how soon architects will use them in designing new buildings.

The *IPC* specifies acceptable material surrounding a urinal. Although a builder usually installs this, you should know what the code approves and whether the builder is using the correct material. You want to make sure you don't have to take down a fixture so the builder can replace surrounding material with the proper material. This will save you time and money! *IPC* Section 419.3 says you must waterproof each side of a urinal with a smooth, non-absorbent, readily-cleanable material (laminated formica, tile or other approved material). Cover the space to a point 2 feet in front of the urinal lip, 2 feet on each side of the urinal, and at least 4 feet above the floor.

## Bidets

Bidets are quite common in many residential units. To guard against any cross-connection possibility, code requires that you install a backflow preventer or provide an air gap. The waste piping is usually $1^1/_4$ inches in diameter and must be attached to a trap of like size. See *IPC* Section 408.2. See Figure 14-4 for details.

## Floor Drains

You may not realize it, but the code recognizes a floor drain as a plumbing fixture. And it's very specific about how it's installed and used. We'll review code requirements in this section, and you may want to review Chapter 8 for additional information.

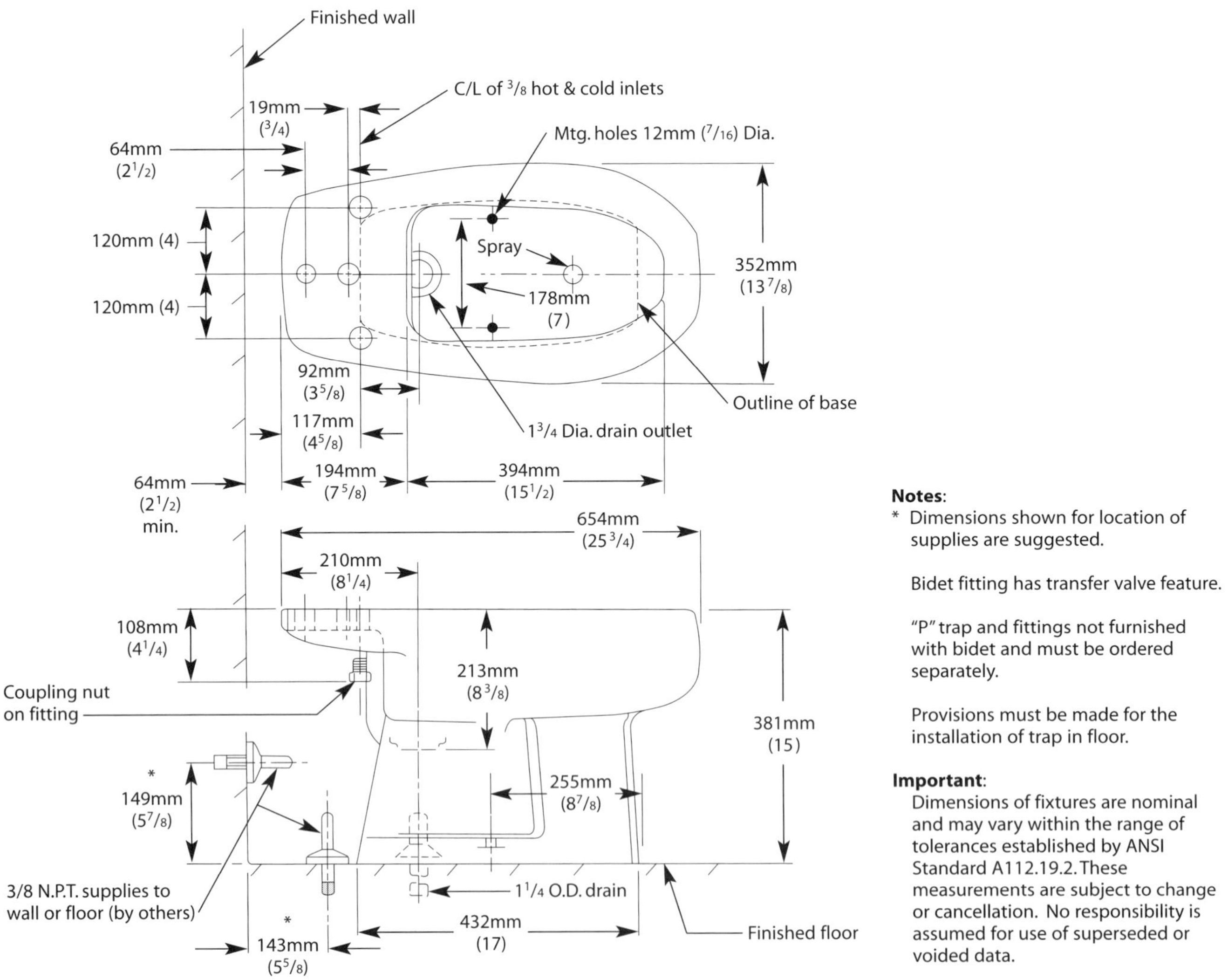


Courtesy: American Standard

**Figure 14-4**

*Roughing-in measurements for a bidet*

*IPC* Section 412 points out just a few simple requirements for floor drains. The traps must be accessible for cleaning. The area of the strainer must equal the area of the drain outlet, and be removable.

When you put in the drain outlet for a floor drain, be aware that it can't be smaller than 2 inches in diameter. If you install it in a public laundry or a central washing facility for a multiple-family dwelling, your minimum drain outlet must be 3 inches. For public washing facilities, the code also requires that the floor drain be located so, in case of an accident, it will readily drain the entire floor area.

## Lavatories and Bathtubs

Wall-hung lavatories installed in commercial buildings are generally secured to a concealed metal carrier. It's important to prevent strain on the fixture connection because commercial lavatories can get rather brutal use. It's a different story with wall-hung lavatories for residential use. You can install a wood backing fastened to the bathroom partition studs. When the wall is finished, secure the metal bracket with long screws to the backing to support the lavatory. Grout the surface where a wall-hung lavatory contacts a finished wall with white cement or other approved material.

Fasten cabinet-mounted commercial or residential lavatories securely to a countertop with special rim clips. Where a lavatory edge rests on a countertop, use a caulking compound or other approved adhesive to make the union watertight. Support other wall-mounted fixtures according to the manufacturer's recommendations, and grout where a fixture meets wall surfaces. Waste outlets for lavatories must be at least $1^1/_4$ inches in diameter. Always provide a strainer, pop-up stopper, crossbar or other acceptable device to hold water in the basin and keep out any items that might stop up the drainage system. See *IPC* Section 416.3.

Manufacturing plants and similar establishments often use multiple wash sinks. These may be circular, semi-circular, linear or square. Regardless of the shape, *IPC* Section 416.1 says you must place a faucet every 20 inches.

*IPC* Table 604.4 says you must use water-conservation lavatory faucets. Check your water flow rates. Residential lavatories must have aerators and a maximum 2.2 gpm water flow rate. Metering-type commercial lavatory faucets require a water flow rate of 0.25 gpm. Non-metering commercial lavatory faucets require a water flow rate of 0.5 gpm. *IPC* Section 405.3.2 says that in public toilet rooms, any required lavatory must be installed in the same room as the required water closet.

### *Bathtubs*

Bathtubs are manufactured in enameled pressed steel, enameled cast iron, or in gel-coated fiberglass. Although the code doesn't demand slip-resistant bathtubs and showers, most manufacturers provide them. See Figure 14-5. The code says that the finished coating be easily cleaned and generally acid-resistant.

When you install a bathtub that's recessed into a tile or other finished wall, be sure you waterproof the joints by grouting the seams with white cement or other approved material. The *IPC* Section 417.4.1 requires that the walls above a built-in tub be of smooth, non-corrosive, non-absorbent waterproof material and extend at least 6 feet above the bathroom floor.

*IPC* Section 407.3 requires that windows and doors within bathtub enclosures conform to the safety glazing standards in the *International Building Code*. *IPC* Section 417.2 says that any water supply riser to a showerhead must be firmly attached to the building structure. This also applies to bathtub showers. Tub and shower valves must deliver 3 gpm to conserve water.

Bathtub waste outlets must be at least $1^1/_2$ inches in diameter. *IPC* Section 407.2 says that a waste outlet must have an approved stopper. It's important to select stoppers that are trouble-free. You'll be expected to service them the first year at your expense. It's surprising what the advantages and

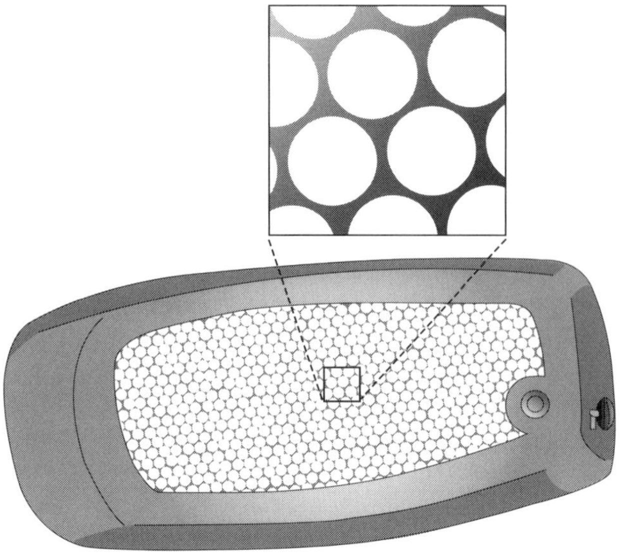

**Figure 14-5**
*Slip-resistant surface for a bathtub*

disadvantages can be in such a simple piece of equipment. Let's look at three types of stoppers.

*Tip-toe waste and overflow (Figure 14-6 A):* This is the most trouble-free of the three types shown. There are no internal moving parts to wear out or break. Usually, accumulated hair or other debris won't cause a stoppage. Drain plugs for this fixture come in two types. The tip-toe drain plug gets its name from the way it works. To close the drain, you can press the stopper with a toe. To open the drain, press the stopper again. The other type of drain plug works like the basket strainer in a kitchen sink. To close the drain, turn it and let it drop. To open the drain, lift and turn to lock it in an open position. If the stopper ever needs replacing, unscrew the stopper and replace it with a new stopper of the same type.

*Trip waste and overflow (Figure 14-6 B):* Some local codes no longer permit this stopper. Be sure to check before you set a tub with this particular type

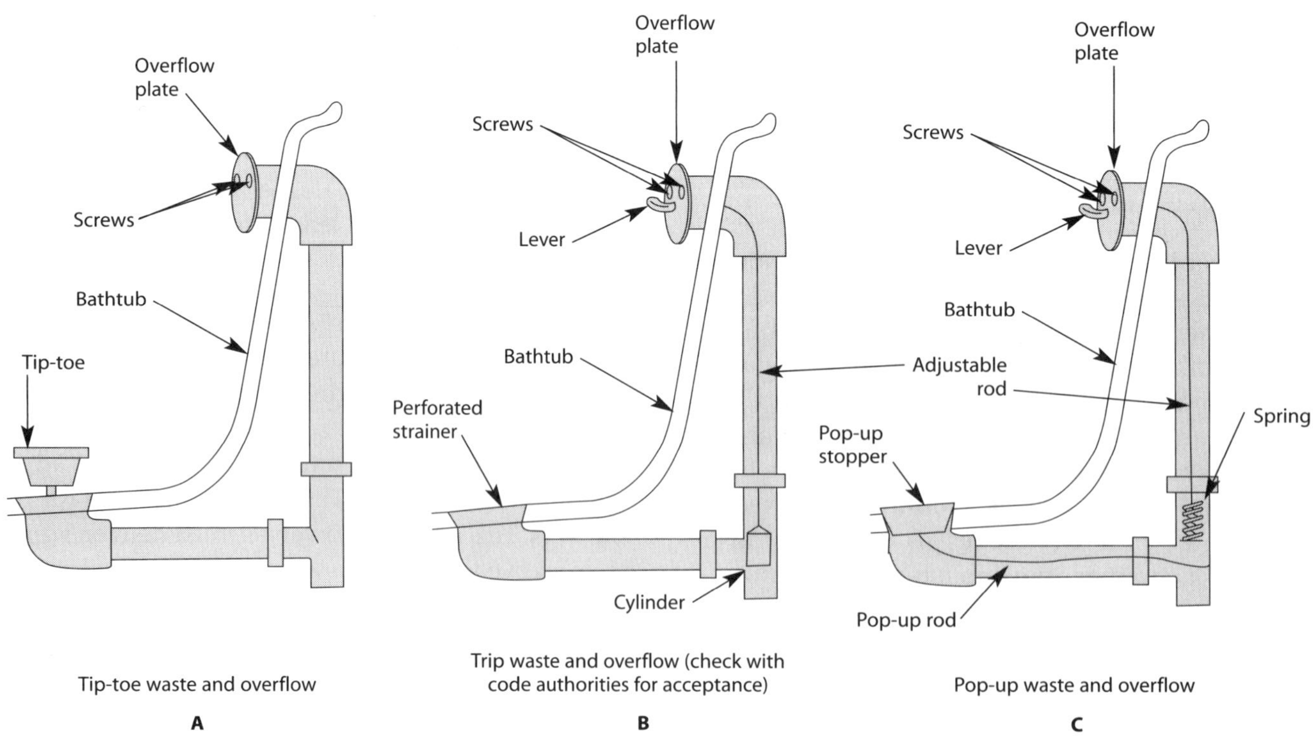


**Figure 14-6**
*Bathtub waste and overflow fixtures*

of waste and overflow. The water is controlled with a heavy brass cylinder attached to a small adjustable rod. The upper part of the rod is attached to the overflow plate. A perforated strainer covers the waste opening at the bottom of the tub. When the lever is pressed down, the rod lifts the cylinder and the wastewater drains away. When the lever is pulled up, the cylinder is lowered to a ground seat, closing the drain. This stopper has its disadvantages. It's notorious for collecting hair, lint and other debris on the ground seat and the cylinder surface. Then the cylinder can't make a watertight seal against the ground seat and won't hold water in the tub. It's also difficult to adjust this stopper. If you adjust the cylinder down a little too much, it'll partially block the drain and water won't drain as it should. If you adjust the cylinder up a little too much, the tub won't hold water properly. And the adjusting rod is only about 1/8 inch in diameter and it tends to wear out after a few years. If the rod breaks, it's not easy to remove unless you can get at the waste and overflow pipe. Now, *IPC* Section 405.8 requires that access panels be provided for fixtures that have concealed slip-joint connections. If you have to replace a cylinder and there's no access panel, you may have to break out an opening in the wall. Or if you have lots of time, you can remove the overflow plate and fish for the cylinder with the hook end of a clothes hanger.

*Pop-up waste and overflow (Figure 14-6 C):* This is the most popular waste and overflow fixture and the one you'll install most often. The pop-up stopper operates just like the trip waste stopper. Each has a lever attached to an adjustable rod, which in turn attaches to the overflow plate. Turn the lever to the left to close the drain and to the right to open it. Instead of a cylinder, the pop-up has a coiled spring attached to the end of the adjusting rod. You don't have to make any precise adjustments to get this stopper to work. It does collect hair, lint and other debris, but it's easy to clean.

*Chain and stopper waste and overflow*: You may come across an old-fashioned type waste and overflow not shown in Figure 14-6. These antiquated items are usually found on the old style leg tub. If it's in decent shape, it has a stopper attached by chain to the overflow plate. If you want to hold water, you put the stopper in the waste opening. If you want to let water out, take the stopper out. As a rule, the stopper is the only part that needs replacing and you can purchase one at most any hardware store. If you ever need to replace a whole waste and overflow fixture, use one of the waste and overflows described above and shown in Figure 14-6.

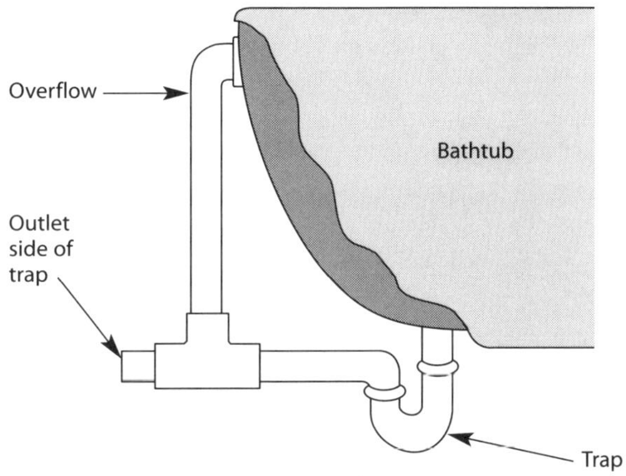


**Figure 14-7**
*Prohibited overflow connection*

*IPC* Section 405.7.1 says overflow must discharge into the drainage system on the fixture side of the trap. See Figure 14-6. Present-day plumbing regulations prohibit the use of certain overflow connections, such as the one in Figure 14-7, which locates the overflow between the trap and the sanitary drainage system. This will allow sewer gases to pass into the overflow and into the building — definitely taboo.

## Showers

There are two types of showers available — manufactured and tiled. Manufactured shower stalls are made of fiberglass, porcelain-coated steel or other approved materials. A porcelain-steel shower stall has three sides and a base that you assemble at the job site. It's usually a very small unit, barely meeting minimum code space requirements. Its sides

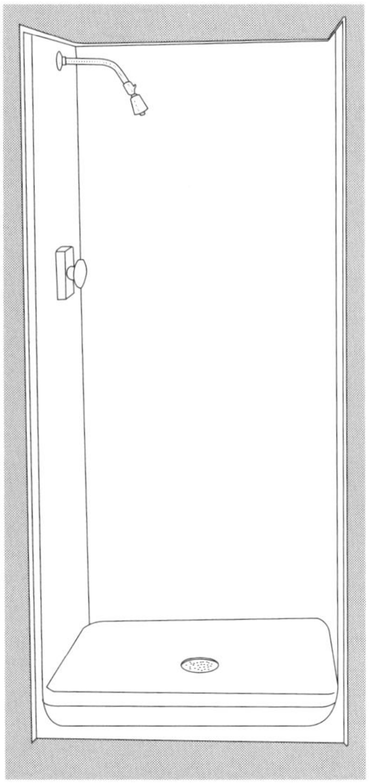

**Figure 14-8**
*Fiberglass shower stall*

are grooved to interlock and make a watertight joint. Although each unit is coated with porcelain, some water and dampness can collect in or around the joints. Over a period of time the joints tend to rust, making it difficult to keep them clean. The base may be made of heavier porcelain-steel, precast concrete or other approved material. A fiberglass shower stall is manufactured in one piece in various sizes and colors. A fiberglass unit has smooth, watertight, easy-to-clean surfaces. See Figure 14-8.

Manufactured shower stall bases usually come with the shower drain built into the base. The base is installed after the floor is poured and partitions are in place. That leaves no wiggle room for roughing-in mistakes. Accuracy is critical for a perfect fit. A manufactured shower stall takes up less space and uses less water than a bathtub.

The most commonly-used shower is built at the job site with three sides and the floor constructed of tile, marble, or other approved material. It can be sized to fit any situation. It can be built on the same level as the existing bathroom floor or recessed to a usable depth. It's OK to install the waste opening in any location within the shower compartment as long as it's low enough so the shower floor slopes toward the drain from all angles. You won't install the tile or marble — that's the tile setter's job. The plumbing or building code regulates materials and installation methods.

## Code Requirements for Showers

Let's review some of the main code requirements for showers:

Waste outlets for shower compartments must be at least $1^1/_2$ inches in diameter. A shower must have a strainer that's at least 3 inches in diameter. The strainer openings must be at least 0.25 inches in diameter. The strainer must be removable for cleaning. A shower trap can't be smaller than the waste outlet pipe ($1^1/_2$ inches) in the shower compartment. See *IPC* Section 417.3. In gang showers or shower spaces that don't have individual drains, you must locate the waste outlets (one or more) and arrange the floor pitch so waste from one shower can't drain over the floor area of another shower. Be sure the shower drains are secured to the waste pipe in an approved manner. See *IPC* Section 417.3.

*IPC* Section 417.4 says a shower compartment must have at least 900 square inches of floor area and at least a 30-inch span between walls. This may be slightly less than code requirements you're used to. It's up to the building contractor to make a smooth and structurally-sound base to receive the shower base or floor. See *IPC* Section 417.5.1. Line floors under shower compartments (other than prefabricated units) and make them watertight using an approved material.

## Shower Floors and Pans

Before you install a shower pan, the carpenter must install the wall framing and curb. If the building has wood floors, the carpenter must provide a solid base of sub-flooring for the shower pan. In buildings with concrete floors, a layer of 15-pound asphalt felt (some local codes permit a $^1/_2$-inch layer of fine sand) is required under the pan. This provides protection for the pan against rough surfaces that could puncture the lining. See *IPC* Section 417.5.2.4.

Shower pan material needs support along its sides until it's tiled. Place backing between the studs. Make sure the pan turn-up is at least 2 inches above the finished threshold level. It's OK to use nails to secure the shower lining to the backing, but never less than 1 inch above the finished threshold. See *IPC* Section 417.5.2. Figure 14-9 shows a shower pan installation.

A shower floor must have a 2 percent slope toward its drain. Shower strainers are equipped with a clamping ring (upper flange). The shower liner is clamped between the two pieces making up the clamping ring, which makes a watertight joint. Again, see *IPC* Section 417.5.2. Figure 14-10 shows the details of a shower pan installation.

Whatever material a shower pan is made of, it's very important that you follow the manufacturer's installation instructions and code requirements for installing the material. PVC plasticized sheets are commonly used but CPE non-plasticized sheets are also acceptable. The sheets must be 0.040 inches thick. Join the seams by solvent welding. Any sheet lead you use for shower pans must weigh at least 4 pounds per square foot. Coat the pan with asphalt paint inside and outside to protect it from corrosion where it contacts concrete or mortar. Join the lead pans by burning. See *IPC* Sections 417.5.2.1 and 417.5.2.3.

Any sheet copper you use for shower pans must weigh no less than 12 ounces per square foot. Although not spelled out in the code, you should paint the inside and outside of a pan as a protection from concrete/mortar corrosion. Join the sheets by brazing or soldering. See *IPC* Section 417.5.2.4.

## Kitchen and Laundry Sinks

Kitchen sinks are usually made of stainless steel, enameled cast iron or enameled pressed steel, in various colors, designs and lengths, with one, two

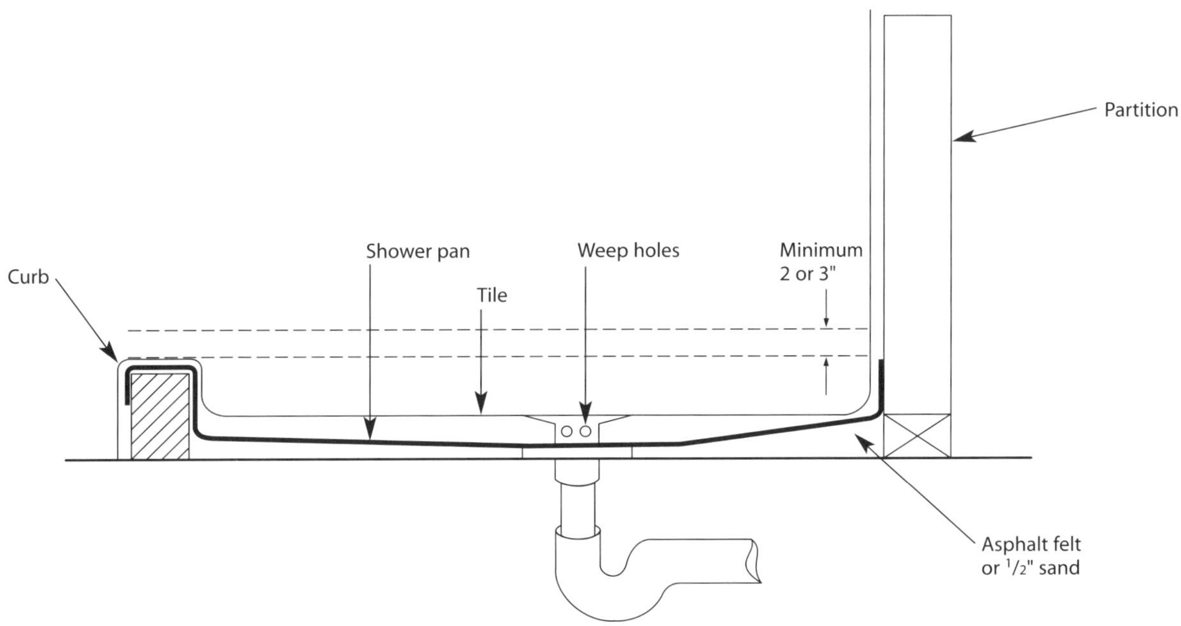


**Figure 14-9**
*Shower pan installation*

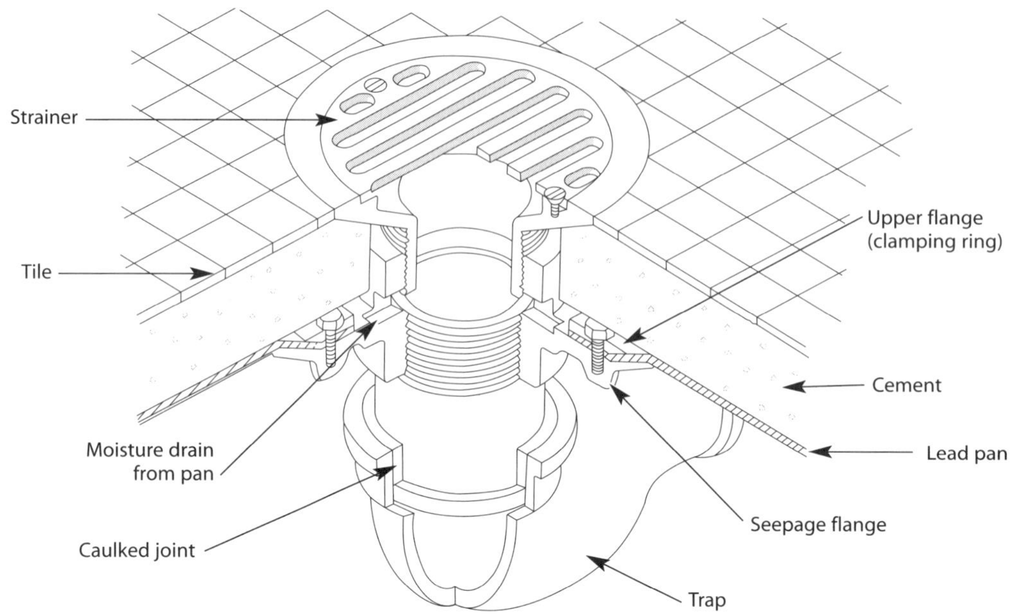


**Figure 14-10**
*Shower pan installation details*

or three bowls. Figure 14-11 shows a single-bowl set in a countertop using a rim and clamps. Figure 14-12 shows a double-bowl, self-rimming type of sink.

The code requires the sink trap, tail pieces, and continuous waste pipe to be at least $1^1/_2$ inches in diameter. The code also requires a $3^1/_2$-inch opening in a domestic sink for a standard size sink strainer and to install a food waste grinder. Provide a strainer or crossbar over the waste outlet to keep large food items out of the drainage system. See *IPC* Section 418.2.

Connect a domestic food waste grinder to a drain that's at least $1^1/_2$ inches in diameter. You can use the same sink compartment and trap. Connect a commercial food waste grinder to a drain that's at least 2 inches in diameter. Use a separate sink compartment and trap for a commercial unit. Make sure a commercial unit always has a supply of cold water available. See *IPC* Section 413. *IPC* Section 413.1 says you don't have to add any dfu for a food waste grinder.

Laundry sinks are made of concrete, enameled pressed steel, enameled cast iron, plastics, vitreous china or other approved material in single- and double-compartment styles. They may be mounted on a cabinet or supported by brackets or a metal stand. Figure 14-13 shows a countertop installation. Figure 14-14 shows two laundry sinks supported by a metal stand.

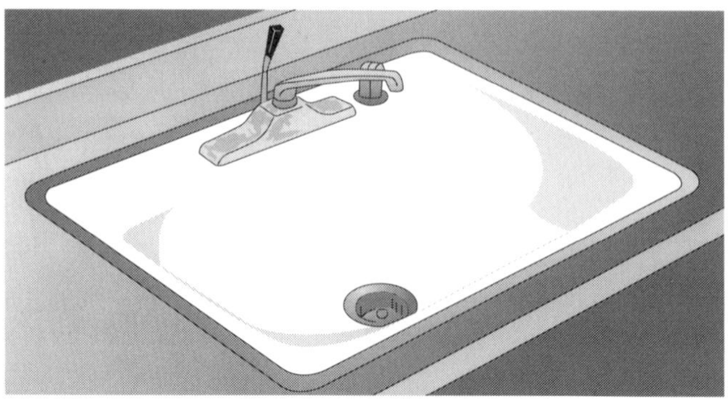

**Figure 14-11**
*Countertop sink secured by rim*

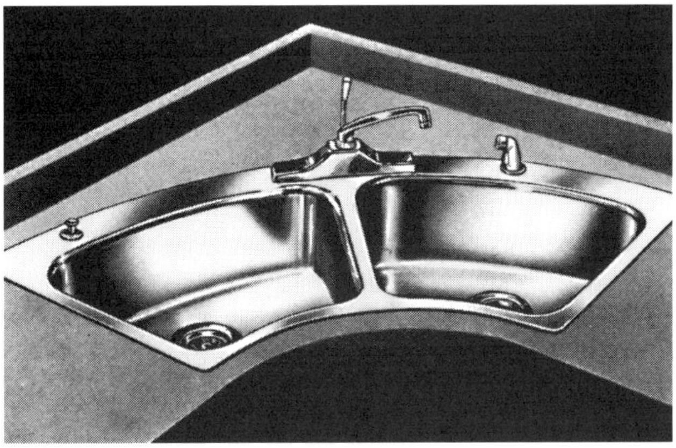

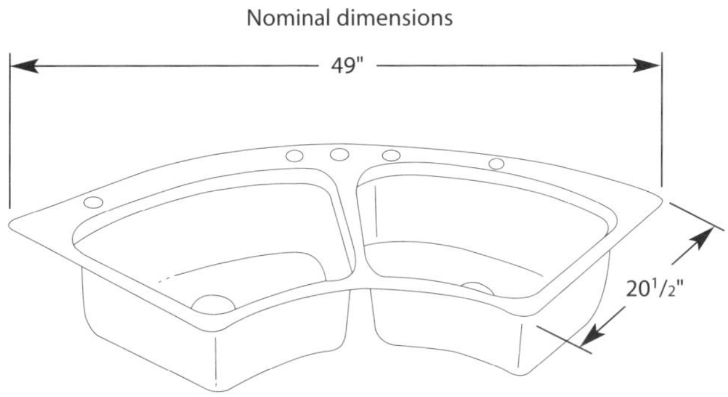


**Figure 14-12**
*Corner stainless steel double bowl self-rimming countertop sink*

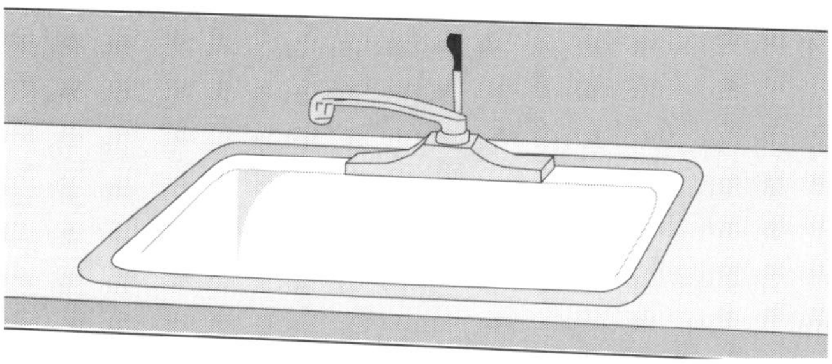

**Figure 14-13**
*Laundry sink supported by countertop*

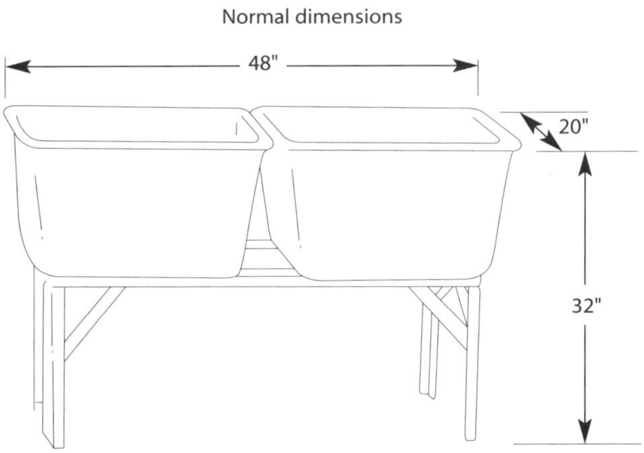


**Figure 14-14**
*Double laundry sink supported by cast iron stand*

Any waste pipe for a laundry sink must be at least $1^1/_2$ inches in diameter. This includes traps, tailpieces, and continuous waste pipe. The waste outlet must have a strainer or crossbar to keep objectionable objects out of the drainage system. See *IPC* Section 415.2.

A laundry sink faucet is usually a swing spout type and may or may not have threads for a hose attachment. If there are threads, be sure the faucet has a backflow prevention device. Use only water conservation faucets in kitchen or laundry sinks.

## Dishwashing and Clothes Washing Machines

In new construction, the cabinetmaker on the job will make the proper opening for dishwasher installation. But it's your responsibility to connect the water and waste to conform to code requirements. As a rule, you'll connect a dishwasher to the hot water supply for the kitchen sink. You can use any approved water piping material, but most plumbers prefer hard or soft copper with brass or copper fittings because it's easier to work with and has a long life. Be sure to install an air gap or backflow preventer per *IPC* Section 608. Here are some guidelines for installing a hot water supply line:

- Install a hand shut-off valve that's easy to get at under the kitchen sink cabinet near the hot water supply pipe.
- Make certain the water supply line's inside diameter is at least $^3/_8$ inch.
- Use a compression fitting or union to make the final connection of any dishwasher factory-installed water fill valve. Don't use a solder fitting because the heat to make the joint can damage a fill valve.

There are specific code requirements for a dishwasher waste connection. The waste piping must be at least as high as the underside of the dishwasher top. *IPC* Section 802.1.6 says you can fasten a waste line securely to the underside of the sink rim or counter, as shown in Figure 14-15. Or you can connect a waste line to a deck-mounted air gap fitting, as shown in Figures 14-16 and 14-17. Then you connect the waste discharge pipe to the sink waste pipe. Make the connection through a dishwasher wye branch, as shown in Figure 14-15, a directional tee as shown in Figure 14-16, or a food waste disposer, shown in Figure 14-17.

The water for a dishwasher must be hot, usually about 150 degrees, the same as all other fixtures in the residence. Every commercial dishwasher needs to have a booster heater to make sure the water stays hot enough to kill germs — about 180 degrees. However, the local health department regulates the water temperature for commercial dishwashers. Also, according to *IPC* Section 802.1.6, commercial dishwasher waste must discharge through an air gap or air break into a standpipe or receptor. See Chapter 8 for more information on indirect waste piping.

Always protect the water supply connected to an automatic clothes washing machine against backflow. See *IPC* Section 406.2. Check with your local authority to find out what's acceptable. The waste must discharge into a washing machine standpipe, somewhere between 18 and 42 inches above the trap. See *IPC* Section 802.4. The waste may also be approved to discharge into a laundry sink. See *IPC* Section 406.3.

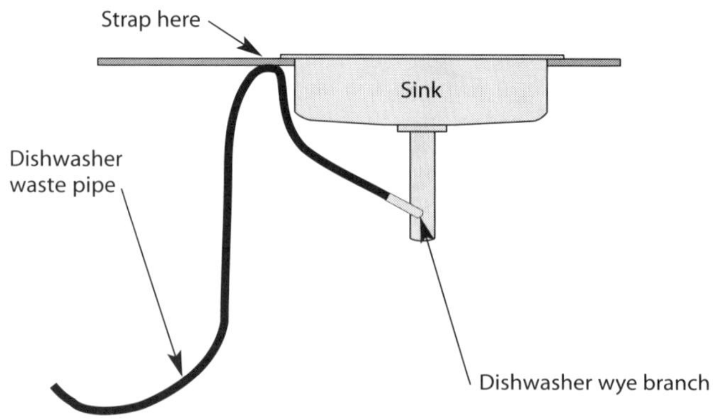


**Figure 14-15**
*Dishwasher connection through wye branch*

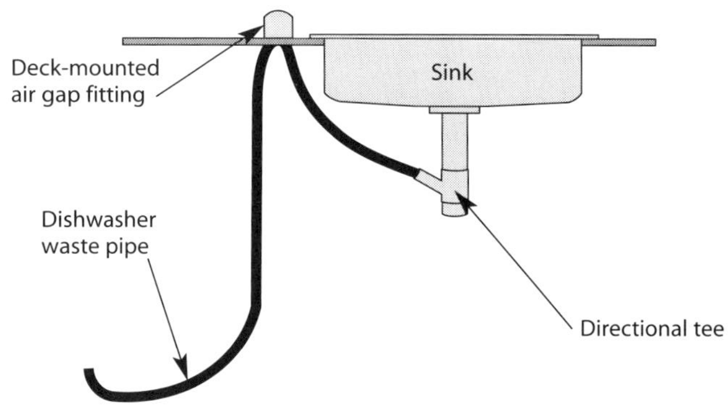


**Figure 14-16**
*Dishwasher connection through directional tee with air gap fitting*

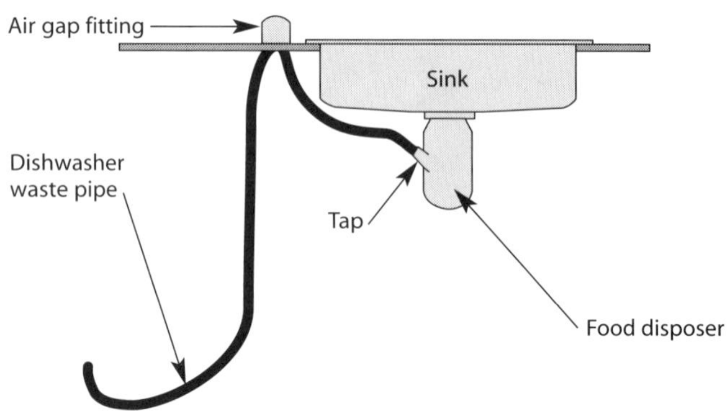


**Figure 14-17**
*Dishwasher connection through food disposer with air gap fitting*

## Special Fixtures

If your work includes specialty plumbing fixtures (baptisteries, ornamental and lily pools, aquariums, ornamental fountain basins, swimming and wading pools, and similar constructions), your water connection will have to have an approved backflow preventer device. Plans for installations of this type must be submitted to your local authority for approval.

Whirlpool bathtubs are quite common in today's plumbing. *IPC* Section 421 lays out the requirements for such installations. It's essential that you install and test it according to the manufacturer's instructions. You must locate the circulating pump above the weir of its trap and be sure it's accessible for repair or replacement.

In plants that manufacture products that can harm skin or eyes, emergency showers and eyewash stations must be provided. A supply of cold water as required by the manufacturer must be installed. These stations don't require a waste connection. See *IPC* Sections 411.1 and 411.2.

### *Garbage Can Washers*

You won't often find a garbage can washer in restaurants today. However, in *IPC* Section 414, you'll find everything you need to know if you should have to install one. The code is greatly concerned about cross-connection. When installing this type of system, make sure to valve both the hot and cold water supply. Provide an approved means to prevent backflow. Figure 14-18 shows a simple installation diagram.

Garbage can washer waste must go into and through a receptacle — usually a floor drain. To prevent large items from entering the drainage system, trap the drain separately and install a removable basket or strainer. Check with local authorities to find out if the wastewater can go into the regular building drainage system or if it must go through the greasy waste system.

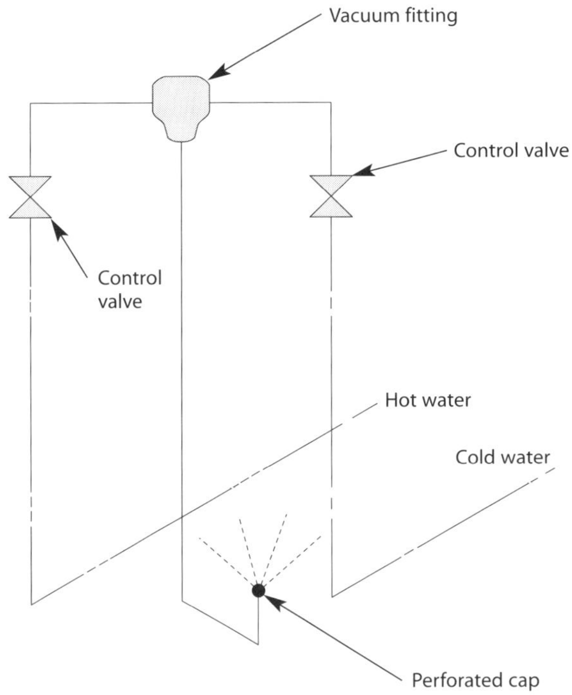


**Figure 14-18**
*Garbage can washer piping diagram*

## Plumbing Fixture Clearances

Install fixtures so they're spaced for easy access for cleaning, repair and intended use. When you install plumbing fixtures in a room, be sure the room is properly lighted and has some form of ventilation. If there's no window for natural ventilation, install an exhaust fan. See *IPC* Section 310.1. Figure 14-19 shows the minimum code clearance for fixtures. Here are code requirements from *IPC* Section 405.

- Each water closet must have at least a 15-inch clearance from its center to any finished wall or partition, and at least a 15-inch clearance from the center of the bowl to the outside edge of a tub apron. There must be at least a 21-inch clearance from the front of the bowl to any finished wall, door or other plumbing fixture. A water closet compartment must be at least 30 inches wide and 60 inches deep. Water closets set in battery installations must have at least a 30-inch center-to-center clearance.

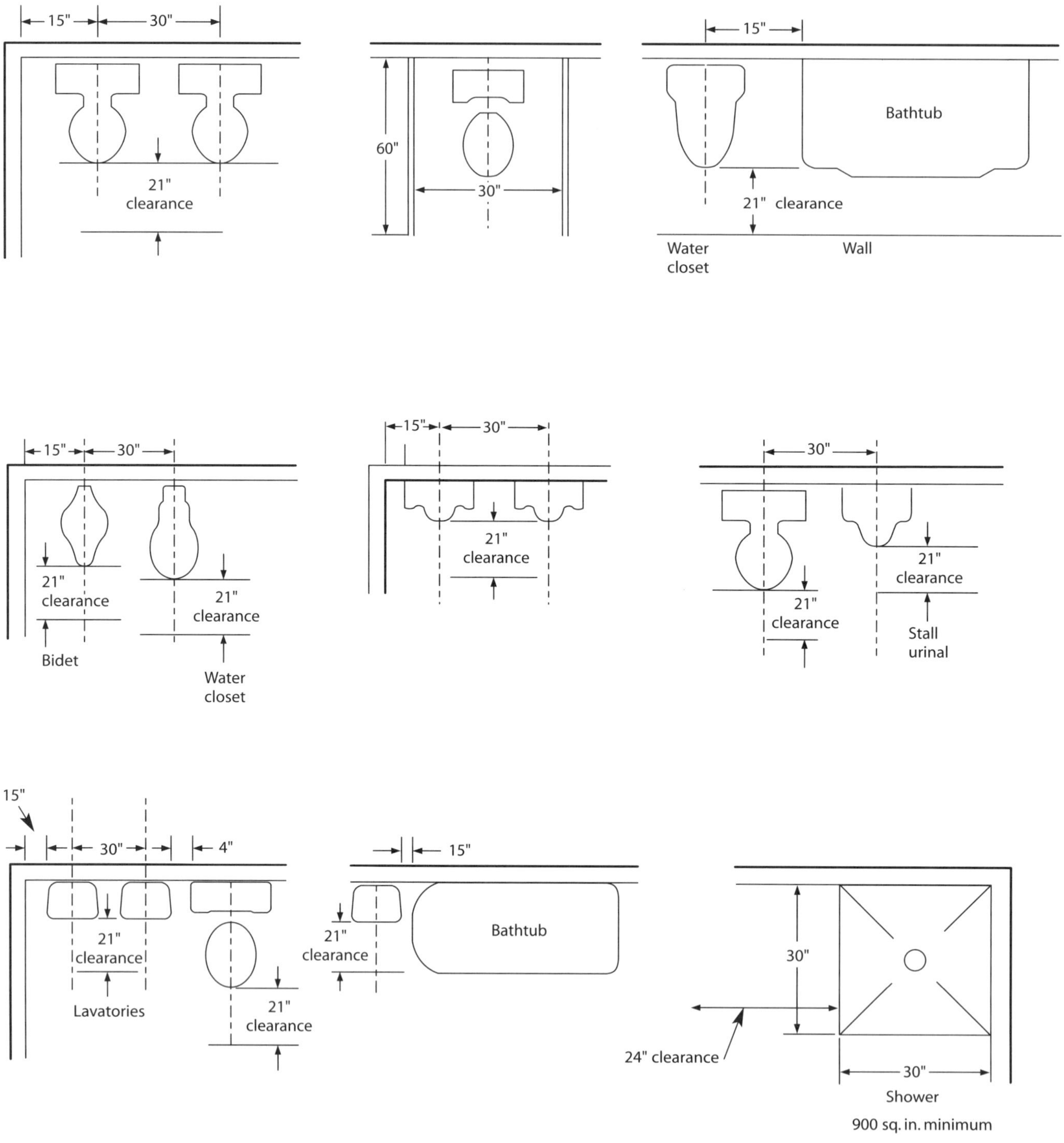


**Figure 14-19**
*Fixture clearances*

- A lavatory must have at least a 30-inch center-to-center clearance and a 15-inch clearance from any finished wall or other plumbing fixture. In front of each lavatory, there must be at least a 21-inch clearance from any wall, door or another fixture.
- A urinal must have at least a 15-inch clearance from the center of the urinal to any finished wall. There must be at least a 21-inch clearance from the front of the urinal to any finished wall, door or other plumbing fixture. Urinals set in a battery installation must have at least a 30-inch center-to-center clearance.
- A bidet must have at least a 15-inch clearance from its center to any finished wall and a 30-inch clearance from its center to the center of a water closet. There must be a 21-inch clearance from the front of the bidet to any finished wall, door or other plumbing fixture.
- According to some local codes, a shower stall must have at least a 24-inch clearance from its opening entrance to any finished wall, door or other plumbing fixture. Shower compartment interior dimensions must have a minimum clearance of 30 inches and at least 900 square inches of interior floor space.

## Minimum Plumbing Facilities

Whenever the code requires plumbing fixtures, it regulates the type and minimum number of each. This includes all premises intended for human occupancy or use. In *IPC* Table 403.1, shown in Figure 14-20, you'll find the occupancy classification, number of occupants, number and type of fixtures. For occupancies not listed in this table, consult your local authority. The *International Building Code* has jurisdiction in determining the number of occupants and the occupancy classification for various building uses. Minimum requirements for residences are:

- Single-family residence: one kitchen sink, one water closet, one lavatory and one bathtub or shower unit. Also, you must provide waste and hot/cold water outlets for a clothes washer.
- Duplex residential units (per unit): one kitchen sink, one water closet, one lavatory and one bathtub or shower unit. Also, you must provide waste and hot/cold water outlets for each dwelling unit clothes washer.
- Apartment units (per unit): one kitchen sink, one water closet, one lavatory and one bathtub or shower unit. Also, you must provide one automatic clothes washer connection for each 20-apartment unit. For example, an apartment building with 50 dwelling units would need connections provided for three clothes washing machines.

### *Places of Employment*

The number of toilet fixtures in manufacturing plants, heavy industry, warehouses and similar establishments is based on the number of employees. So if toilet facilities are needed for a medium-sized manufacturing plant employing 200 persons, *IPC* Section 403.3 says that the occupant load must be considered to be 50 percent male and 50 percent female. If there will be 100 men and 100 women working the plant, that occupant load will work. But, if this isn't the case, the owner can present statistical data showing a different number of males and females will be employed in the plant. Then the plumbing official can determine the correct number and type of plumbing fixtures required.

Let's work out a particular example using *IPC* Table 403.1 (our Figure 14-20). The *IPC* lists this type of facility in the table under the heading "Factory and Industrial." Let's say the owner provides information that there would be 150 male and 50 female employees. Then there must be two water closets and two lavatories for the male employees. For female employees, there must be one water closet and one lavatory. The code also says that one drinking fountain and one service sink are required. Any manufacturing plant that might possibly subject its employees to excessive heat, infection or irritating materials must also provide emergency showers and eyewash stations. See *IPC* Section 411.

In any commercial building where plumbing fixtures are required, separate facilities must be provided for each sex. However, IPC Section 403.2 says that separate toilet facilities aren't required in private buildings, buildings where fewer than 15 employees work, or buildings where the total occupant load isn't more than 15.

IPC Section 403.1.1 says that fixtures located within toilet and bathing rooms for unisex use may be included to determine the minimum required number of fixtures for assembly and mercantile establishments. In mercantile and assembly occupancies, toilet facilities must be in the employees' regular work area. These toilet facilities may be separate for employees only, or they may be combined for employees and public customer use. However, IPC Section 403.4 makes an exception for toilet facilities for employees in kiosks and storage structures (generally located in shopping malls under the same ownership, lease or control) located in adjacent structures. The toilet facilities can be no more than 500 feet away from the regular work area.

Toilet facilities in other types of public buildings must be no more than 500 feet from each work area. In multi-story buildings, they can be no more than one story above or below the area. *IPC* Section 403.4.1 makes an exception in that the travel distance to toilet facilities in factories and industrial buildings may be more than 500 feet if the local plumbing official approves it.

Owners of restaurants, nightclubs, public assembly places and mercantile occupancies must provide toilet facilities for their employees. They may be either separate or combined with their public customer toilet facilities. If the building has more than one floor level, the required toilet facilities must not be located more than one story above or below the employees' regular work area. As a rule, the travel distance can't be more than 500 feet. If the business is located in a covered mall, the travel distance must not be more than 300 feet. *IPC* Section 403.5 has an exception that tenants don't have to provide employee toilet facilities if their space is in a building with central toilet facilities where the travel distance is not more than 300 feet and not more than one story above or below the tenant space.

**TABLE 403.1**
**MINIMUM NUMBER OF REQUIRED PLUMBING FIXTURES**
**(See Sections 403.2 and 403.3)**

| NO. | CLASSIFICATION | USE GROUP | DESCRIPTION | WATER CLOSETS (URINALS SEE SECTION 419.2) | | LAVATORIES | | BATHTUBS/ SHOWERS | DRINKING FOUNTAIN (SEE SECTION 410.1) | OTHER |
|---|---|---|---|---|---|---|---|---|---|---|
| | | | | MALE | FEMALE | MALE | FEMALE | | | |
| 1 | Assembly (see Sections 403.2, 403.5 and 403.6) | A-1 | Theaters usually with fixed seats and other buildings for the performing arts and motion pictures | 1 per 125 | 1 per 65 | 1 per 200 | | ——— | 1 per 500 | 1 service sink |
| | | A-2 | Nightclubs, bars, taverns, dance halls and buildings for similar purposes | 1 per 40 | 1 per 40 | 1 per 75 | | ——— | 1 per 500 | 1 service sink |
| | | | Restaurants, banquet halls and food courts | 1 per 75 | 1 pe r 75 | 1 per 200 | | ——— | 1 per 500 | 1 service sink |
| | | A-3 | Auditoriums without permanent seating, art galleries, exhibition halls, museums, lecture halls, libraries, arcades and gymnasiums | 1 per 125 | 1 per 65 | 1 per 200 | | ——— | 1 per 500 | 1 service sink |

**Figure 14-20**
*Minimum number of required plumbing fixtures*

**TABLE 403.1 (continued)**
**MINIMUM NUMBER OF REQUIRED PLUMBING FIXTURES**
**(See Sections 403.2 and 403.3)**

| NO. | CLASSIFICATION | USE GROUP | DESCRIPTION | WATER CLOSETS (URINALS SEE SECTION 419.2 | | LAVATORIES | | BATHTUBS/ SHOWERS | DRINKING FOUNTAIN (SEE SECTION 410.1) | OTHER |
|---|---|---|---|---|---|---|---|---|---|---|
| | | | | MALE | FEMALE | MALE | FEMALE | | | |
| | | | Passenger terminals and transportation facilities | 1 per 500 | 1 per 500 | 1 per 750 | | ——— | 1 per 1,000 | 1 service sink |
| | | | Places of worship and other religious services. Churches without assembly halls | 1 per 150 | 1 per 75 | 1 per 200 | | ——— | 1 per 1,000 | 1 service sink |
| | | A-4 | Coliseums, arenas, skating rinks, pools and tennis courts for indoor sporting events and activities | 1 per 75 for the first 1,500 and 1 per 120 for the remainder exceeding 1,500 | 1 per 40 for the first 1,500 and 1 per 60 for the remainer exceeding 1,500 | 1 per 200 | 1 per 150 | ——— | 1 per 1,000 | 1 service sink |
| | | A-5 | Stadiums, amusement parks, bleachers and grandstands for outdoor sporting events and activities. | 1 per 75 for the first 1,500 and 1 per 120 for the remainder exceeding 1,500 | 1 per 40 for the first 1,500 and 1 per 60 for the remainder exceeding 1,500 | 1 per 200 | 1 per 150 | ——— | 1 per 1,000 | 1 service sink |
| 2 | Business (see Sections 403.2, 403.4 and 403.6) | B | Buildings for the transaction of business, professional services, other services involving merchandise, office buildings, banks, light industrial and similar uses. | 1 per 25 for the first 50 and 1 per 50 for the remainder exceeding 50 | | 1 per 40 for the first 50 and 1 per 80 for the remainder exceeding 50 | | ——— | 1 per 100 | 1 service sink |
| 3 | Educational | E | Educational facilities | 1 per 50 | | 1 per 50 | | ——— | 1 per 100 | 1 service sink |
| 4 | Factory and industrial | F-1 and F-2 | Structures in which occupants are engaged in work fabricating, assembly or processing of products or materials. | 1 per 100 | | 1 per 100 | | (see Section 411) | 1 per 400 | 1 service sink |
| 5 | Institutional | I-1 | Residential Care | 1 per 10 | | 1 per 10 | | 1 per 8 | 1 per 100 | 1 service sink |
| | | I-2 | Hospitals, ambulatory nursing home patients [b] | 1 per room [c] | | 1 per room [c] | | 1 per 15 | 1 per 100 | 1 service sink per floor |
| | | | Employees, other than residential care [b] | 1 per 25 | | 1 per 35 | | ——— | 1 per 100 | ——— |
| | | | Visitors, other than residential care | 1 per 75 | | 1 per 100 | | ——— | 1 per 500 | ——— |
| | | I-3 | Prisons [b] | 1 per cell | | 1 per cell | | 1 per 15 | 1 per 100 | 1 service sink |
| | | I-3 | Reformitories, detention centers, and correctional centers, [b] | 1 per 15 | | 1 per 15 | | 1 per 15 | 1 per 100 | 1 service sink |

**Figure 14-20 (continued)**

*Minimum number of required plumbing fixtures*

**TABLE 403.1 (continued)**
**MINIMUM NUMBER OF REQUIRED PLUMBING FIXTURES**
**(See Sections 403.2 and 403.3)**

| NO. | CLASSIFICATION | USE GROUP | DESCRIPTION | WATER CLOSETS (URINALS SEE SECTION 419.2) MALE | WATER CLOSETS FEMALE | LAVATORIES MALE | LAVATORIES FEMALE | BATHTUBS/ SHOWERS | DRINKING FOUNTAIN (SEE SECTION 410.1) | OTHER |
|---|---|---|---|---|---|---|---|---|---|---|
| | | I-4 | Adult daycare and childcare. [b] | 1 per 15 | | 1 per 15 | | 1 per 15 [d] | 1 per 100 | 1 service sink |
| 6 | Mercantile (see Sections 403.2, 403.5 and 403.6) | M | Retail stores, service stations, shops, salesrooms, markets and shopping centers | 1 per 500 | | 1 per 750 | | ——— | 1 per 1,000 | 1 service sink |
| 7 | Residential | R-1 | Hotels, motels, boarding houses (transient) | 1 per guestroom | | 1 per guestroom | | 1 per guestroom | ——— | 1 service sin |
| | | R-2 | Domritories, Fraternities, Sororities and Boarding houses (not transient) | 1 per 10 | | 1 per 10 | | 1 per 8 | 1 per 100 | 1 service sink |
| | | R-2 | Apartment House | 1 per dwelling unit | | 1 per dwelling unit | | 1 per dwelling unit | ——— | 1 kitchen sink per dwelling unit; 1 automatic clothes washer connection per 20 dwelling units [e] |
| | | R-4 | Residential Care/Assisted Living Facilities | 1 per 10 | | 1 per 10 | | 1 per 8 | 1 per 100 | 1 service sink |
| 8 | Storage (see Sections 403.2 and 403.4) | S-1<br>S-2 | Structures for the storage of goods, warehouses, storehouse and freight depots. Low and Moderate Hazard. | 1 per 100 | | 1 per 100 | | 1 per 1,000 | See Section 411 | 1 service sink |

a. The fixtures shown are based on one fixture being the minimum required for the number of persons indicated or any fraction of the number of persons indicated. The number of occupants shall be determined by the International Building Code.

b. Toilet facilities for employees shall be separate from facilities for inmates or patients.

c. A single-occupant toilet room with one water closet and one lavatory serving not more than two adjacent patient rooms shall be permitted where such room is provided with direct access from each patient room and with provisions for privacy.

d. For day nurseries, a maximum of one bathtub shall be required.

e. For attached one-and two-family dwellings, one automatic clothes washer connection shall be required per 20 dwelling units.



**Figure 14-20 (continued)**
*Minimum number of required plumbing fixtures*

Businesses located in shopping centers and establishments that are frequented by the public must provide toilet facilities for the number of employees and the public that's reasonably anticipated. The travel distance for central locations can't be more than 500 feet (300 feet in covered malls). Facilities can't be located more than one story above or below the tenant space.

There are usually two classifications of public use which determine the number and type of plumbing fixtures required:

*Classification #1:* Establishments that provide countable seating capacity, such as churches, theaters, stadiums and restaurants. To determine the number of plumbing fixtures, look at Figure 14-20

(*IPC* Table 403.1). Find the correct occupancy and you have the number of plumbing fixtures required.

*Classification #2:* Establishments such as in retail stores, office buildings, and similar establishments that have no countable seating capacity. In such cases, *IPC* Section 403.1 says that the *International Building Code* must determine the number of occupants. This tells me that the architect, not the plumbing contractor, must provide the necessary information to determine the number and types of plumbing fixtures required by code. *IPC* Section 403.6.2 says required public toilet facilities must be free. If public facilities are installed that charge a fee, those facilities must be in addition to the facilities required by code.

## Unisex Toilet Facilities

Unisex toilet facilities are outlined in *IPC* Section [B]403.1.1. Things may change in the future, but most local authorities now accept rather minimum unisex toilet facilities. The requirements must comply with *IPC* section 404.

## Fixtures for the Handicapped

Federal and state laws now require toilet facilities for the physically handicapped in public buildings and privately-owned multi-story buildings. All required toilet rooms in public buildings must provide handicapped facilities for each sex on each floor. Accessible routes to toilet rooms must be readily available. Hallways and toilet room doors must provide readily visible direction signs. Single-family residences and buildings that are considered hazardous are exempt from this requirement. *IPC* Section 404 has some information on fixtures for the handicapped, but it's dealt with more fully in the *International Building Code*. Here's some basic information for common plumbing fixtures:

- *Water closets:* Toilet rooms must have at least one toilet compartment that's wheelchair-accessible. It must be at least 3 feet wide and 4 feet 8 inches deep. The compartment must have handrails on each side, parallel and 33 inches above the finished floor. Handrails should be 1$^1/_2$ inches in outside diameter, with a 1$^1/_2$ inch clearance between the rail and compartment wall. Handrails must be securely fastened at each end and at the center. The compartment door must swing out and be at least 32 inches wide. A water closet must be manufactured and designed for the handicapped and the seating area must be at least 17 to 19 inches above the finished floor.
- *Lavatories:* Toilet rooms must have at least one lavatory manufactured for the handicapped. It must have a narrow apron which, when mounted at standard height, is accessible by a person in a wheelchair. You may mount it higher than the standard height, when the particular design demands, as long as it's accessible to persons in wheelchairs. The code requires that drain pipes and hot water pipes be insulated to prevent burns to anyone in a wheelchair who has no feeling in his or her limbs.
- *Urinals:* Toilet rooms for men must have at least one wall-mounted urinal useable by anyone that's handicapped. It must be installed where the basin opening is no more than 19 inches above the toilet room floor. If your local code permits floor-mounted urinals, the basin must be level with the finished toilet room floor.
- *Drinking fountains:* At least 50 percent of all drinking fountains in a building and at least one on each floor must be accessible to the handicapped. They can be hand-operated or hand-and foot-operated, with up-front spouts and controls. If a drinking fountain is wall mounted no more than 36 inches from the floor and is hand-operated, it can be used by the able-bodied as well as the physically handicapped. You can't fully recess drinking fountains that will serve the handicapped unless the alcove is at least 32 inches wide.

I've provided floor plans showing several acceptable layouts for handicapped toilet rooms in Figures 14-21 through 14-29.

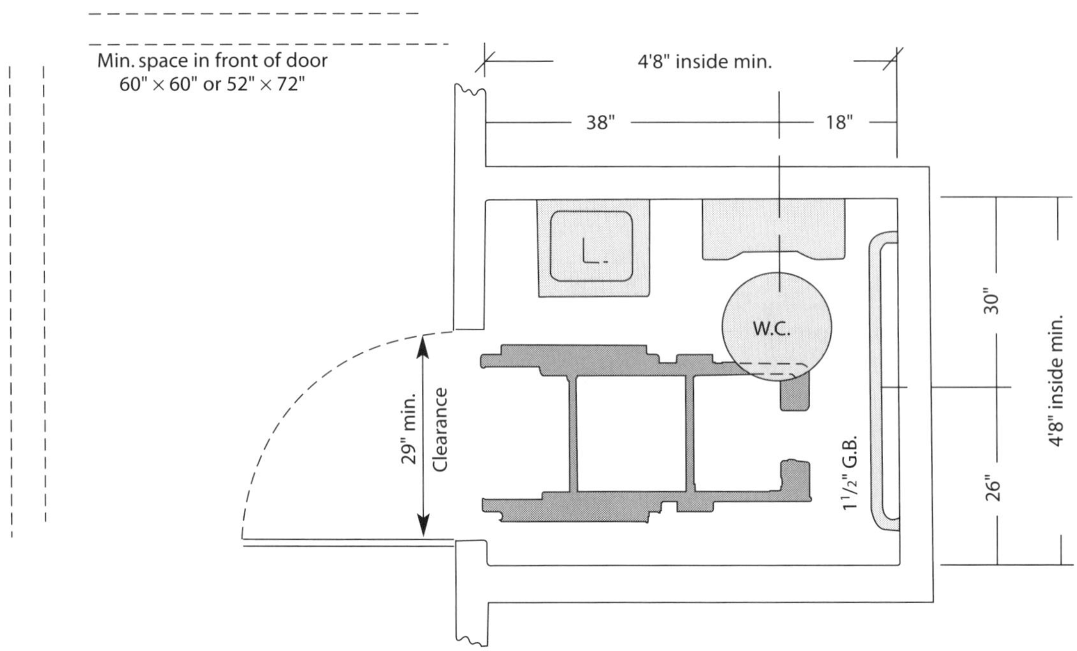


**Figure 14-21**

*Toilet room with side approach to water closet*

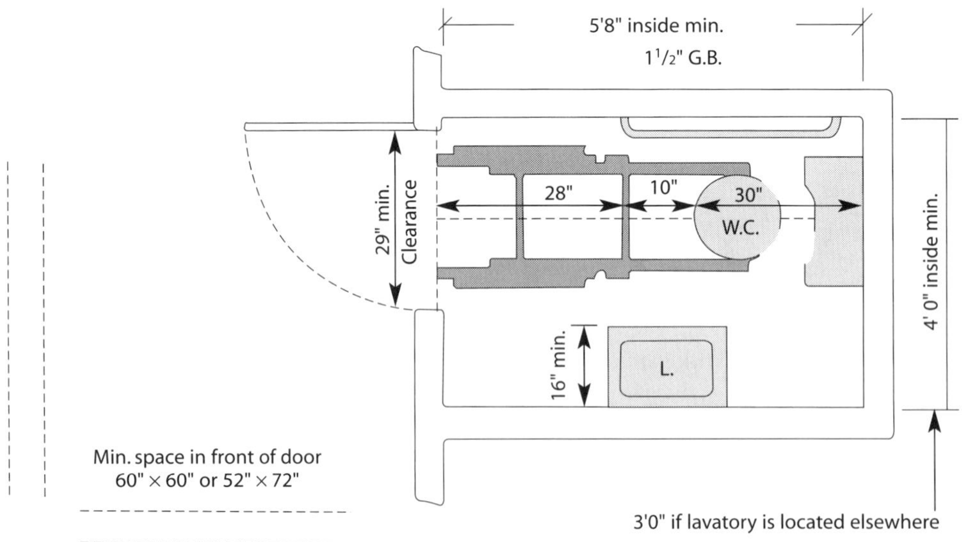


**Figure 14-22**

*Toilet room with front approach to water closet*

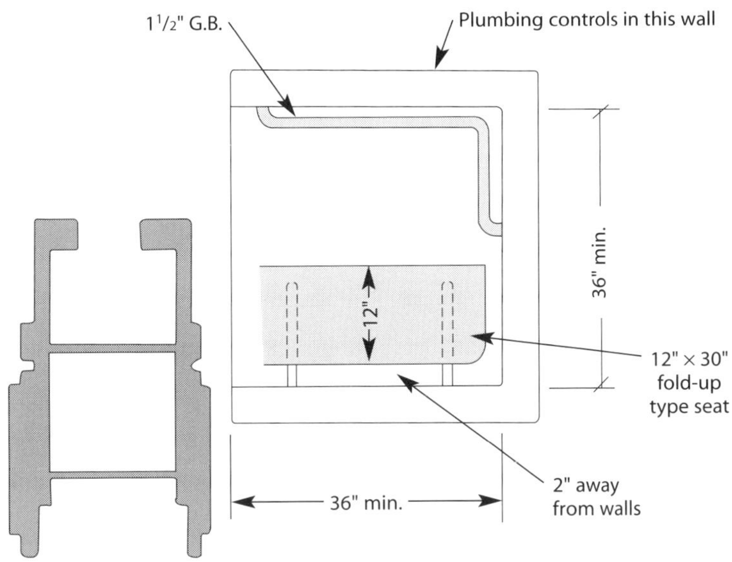


**Figure 14-23**
*Minimum size shower for a handicapped person*

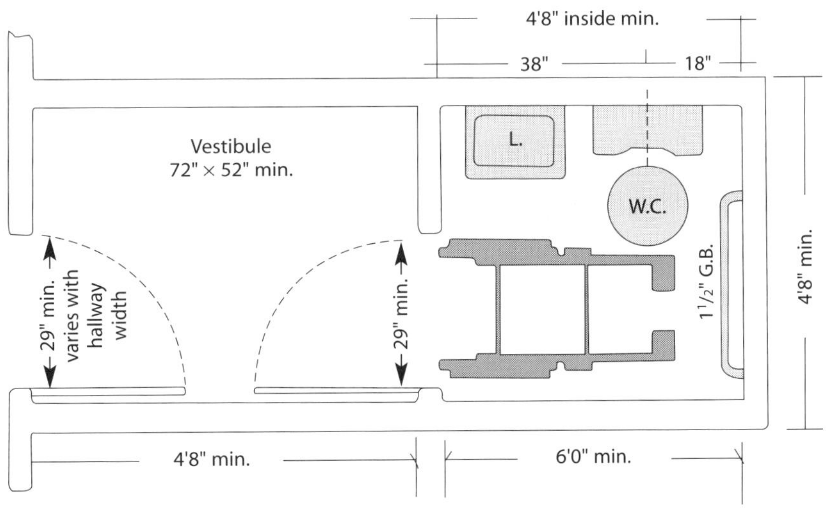


**Figure 14-24**
*Toilet room with minimum size vestibule*

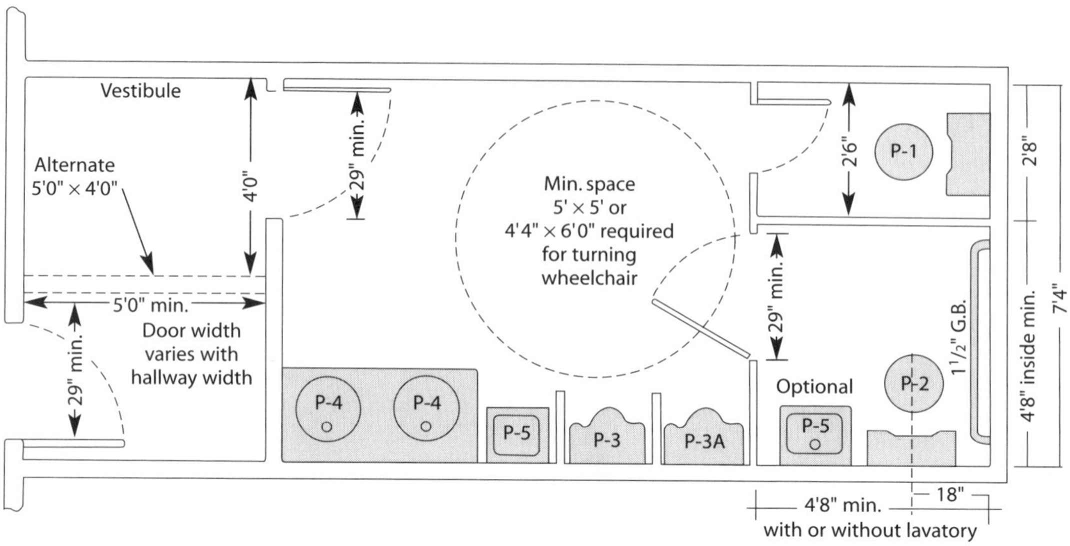


**Figure 14-25**

*Public toilet room with alternate size vestibule and minimum space for turning wheelchair*

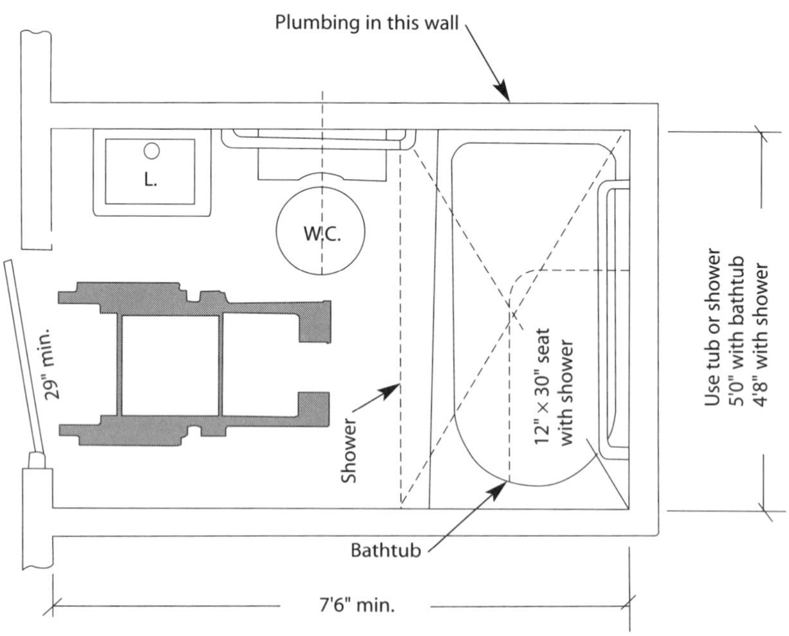


**Figure 14-26**

*Bathroom layout for hotels, motels and apartments, acceptable for handicapped use*

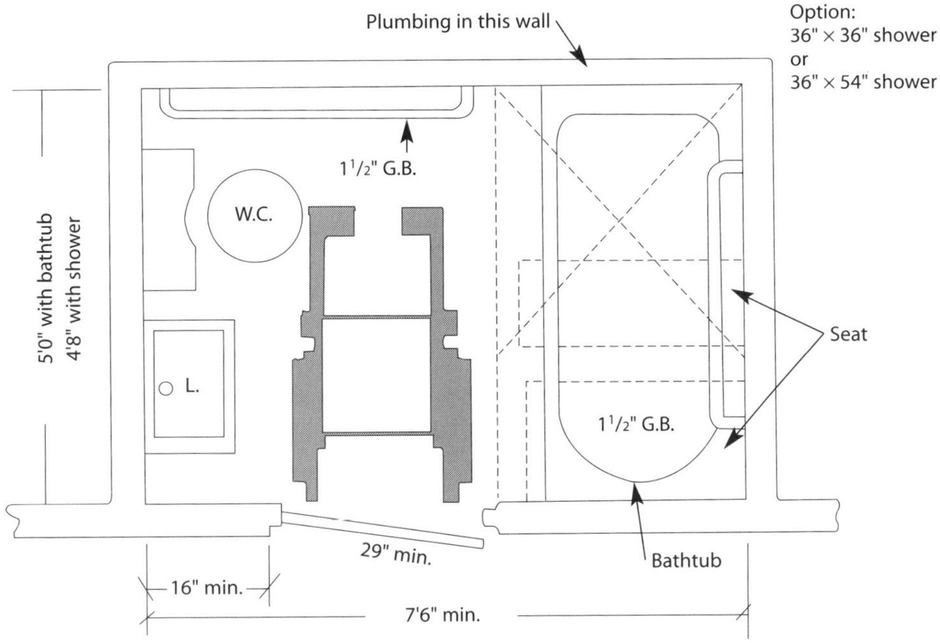


**Figure 14-27**

*Alternate bathroom layout for hotels, motels and apartments, acceptable for handicapped use*

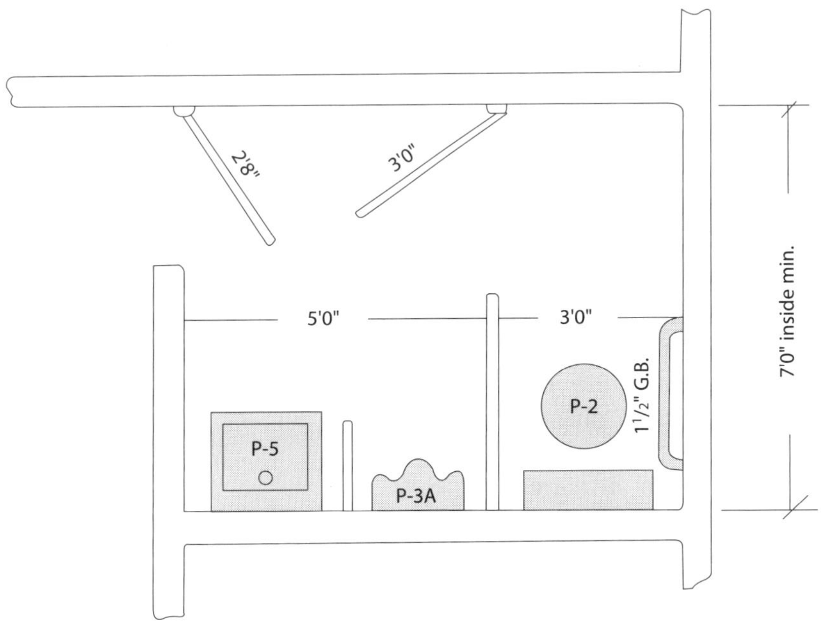


**Figure 14-28**

*Minimum size toilet room designed with three fixtures*

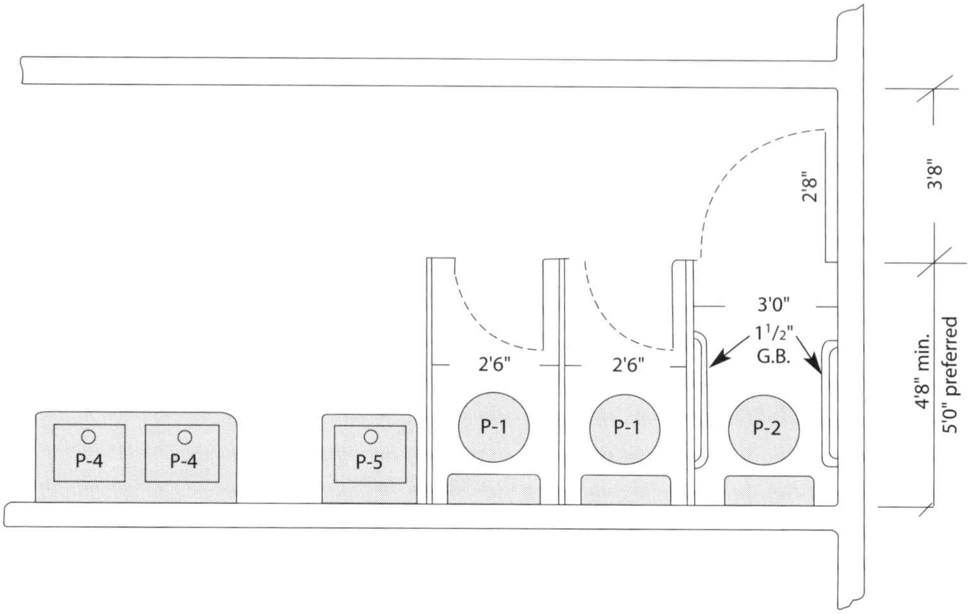


**Figure 14-29**

*Multiple fixtured toilet room layout with minimum size*

## Healthcare Plumbing Fixtures and Equipment

Other than supplying and installing approved special plumbing fixtures, most of *IPC* Section 422 is about the mechanical code and trade. Let me lift out the small areas of concern to a plumber. First, these special plumbing fixtures, equipment, devices and apparatuses are in nursing homes, homes for the aged, orphanages, infirmaries, first aid stations, psychiatric facilities, clinics, dentist and doctor offices, mortuaries, educational facilities, surgery, research and testing laboratories, establishments that manufacture pharmaceutical drugs and medicines, and similar establishments. Your main job here is to protect the water supply or drainage system from backflow, flooding, fouling or contamination. Fixtures used must be designed for therapy, special cleansing or disposal of waste material, combinations of such purposes, or any other special purpose. The fixtures must be of smooth, impervious, corrosion-resistant materials and able to withstand temperatures greater than 180 degrees Fahrenheit.

Clinical sinks must have integral traps with visible trap seals. They must be either the siphonic or blowout type and capable of resealing the trap. They must be designed with a flushing rim for cleaning the interior surface. The fixture must be designed to have the flushing and cleansing characteristics of a water closet.

# Chapter 14 — Questions and Answers

**14-1 For general applications using sheet copper, the *IPC* requires that you use copper weighing at least _____ per square foot.**

(A) 12 ounces

(B) 14 ounces

(C) 18 ounces

(D) 20 ounces

***Answer:* *A***

*Code response: Section 402.3*
*This book, page 223*

**14-2 Sheet lead you use for pans must weigh at least ____ per square foot.**

(A) 2.5 pounds

(B) 3 pounds

(C) 3.5 pounds

(D) 4 pounds

***Answer:* *D***

*Code response: Section 402.4*
*This book, page 223*

**14-3 According to the *IPC*, unless there is data showing a building will be occupied by more of one sex than the other, the ratio of all fixtures for an office building must be**

(A) determined by the code official.

(B) 40% male and 60% female.

(C) 50% male and 50% female.

(D) 60% male and 40% female.

***Answer:* *C***

*Code response: Section 403.3*
*This book, page 239*

**14-4 The *IPC* requires that toilet seats for public use be**

(A) the round closed-front type.

(B) the elongated closed-front type.

(C) the round open-front type.

(D) the elongated open-front type.

***Answer:* *D***

*Code response: Section 420.3*
*This book, page 226*

**14-5 According to the *IPC*, a residential tank-type water closet must use no more than _____ gallons of water per flush.**

(A) 1.6

(B) 1.8

(C) 2.0

(D) 2.2

***Answer: A***

*Code response: Table 604.4*
*This book, page 227*

**14-6 In public toilet rooms, the *IPC* allows you to substitute _____ per cent of the required water closets with urinals.**

(A) 33

(B) 50

(C) 60

(D) 67

***Answer: D***

*Code response: Section 419.2*
*This book, page 227*

**14-7 The *IPC* states that when you install a bidet, you must install it using _______ to protect the water supply against cross-connection.**

(A) an air break connection

(B) an air gap

(C) a soft-seat check valve

(D) a direct connection

***Answer: B***

*Code response: Section 408.2*
*This book, page 228*

**14-8 The *IPC* requires that a shower compartment have at least ______ square inches of floor space.**

(A) 900

(B) 1,000

(C) 1,100

(D) 1,200

***Answer: A***

*Code response: Section 417.4*
*This book, page 232*

**14-9 When urinals are set in a battery installation, the *IPC* requires at least a(n) _______ center-to-center clearance.**

(A) 18-inch

(B) 24-inch

(C) 30-inch

(D) 36-inch

***Answer:* C**

*Code response: Figure 405.3.1*
*This book, page 239 and Figure 14-19*

**14-10 The *IPC* requires that you install plumbing fixtures so they're easy to**

(A) clean.

(B) inspect.

(C) use.

(D) replace.

***Answer:* A**

*Code response: Section 405.2*
*This book, page 237*

◆ *Chapter 15* ◆

# Let's Talk About the Gas Code

The *International Fuel Gas Code (IFGC)* doesn't leave much to chance when it comes to outlining the piping materials and installation methods you must use when working with gas. Improperly-installed water-supply systems can flood buildings and do a whole lot of damage. But it's rare that they'll kill anyone. Not so with gas. While natural gas isn't in itself poisonous, it can replace the oxygen in a closed space and cause suffocation. And under certain conditions, it's a powerful explosive!

Your plumbing license qualifies you to size and install gas systems, and you'll find that you won't be a journeyman plumber for long before you have a job that involves gas piping. It's therefore important that you have at least a basic understanding of just what gas is and what are its properties. Besides, you just might get a question or two on your next plumbers' exam.

Every substance known to man is in one of three states — a solid, a liquid, or a gas. Although gases are lighter than the other two, they do have weight. Gas is made of constantly-moving atoms, with no fixed shape or volume. When these atoms are forced into a container, they take on the container's shape but occupy only about one thousandth of the container's interior space. Actually, the spaces between the atoms are empty, making it possible to compress a lot of gas into a very small space.

History credits the Chinese with being the first to use natural gas for industrial purposes, nearly 3,000 years ago. They used hollow bamboo to pipe natural gas for use in evaporating brine to make salt.

In America, in 1821, Fredonia, New York was the first city to use natural gas on a commercial scale. By 1850, many cities and towns were using natural gas for lighting purposes. The raw gas piped directly to these lighting outlets was yellow and gave off a poor light. Later, scientists learned to mix air with the gas to produce a better flame.

## Kinds of Gas

The three kinds of gas widely used in homes and businesses today are natural, manufactured, and liquefied petroleum. Natural gas is classified according to its chemical formation, and according to the chemical impurities it contains. There are four different types of natural gas. Wet gas is natural gas that contains the heavier hydrocarbons such as propane and butane. Dry gas is natural gas that all hydrocarbons heavier than methane and ethane have been removed from. Sour gas is natural gas that contains varying amounts of hydrogen sulfide. It's used primarily to produce carbon black, which is an important ingredient in all rubber products and black printing ink. Sweet gas is natural gas with little or no hydrogen sulfide. This gas is used chiefly as fuel in homes, factories, and commercial institutions.

The natural gas that millions of consumers use as fuel in homes and industries is known as dry, or sweet, gas. It's clean and has no odor. However, because of these properties, a gas leak could go undetected until it caused an explosion. Consequently, gas producers add a chemical scent to gas before it enters a pipeline. This odor is very noticeable and gives warning of escaping gas before the concentration can reach a dangerous level.

Manufactured gas is produced chiefly from coal and coke. It's generally added to other fuels to increase heating capacity. Manufactured gas is also used as fuel in homes and industries. It can be poisonous, since it contains carbon monoxide, and it's explosive under certain conditions.

Liquefied petroleum gas is known in the plumbing trade as LP or bottle gas. It's basically butane or propane, or a mixture of the two. See *IFGC* Section 402.6.1. It's produced primarily in plants that process natural gas. Although LP gas is still a gas after processing, it can become a liquid under moderate pressure. In this liquid state, it's easy and economical to transport and store in pressure containers. The liquid returns to its original gaseous state at normal atmospheric pressure and temperature. LP gas is colorless and nonpoisonous. A chemical scent is added to LP gas to detect leaks in gas systems. LP gas is a convenient fuel for homes and businesses in remote areas. No place is too remote to be served by LP gas.

## Understanding Gas Piping Installation

Your responsibilities in sizing and installing a building's gas supply system will be a little different from those required for doing a water supply system. For water service piping, you must take it from the meter (essentially the property line) to the building, as well as all the work required inside. But with gas service piping, the gas supplier is responsible for running the gas line from the property line to the meter (usually located at the building's exterior wall). Your responsibilities begin at the meter location and include all the piping within the building.

The *IFGC* is a separate book with a complexity all its own. Regardless of your experience level, don't expect compliance with the gas code to be easy. Although a gas system is usually installed by a gas fitter and/or installer, as a plumber you must know how to do this work in accordance with the gas code.

This chapter will give you the relevant information you need from the *IFGC*, with some practical examples. This chapter and the next will give you a working knowledge of gas code essentials. That isn't to say that the information here replaces the gas code. The *IFGC* is always your final authority.

## Sizing Gas Systems

To properly size a building's gas main and branch lines, you have to know (1) the maximum gas demand (according to the manufacturer's input rating) required by each appliance and equipment outlet, and (2) the length of piping required to reach the most remote outlet. There are a few other variables, like pressure loss and specific gravity that need to be taken into consideration. Thank goodness, we don't have to fret over such things because the *IFGC* includes them in its tables.

Every gas appliance will have a metal plate in a visible location showing the maximum gas demand (Btu input rate). Before you can size distribution piping, you must convert each Btu input rating to cubic feet of gas. This may sound a bit complicated, but actually it's fairly simple. One cubic foot of natural gas releases about 1,000 Btu per hour. Manufactured gas releases about 550 Btu per hour. In our example, we'll use the formula for natural gas. In a geographic area where manufactured gas is used, you'll use the same formula we'll use for natural gas (substituting 550 Btu) to find the demand in cubic feet per hour. For example, let's say you're sizing pipe for a gas range that has a maximum demand of 70,000 Btu per hour. Since 1 cubic foot of natural gas releases 1,000 Btu, you divide 70,000 Btu by 1,000 to get 70 cubic feet per hour. If you're using manufactured gas, you'd divide 70,000 Btu by 550 to get 127.27.

Sometimes you may have to install a used appliance, where the Btu rating plate is missing or illegible. Here's how you can make a safe installation. If the appliance inlet pipe is $^3/_4$ inch, install a $^3/_4$ inch supply pipe. You could use a larger supply pipe without violating the gas code, but it won't work any better. Just don't make the supply pipe smaller than the appliance's inlet pipe.

Sometimes you have to install an additional gas appliance in an existing system. According to *IFGC* Section 401.4, it's your responsibility to check the existing piping to make sure it can serve all the appliances. If you can't do this, install a new gas line sized to serve the additional appliance.

There are some 34 tables listed in *IFGC* Section 402. The table we'll use for our examples is *IFGC* Table 402.4(2), shown as Figure 15-1. This is the table you'll use most often. It shows the capacities of different sizes and lengths of pipe for gas at pressures of 0.5 psi or less in cubic feet per hour of 0.60 specific-gravity gas. If you want to review the description of other individual tables and their particular use, you'll find explanations in *IFGC* Appendix A, Subsections A.2.1, A.2.2, A.2.3, A.2.4, and A.2.5.

Let's run through a couple of examples using *IFGC* Table 402.4(2). I'll show you how to use this table to size the piping arrangements, first for a commercial building, then for a residence.

**TABLE 402.4(2)**
**SCHEDULE 40 METALLIC PIPE**

| | |
|---|---|
| **Gas** | Natural |
| **Inlet Pressure** | 0.50 psi or less |
| **Pressure Drop** | 0.5 inch WC |
| **Specific Gravity** | 0.60 |

| | PIPE SIZE (in.) | | | | | | | | | | |
|---|---|---|---|---|---|---|---|---|---|---|---|
| **Nominal** | $^1/_4$ | $^3/_8$ | $^1/_2$ | $^3/_4$ | 1 | $1^1/_4$ | $1^1/_2$ | 2 | $2^1/_2$ | 3 | 4 |
| **Actual ID** | 0.364 | 0.493 | 0.622 | 0.824 | 1.049 | 1.380 | 1.610 | 2.067 | 2.469 | 3.068 | 4.026 |
| **Length (ft)** | **Maximum Capacity in Cubic Feet of Gas per Hour** | | | | | | | | | | |
| 10 | 43 | 95 | 175 | 360 | 680 | 1,400 | 2,100 | 3,950 | 6,300 | 11,000 | 23,000 |
| 20 | 29 | 65 | 120 | 250 | 465 | 950 | 1,460 | 2,750 | 4,350 | 7,700 | 15,800 |
| 30 | 24 | 52 | 97 | 200 | 375 | 770 | 1,180 | 2,200 | 3,520 | 6,250 | 12,800 |
| 40 | 20 | 45 | 82 | 170 | 320 | 660 | 990 | 1,900 | 3,000 | 5,300 | 10,900 |
| 50 | 18 | 40 | 73 | 151 | 285 | 580 | 900 | 1,680 | 2,650 | 4,750 | 9,700 |
| 60 | 16 | 36 | 66 | 138 | 260 | 530 | 810 | 1,520 | 2,400 | 4,300 | 8,800 |
| 70 | 15 | 33 | 61 | 125 | 240 | 490 | 750 | 1,400 | 2,250 | 3,900 | 8,100 |
| 80 | 14 | 31 | 57 | 118 | 220 | 460 | 690 | 1,300 | 2,050 | 3,700 | 7,500 |
| 90 | 13 | 29 | 53 | 110 | 205 | 430 | 650 | 1,220 | 1,950 | 3,450 | 7,200 |
| 100 | 12 | 27 | 50 | 103 | 195 | 400 | 620 | 1,150 | 1,850 | 3,250 | 6,700 |
| 125 | 11 | 24 | 44 | 93 | 175 | 360 | 550 | 1,020 | 1,650 | 2,950 | 6,000 |
| 150 | 10 | 22 | 40 | 84 | 160 | 325 | 500 | 950 | 1,500 | 2,650 | 5,500 |
| 175 | 9 | 20 | 37 | 77 | 145 | 300 | 460 | 850 | 1,370 | 2,450 | 5,000 |
| 200 | 8 | 19 | 35 | 72 | 135 | 280 | 430 | 800 | 1,280 | 2,280 | 4,600 |

For SI: 1 inch = 25.4 mm, 1 foot = 304.8 mm, 1 cubic foot per hour = 0.0283 $m^3/h$, 1 pound per square inch = 6.895 kPa, 1-inch water column = 0.2488 kPa.



**Figure 15-1**

*Maximum capacity of pipe in cubic feet of gas per hour for gas pressures of 0.5 psi or less and a pressure drop of 0.5-inch water column*

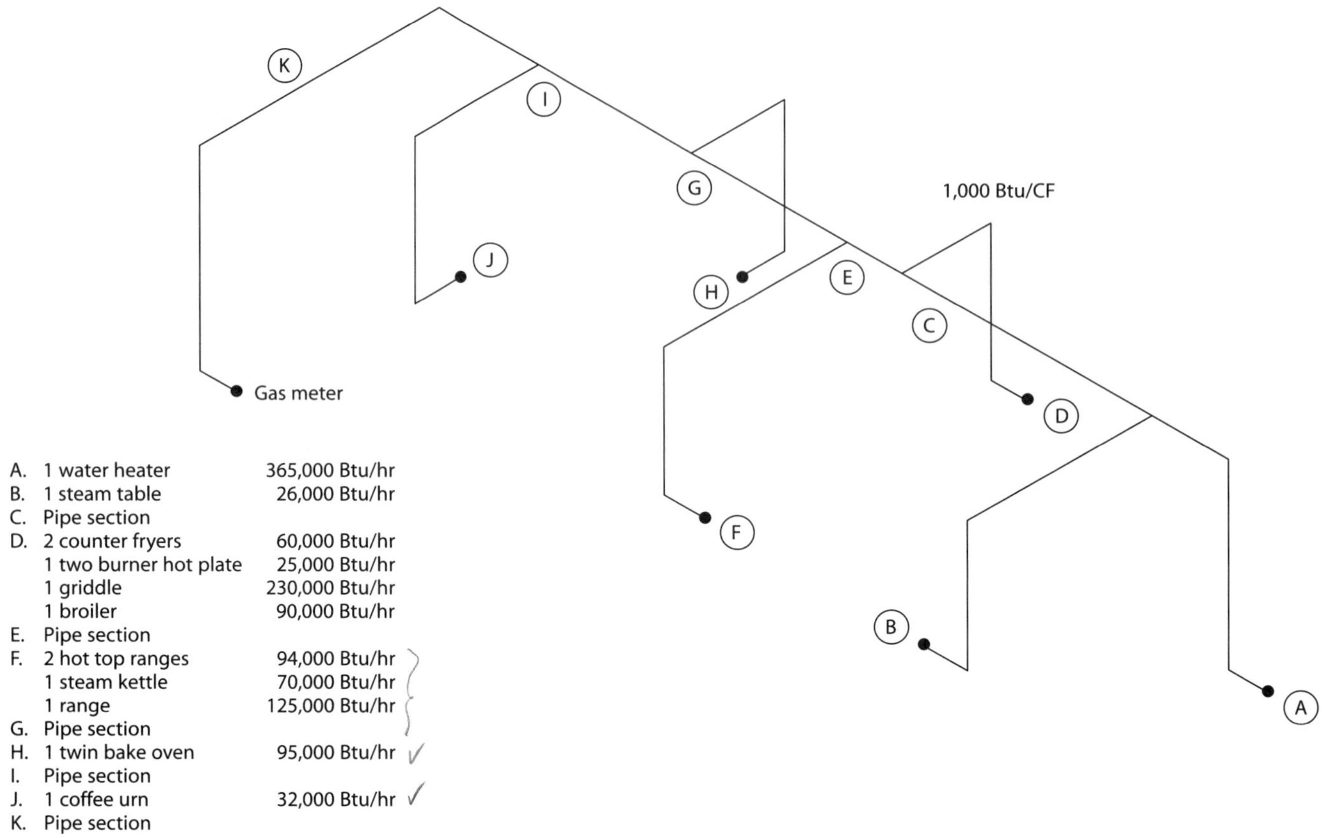


**Figure 15-2**
*Natural gas installation for a commercial kitchen*

## Commercial Gas Sizing Example

Figure 15-2 shows a typical gas piping arrangement for a restaurant. It's similar to what you might find on some plumber's exams. Here's how to size the system.

Assume the total developed length of gas piping in Figure 15-2 is 147 feet from the gas meter to Outlet A. Remember, the total developed length is the only distance you use to calculate the size of any section of the gas piping. Now find the correct distance row in *IFGC* Table 402.4(2). The 125-foot row is too small, so you need to use the 150-foot row.

At the left of Figure 15-2, there's a breakdown showing each type of equipment used and its Btu-per-hour rating. See Section 402.2. The maximum gas demand of Outlet A is 365,000 Btu per hour, or 365 cubic feet per hour, assuming 1,000 Btu per cubic foot. As you scan across the 150-foot column in Table 402.4(2), the first figure over 365 cubic feet per hour is in the column for $1^1/_2$-inch pipe, which has a capacity of 500 cubic feet per hour. This $1^1/_2$-inch pipe is the correct pipe to use.

Use the same procedure to size each remaining section. The maximum gas demand for Section B is 26 cubic feet per hour, so the pipe size is $^1/_2$ inch. The maximum gas demand for Section C includes the combined demand of Sections A and B, for a total of 391 cubic feet per hour. Again, you'd use $1^1/_2$-inch pipe. The maximum demand for section D includes the combined demand of five appliances:

*30,000 + 30,000 + 25,000 + 230,000 + 90,000 = 405,000 Btu*

Converting to cubic feet gives 405 cubic feet per hour. Here, too, you'd use $1^1/_2$-inch pipe. The maximum gas demand for Section E includes the combined cubic feet per hour ratings of Sections A, B, C, and D, for a total of 796 cubic feet per hour. Now you have to go to the column for 2-inch pipe to find a capacity over 796 cubic feet per hour.

The maximum gas demand for Section F is 289 cubic feet per hour. Note that the two hot top ranges have a combined Btu rating of 94,000, plus the steam kettle and range. The correct pipe size for Section F is $1^1/_4$ inches. The maximum gas demand for Section G includes the combined demand for Sections A, B, C, D, E, and F, for a total of 1,085 cubic feet per hour. The correct pipe size is $2^1/_2$ inches, which has a capacity of 1,500 cubic feet per hour.

Continuing sizing the various sections of pipes toward the meter: Section H maximum gas demand is 95 cubic feet per hour. The correct pipe size is 1 inch. The maximum gas demand for Section I is the combined demand of all previous sections, for a total of 1,180 cubic feet per hour. Since this total isn't over 1,500 cubic feet per hour, the correct pipe size for Section I is $2^1/_2$ inches. The maximum gas demand for Section J is 32 cubic feet per hour. The correct pipe size is $^1/_2$ inch. The maximum demand for Section K is the cumulative total for the entire restaurant, or 1,212 cubic feet per hour. Since this total isn't over 1,500 cubic feet per hour, you can use a $2^1/_2$-inch pipe for the job.

How about it! You've just sized the entire gas piping system for a commercial restaurant. However, if you still feel like a "greenhorn," your friendly plumbing inspector may be willing to check your figures and offer you a sense of security.

## Residential Gas Sizing Example

Obviously, what works for a commercial job will work for a single-family residence. You start with the total developed length of the gas piping from the gas meter to the most remote outlet. Then you size the pipes, section by section, beginning at the most remote outlet and working your way backward toward the gas meter. Very likely, the total developed length for a residence will be less, and the Btu rating of the appliances will be lower than for most commercial jobs.

Figure 15-3 is a simple gas piping layout similar to the one you'll find in most single-family residences. The distance of each section of pipe is shown, along

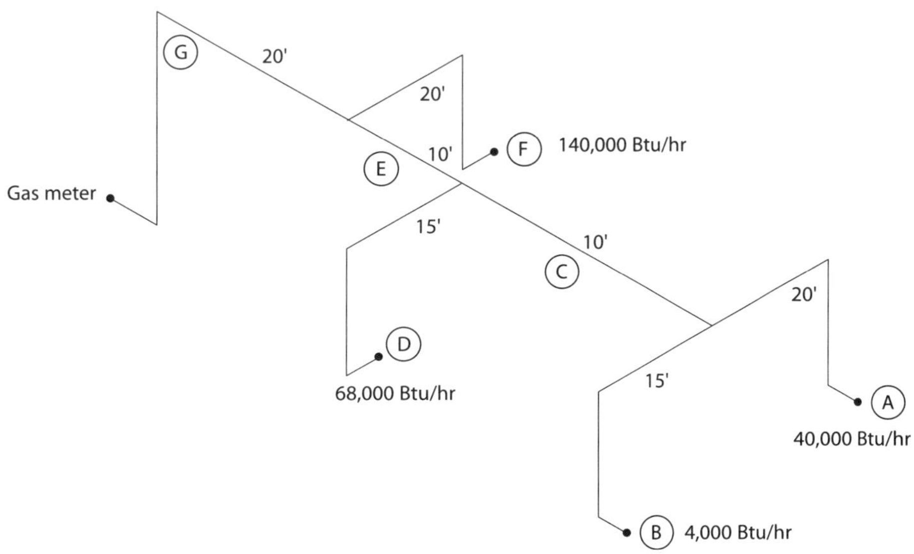


**Figure 15-3**
*Natural gas installation for a residence*

with the Btu per hour demand for each appliance. Let's size the gas piping for this single-family home. The developed length of the gas piping, measured from the gas meter to the most remote outlet, Outlet A, is 60 feet. In *IFGC* Table 402.4 (2), you use the 60-foot row to find the capacity of each pipe size.

Starting with Section A, the demand is 40,000 Btu per hour, or 40 cubic feet per hour. The correct pipe size is 1/2 inch. The maximum demand of Outlet B is 4,000 Btu, or 4 cubic feet per hour. A 1/4-inch pipe will do at this point. The combined maximum demand for pipe Section C is 44 cubic feet per hour. A 1/2-inch pipe will handle this.

Maximum demand for Outlet D is 68,000 Btu per hour or 68 cubic feet per hour. The correct pipe size is 3/4 inch. Looking at Section E, the combined maximum demand is 112 cubic feet per hour, so a 3/4-inch pipe will still work. The maximum demand for Outlet F is 140,000 Btu per hour, or 140 cubic feet per hour. At this point, you need 1-inch piping. Moving on to Section G, the combined maximum gas demand for the entire single-family residence is 252,000 Btu or 252 cubic feet per hour. The correct pipe size is 1 inch.

So much for sizing a gas supply system. If you follow the proper table and pay attention to detail, you should have no difficulty sizing gas systems. In the next chapter you'll learn how to install them.

# *Chapter 15 — Questions and Answers*

**15-1 LP gas may be a mixture of butane and ________________ gas.**

(A) heavy

(B) sour

(C) dry

(D) propane

***Answer: D***

*Code response: Section 402.6.1*
*This book, page 254*

**15-2 The maximum gas demand required by each appliance is determined by the**

(A) *International Mechanical Code.*

(B) *International Fire Code.*

(C) manufacturer's input rating.

(D) local gas supplier.

***Answer: C***

*Code response: Section 402.2*
*This book, page 254*

**15-3 To properly size each section of gas piping you must know the**

(A) type of gas being used.

(B) distance to the most remote outlet.

(C) pressure loss.

(D) specific gravity.

***Answer: B***

*Code response: Section 402.4.1*
*This book, page 254*

**15-4 When installing additional appliances to existing piping, it's the ________________ responsibility to check the existing gas piping to determine its adequacy.**

(A) gas supplier's

(B) plumbing plans examiner's

(C) plumbing contractor's

(D) code official's

***Answer: C***

*Code response: Section 401.4*
*This book, page 255*

**15-5 When you are to install an additional gas appliance in an existing system, the code states you must first check the existing piping**

(A) to determine if it has adequate capacity.

(B) to determine if it is properly identified.

(C) for defects.

(D) for the inspector's stamp.

***Answer:* *A***

*Code response:* *Section 401.4*
*This book, page 255*

**15-6 Each appliance must have a permanent factory-applied nameplate affixed, showing each of the following, except the**

(A) total number of cubic feet of gas.

(B) minimum clearance requirements.

(C) Btu input rate.

(D) manufacturer's name.

***Answer:* *A***

*Code response:* *Section 301.5*
*This book, page 254*

**15-7 Of the following gases, _______________ gas can become liquefied under only moderate pressure.**

(A) dry

(B) natural

(C) manufactured

(D) LP

***Answer:* *D***

*Code response:* *Definitions: page 13*
*This book, page 254*

**15-8 The maximum amount of gas input for an appliance is usually expressed**

(A) by the *International Fuel Gas Code*.

(B) by the gas supplier.

(C) in cubic inches.

(D) in cubic feet.

***Answer:* *D***

*Code response:* *Definitions: page 11*
*This book, page 254*

**15-9 When you have to add another appliance to an existing system that doesn't have the capacity, you'll need to**

(A) get the gas supplier to sign off.

(B) add copper expansion joints to the existing piping.

(C) install a separate line.

(D) get the code official's approval.

***Answer:* C**

*Code response: 401.4*
*This book, page 255*

**15-10 When appliance input rate (Btu) is not indicated (as shown in Figure 15-2) you must**

(A) get the code official's approval.

(B) use Table 402.2.

(C) get the job engineer's approval.

(D) use Table 402.4(2).

***Answer:* B**

*Code response: Section: 402.2*
*This book, page 256*

◆ *Chapter 16* ◆

# Gas Piping Materials and Installation Methods

The *IFGC* regulates the materials and installation methods you can use for gas systems. Some materials are limited to certain installations, and some materials are prohibited entirely. When you select materials for gas supply pipes, tubing, valves, or fittings, you must be certain they conform to the standards and specifications listed in Figure 16-1. If you don't find a specific material on this list, ask your local plumbing official. It's possible the material in question has been approved by the manufacturer and/or plumbing official through independent testing, but not added to the standards and specifications list. See *IFGC* Sections 403.1 and 403.3. If you install used materials, it's your responsibility to check the code requirements. All materials must be free from any foreign substances and be adequate for the installation you have in mind (*IFGC* Section 403.2).

## Materials

There are many materials you can use for gas piping. Some are OK for underground installations, some are OK for above ground only and some are OK for both. See Figure 16-2 for information on where you can use common gas piping materials.

### *Steel and Wrought Iron Pipe*

Steel and wrought iron pipe are widely used in gas systems because of their versatility. They're appropriate for interior, exterior and underground use. However, you can't use them under a concrete slab. Steel and wrought iron pipe must be at least standard weight (Schedule 40). Check *IFGC* Section 403.4.2 and Figure 12-3 for various weights of steel or wrought iron pipe. Remember to protect these pipes with an approved corrosion-resistant material if they're in contact with corrosive material or atmosphere. See *IFGC*, Section 403.8.

*IFGC* Sections 403.7 and 403.9.1 have clear rules about the condition of the piping you may use. When you cut and thread pipe in the field, you must be sure it's free from cutting burrs. Use a pipe reamer or file to remove burrs so gas won't be restricted. Always check the threads. Don't use threads that are chipped, stripped, corroded, or otherwise damaged. Above all, never try to repair defects in pipe, tubing or fittings. Just replace them with good materials.

| Materials | Standards |
|---|---|
| Aluminum alloy pipe | ASTM B 241 (alloy 5456 prohibited) |
| Aluminum alloy tubing | ASTM B 210 or ASTM B 241 |
| Anodeless risers | ASTM D 2513 |
| Copper tubing type K or L | ASTM B88 or ASTM B280 |
| Corrugated stainless steel tubing | ANSI/AGA LC 1 |
| Flanges | ASME B16.1, AMSE B16.20, AWWA C111/A 21.11 or MSS SP-6 |
| L P gas systems | NFPA 58 |
| Metallic pipe threads | ASME B 1.20.1 |
| Plastic pipe, tubing, fittings | ASTM D 2513 |
| Steel tubing | ASTM A 254 or ASTM A 539 |
| Steel and wrought-iron pipe | ASTM B 36.10, 10 M, ASTM A 53; or ASTM A 106 |
| Valves | ASME B 16.33, ANSI Z 21.15 |
| Appliance connectors | ANSI Z 21.69 |

**Note:** Although these are the standards according to the *International Fuel Gas Code*, you may find that pipe, tube or fittings tables may vary in other codes.

**Figure 16-1**

*Gas piping materials, standards and specifications*

| Material | Exterior underground | Interior above ground |
|---|---|---|
| Galvanized steel pipe | X | X |
| Wrought-iron pipe | X | X |
| Copper and brass pipe | X | X |
| Copper tubing | X | X |
| Aluminum alloy pipe | | X |
| Aluminum tubing | | X |
| Plastic pipe | X | |
| Steel tubing | X | X |

**Notes:**
1. Gas piping systems installed underground must have at least 12" of earth cover. Some local authorities require 18" earth cover if piping is plastic.
2. Some local authorities may list materials that don't appear in this chart.
3. Copper, brass, aluminum and steel pipe or tubing can't be used where gas is corrosive.
4. Always check local code requirements.

**Figure 16-2**

*Common gas piping materials and where to use them*

Pipe up to 2 inches is generally threaded. Pipe threads must be standard and tapered, in compliance with *IFGC* Section 403.9 and Figure 12-4. Seal all threaded joints tight with an approved pipe compound that's resistant to the action of liquefied petroleum gas or any other chemicals in the gases carried through the piping. See *IFGC* Section 403.9.3. Your plumbing supplier will stock the approved pipe compound you need for gas piping. Pipe sizes greater than 2 inches are usually flanged or welded.

Be sure your pipe joints can pass muster. They must be tight and able to withstand the pressure, vibration, temperature expansion and contraction, fatigue and the weight of the pipe and its contents. See *IFGC* Section 403.10. Most of the code requirements are the responsibility of the designing engineer, but if he's made a mistake, it's up to you to find it before you install it.

You can use steel tubing as long as the gases it will carry won't corrode it. All metallic piping and fittings must be coated with an approved coating or wrapping to protect them from the action of any corrosive materials with which it may come in contact. See *IFGC* Section 403.8.

For threaded fittings, see *IFGC* Section 403.10.4. Don't use threaded fittings larger than 4 inches without the approval of your local authority. Fittings used with steel or wrought iron pipe must be of steel, brass, bronze, malleable iron, or cast-iron. You can use cast-iron *fittings* in a gas system, but you can't use cast-iron *pipe. IFGC* Section 403.10.4 lists these guidelines for cast-iron fittings:

- Cast-iron flanges are acceptable. Study *IFGC* Sections 403.12.1, 403.12.2 and 403.13 for flange specifics.
- Cast-iron bushings are *not* acceptable.
- Never use cast-iron fittings in a system containing flammable gas/air mixtures.
- Cast-iron fittings 4 inches and larger must be approved by the local authority before used inside a building.
- Cast-iron fittings 6 inches and larger *can't* be used unless first approved by the local authority.

## Aluminum, Copper and Brass Pipe

You must protect aluminum alloy pipe from external corrosion. Use an approved coating where the pipe is in contact with masonry, plaster, or insulation, or where it's constantly wet by water, detergents, or sewage. It's *never* acceptable to install aluminum alloy pipe outside or underground. See *IFGC* Section 403.4.4 and Figure 16-2 for location use. If you use aluminum alloy pipe, you must use aluminum alloy fittings. Threads must *not* form the joint seal, so use approved joint compounds. See *IFGC* Section 403.9.3. You *can't* use zinc-aluminum fittings in systems containing flammable gas-air mixtures. *IFGC* Section 403.10.4 has many more details on this.

You can use seamless copper, aluminum alloy or steel tubing with gases that won't corrode them. Check with local gas supplier for gas type. See *IFGC* Section 403.5. You can also use both copper pipe (Type K or L) and brass pipe in a gas system, unless the gases are corrosive. Gases are classified as corrosive if they contain, on average, more than 0.3 grains of hydrogen sulfide per 100 cubic feet. *Never* bypass checking with the local gas supplier as to the type of gas to be used. See *IFGC* Section 403.5.2.

You can use threaded copper, brass or aluminum alloy in iron pipe sizes (also seamless copper tubing) with gases that won't corrode them. Copper and brass pipe are suitable for interior, exterior and underground gas piping but *never* under a concrete slab. Figure 16-2 shows the acceptable locations and Figure 16-3 shows the proper installation method. See *IFGC* Section 404.11.

If you use copper or brass pipe, the fittings must be manufactured of copper, brass, or bronze. See *IFGC* Section 403.10.4, Subsection 3.

You can make brazed joints in copper pipe, brass pipe and copper tubing. But this kind of gas piping requires special handling, so you're not likely to see it in general construction work. However, if you do:

- Any brazing material you use must have a melting point of more than 1,000 degrees Fahrenheit.
- Any brazing alloy can contain no more than 0.05 percent phosphorous. See *IFGC* Section 403.10.2.

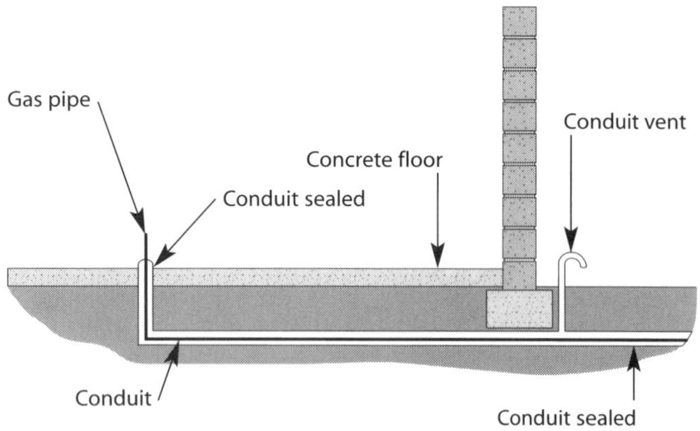


**Figure 16-3**
*Installation beneath concrete floor*

Most local authorities will let you use approved gas flare fittings for joining copper and aluminum tubing where the joints aren't concealed. This type of installation usually includes connecting the tubing from a gas outlet pipe to an appliance pipe inlet. See *IFGC* Section 403.10.3.

## *Plastic Pipe*

Plastic pipe and fittings, when approved for use by the local authority, must be marked "gas" and conform to ASTM (American Society for Testing and Materials) specifications. Corrosive gas or soil won't affect plastic pipe and fittings. Nevertheless, the code approves it only for outside underground installation. Plastic pipe and fittings you use in an LP-gas system must be in accordance with National Fire Protection Association's 58 specifications. See *IFGC* Section 403.6. Be sure to check with the local authority on specifics.

Remember that defects in gas pipe, tubing or fittings can't be repaired. Don't even think about repairing a gas system. It's just too dangerous. Replace defective pipe and fittings with new material. See *IFGC* Section 403.7.

Plastic pipe and fittings must meet the requirements in *IFGC* Sections 403.11 and 404.14.2. Here are some of the highlights:

- You can join plastic pipe and fittings with solvent cement, heat fusion or compression-type mechanical joints.
- Whatever type joint you use, you must make a gastight joint that's as strong as the pipe you're joining.
- Use approved transition fittings to make joints between plastic pipe and metallic piping.
- When you make a compression mechanical joint, the gasket material in the fitting must be compatible with the plastic piping and with the gas in the system.
- Use the joining method recommended by the plastic manufacturer to make all joints.

As to special fittings, I quote *IFGC* Section 403.10.4, Subsection 8:

"Fittings such as couplings, proprietary-type joints, saddle-tees, gland-type compression fittings, and flared, flareless, or compression-type tubing fittings shall be: used within the fitting manufacturer's pressure-temperature recommendations; used within the service conditions anticipated with respect to vibration, fatigue, thermal expansion or contraction; installed or braced to prevent separation of the joint by gas pressure or external physical damage; and shall be approved."

## What You Need to Know to Install Gas Piping

In general, these are the rules you must keep in mind when designing, installing, making modifications to or maintaining gas piping systems:

- Don't install gas piping in any circulating air duct, clothes chute, chimney, gas vent, ventilating duct, dumbwaiter, or elevator shaft. See *IFGC* Section 404.1.

- When you install gas piping in concealed locations, don't use unions, tubing fittings, right and left couplings, bushings, compression couplings and swing joints made by combinations of fittings. See *IFGC* Section 404.3. However, the code permits the following two exceptions: (1) fittings joined to piping by brazing and (2) fittings approved for use in concealed locations. Always check anything questionable with the local authority before you install it.
- Where internal or external crevice corrosion is known to occur, make certain your threaded or socket welded joints are protected with an approved coating or wrapping. See *IFGC* Section 404.8.1.
- Don't install gas piping in the same trench with water, sewer or drainage pipe, unless first approved by the local authority. This is a general rule of thumb.

## Exterior Gas Piping Installation

You must always install horizontal metallic gas piping at least 12 inches below grade. See *IFGC* Section 404.9. Though the *IFGC* doesn't mention it, many local authorities insist on a depth of 18 inches for plastic piping. When installing individual lines to lights, grills or some other gas appliances located outside a building, a trench can be 8 inches below finished grade, as long as you get approval first and the location isn't susceptible to physical damage. See *IFGC* Section 404.9.1. Lay underground gas piping in open trenches that are graded in a way that provides a firm, substantially-continuous bearing all along the bottom of the pipe. See *IFGC* Section 404.10.

If you're laying metallic pipe in corrosive soil, protect it with an approved coating or wrapping. When you join dissimilar metal pipes underground, use an insulating coupling or fitting. Never install metal piping in contact with cinders, as this will corrode it. See *IFGC* Section 404.8.

Here are some important general installation requirements:

- Make sure you support piping in open trenches to prevent sagging and excessive stress during backfilling.
- Use clean and fine material to fill all trenches.
- Where freezing temperatures occur, make the trench bottom below the frost line.
- In cold climates, be sure to insulate the pipe where it enters a building above ground, in crawl spaces, or anywhere it isn't protected from the cold. Many gases contain moisture that can freeze, block or rupture pipe.

## Installing Gas Piping Under a Slab

Occasionally you'll have a job where you have to install gas piping underground, beneath a building floor slab. When this is unavoidable, the *IFGC* permits this type of installation only in accordance with Section 404.11 and Figure 16-3. Here's the scoop:

- Encase the entire length of gas piping (down through, under and back up through the floor) in a conduit of wrought iron, plastic pipe, or steel pipe designed to withstand the superimposed loads. Local authorities usually won't accept less than Schedule 40 pipe.
- Seal the termination of the conduit above the floor to prevent any gas leaking into the building.
- Extend the termination of the conduit at least 4 inches outside the building and seal it tightly to keep out water and insects.
- Extend a vent above the grade and secure it to the conduit. This vent carries any leaking gas to the outside of the building.
- Protect the conduit from corrosion, according to *IFGC* Section 404.8.

## *Installing Gas Piping Outdoors Above Ground*

If you're ever required to install gas piping above grade outside a building, follow these code requirements. Locate the pipe where it can't be physically damaged, and support it securely. If it must pass through a wall constructed of corrosive materials, protect the pipe with an approved coating or wrapping. If you use a protective pipe sleeve, be sure to seal the annular space between the pipe and sleeve. Piping installed above ground or on building roofs must be elevated at least $3^1/_2$ inches above the surface. See *IFGC* Sections 404.7 and 404.4. Check with your local officials on insulation requirements for exposed piping.

## *Installing Interior Gas Piping*

If you're installing piping in concrete floors — obviously, it's more difficult to conceal gas piping if the appliance isn't adjacent to a wall. The *IFGC* Section 404.6 recommends the following installation method:

If you're working with a solid floor, lay the piping in an open channel cut into the concrete floor. It's important that you keep damage to the building to a minimum, so only cut the floor slab as deep as is necessary to conceal the pipe. Protect the pipe by installing a removable grille or a cover that provides access to the piping. Where the piping is exposed to excessive moisture or corrosive substances, it must be protected with an approved material. See Figure 16-4.

In the rare cases where it's unavoidable, the code may let you embed gas piping in freshly-poured concrete floors if you get approval from the local jurisdiction. You must follow these strict guidelines:

- Install gas piping in a casing of Schedule 40 steel, wrought iron, PVC or ABS plastic pipe.
- Tightly seal the ends and joints.
- Extend both ends of the casing at least 2 inches beyond the point where the pipe comes out of the floor. When in doubt, review *IFGC* Section 404.6.

When gas piping must pass through masonry walls, protect the pipe against corrosion by painting or sleeving. If your piping is in vertical masonry walls, be sure to provide adequate chases for protection. See *IFGC* Section 404.2. Figure 16-5 shows two installation examples.

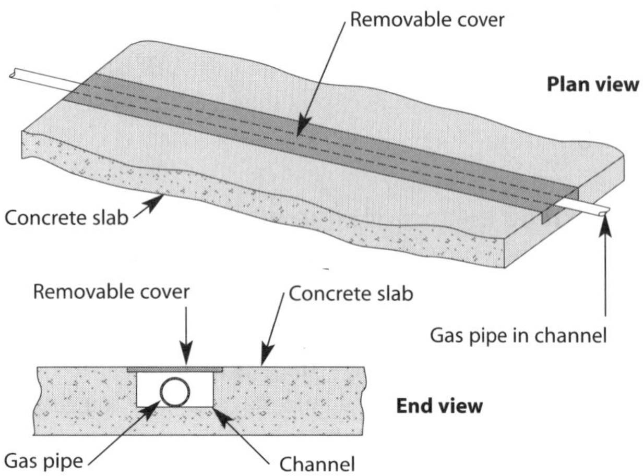


**Figure 16-4**
*Concrete floor open-channel installation*

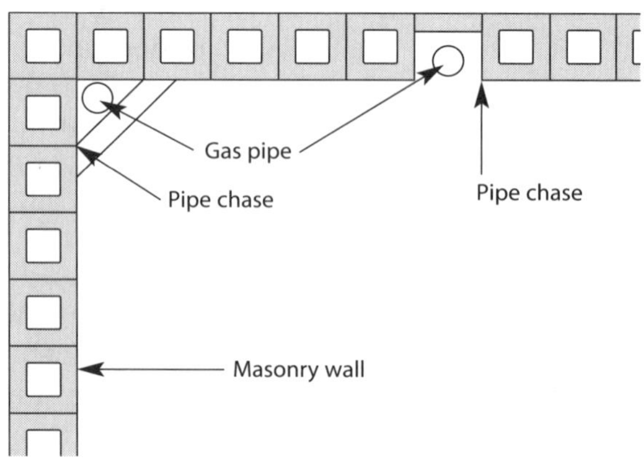


**Figure 16-5**
*Vertical masonry wall chase — plan view*

## *Installing Gas Piping in Wood Studs, Joists or Rafters*

*IFGC* Section [B]302.1 has some very specific instructions about installing gas piping or tubing within a building. First off, and very important, you mustn't weaken a building when you install any gas piping. When gas piping must penetrate the floor or ceiling of a building that requires a fire-resistance rating, you must protect it in accordance with the *International Building Code*. See *IFGC* Section [B]302.2. Check with the general contractor to see if this is your responsibility or his. Always read job specifications carefully. Impairing a building's fire-resistance rating can mean big bucks.

When cutting, notching and boring in wood members:

- If a notch is at the end of a joist, don't make it deeper than one-fourth the joist depth.
- When it's absolutely necessary to bore a hole, make it at least 2 inches from the top and bottom of the joist. Don't make a hole that's more than one-third the depth of the joist.
- Don't bore any hole in the top or bottom of the joist that's more than one-sixth the depth of the joist.
- Never bore a hole in the middle one-third of a span. See *IFGC* Section [B]302.3.2.

Don't make a cut or notch in a wood exterior wall or bearing partition that's more than 25 percent of the depth of the wall or partition. If you're working with studs in nonbearing partitions (supporting no loads), you can cut or notch up to 40 percent of any stud's width. See *IFGC* Section [B]302.3.3.

You can make a bored hole in a wood stud if the hole isn't more than 40 percent of the depth of the stud. In non-bearing partitions, or in any wall where each bored stud is doubled, you can make a bored hole if the hole isn't more than 60 percent of the depth of the stud or partition. But you can't bore in two double studs that are next to each other. Never bore a hole closer than $^5/_8$ inch to the stud's edge. Never bore a hole at the same section where a stud has been cut or notched. See *IFGC* Section [B]302.3.4.

If you have to cut, notch or bore steel members in a multi-story building with cold-formed steel framing, be sure you don't do anything that might weaken the building structure. *IFGC* Sections [B]302.6 and 302.7 suggest you check with the design professional in charge before proceeding. A great idea! In fact, an imperative idea! It's OK to cut, notch and bore flanges and lips of non-structural cold-formed steel wall studs along the centerline of the framing member if you don't make them more than $1^1/_2$ inches in width or 4 inches in length. You can't space any holes less than 24 inches apart or closer than 10 inches from the bearing end. See *IFGC* Section [B]302.7.

Light metal stud partitions are replacing wood partitions in many new buildings. The metal studs are hollow — essentially a rectangular hollow pipe with most of one of the wide sides omitted. Simply install the gas pipe or tubing along the inside of the stud. Wrap the pipe or tubing with an approved material to prevent contact with the metal and corrosion. Secure the pipe with tie wire. Metal studs have sharp edges. Wear gloves.

When installing piping materials other than black or galvanized steel, protect piping from drywall screw and nail damage. If a hole or notch that supports the pipe is less than 1 inch from a stud, joists, rafters or similar members, protect the pipe with shield plates (stud guards). The stud guard must be at least $^1/_{16}$ inch thick and reinforce the damaged wood member. The stud guard must also extend at least 4 inches above sole plates, below top plates and to each side of the wood member. See *IFGC* Section 404.5.

## Gas Outlets

Extend gas piping outlets through finished walls or ceilings at least 1 inch beyond the pipe threads. If the outlet extends through floors, outdoor patios, or slabs, the extension must be at least 2 inches beyond the pipe threads. Make certain that any outlet fitting or pipe is securely supported. Always locate gas outlets in the room where the appliance will be installed. Never, never locate an outlet behind doors. See *IFGC* Section 404.13. Follow the manufacturer's installation instructions if you install any flush-mounted-type quick disconnect outlets that are listed and labeled as gas convenience outlets. See *IFGC* Sections 404.12 and 404.13.

Never place any device inside gas piping, or use fittings that will reduce the cross-sectional area or restrict the free flow of gas. Use only gas filters that are approved by the local authority.

If you install future gas outlets, make sure they're capped gastight. See *IFGC* Section 404.12, and check out the exception.

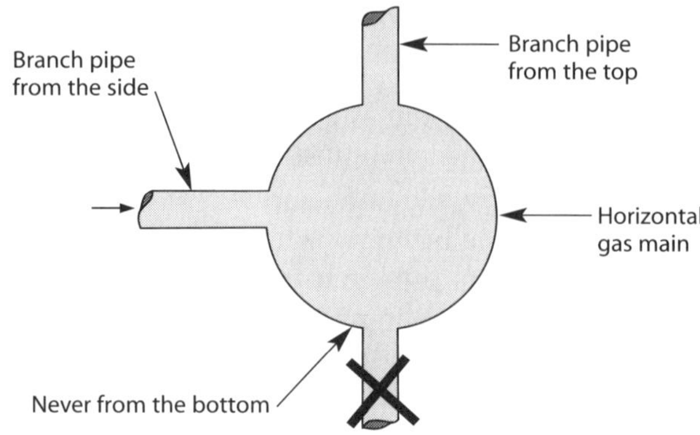


**Figure 16-7**
*Branch pipe connection to horizontal pipe*

## Installing Drip Pipes and Shutoff Valves

Always install gas mains (other than dry gas) so they can drain dry. Use a pitch or slope that's at least 1/4 inch per 15 feet of run and always toward the meter. Place drips wherever condensation may collect. Usually you make a drip pipe at the job constructed with a tee, nipple and cap. Don't make a drip pipe smaller than the pipe it serves and be sure it's accessible for emptying. Protect a drip pipe from freezing. Some local authorities require an 18-inch drip pipe. See Figure 16-6 for an illustration. Install a drip pipe at the outlet of the house meter to provide a trap so that accumulated condensate will shut off the gas flow to the house before the condensate can run back into the meter.

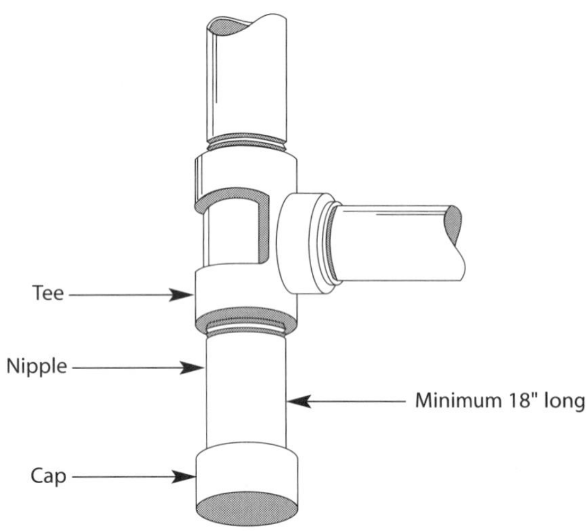


**Figure 16-6**
*Drip pipe*

When the gas utilization equipment doesn't have a sediment trap, you'll have to construct one. Install it as close as possible to the inlet of the equipment. Code doesn't require a sediment trap for illuminating appliances, ranges, clothes dryers and outdoor grills. See *IFGC* Section 408.4. Always connect gas branch pipes to other horizontal pipes at the top or side of the feeder pipe and never from the bottom. This will keep condensate collections from obstructing the flow of gas to the branch line. See Figure 16-7.

Buildings with multiple tenants and multiple gas lines and a master meter must have a separate shutoff valve for each building or tenant. It's a good idea to locate each shutoff valve on the outside of the building in an accessible location near the meter. See *IFGC* Section 409.3.

In residential buildings larger than a single-family or duplex, where there is a common gas main, there must be a shutoff valve for each unit in an accessible location. See *IFGC* Section 409.3.1.

Where a common gas main serves more than one building, you must install an accessible shutoff valve outdoors for each building. See *IFGC* Section 409.3.2.

Be sure each shutoff valve has an identification tag for ready identification. See *IFGC* Section 409.3.3.

The gas supplier must install a shutoff valve on the inlet or supply side of the gas meter. See *IFGC* Section 409.2.

Don't ever install gas shutoff valves in concealed locations or in spaces used as plenums. See *IFGC* Section 409.1.2.

Use only shutoff valves approved for gas systems. The valves must be manufactured of materials that are compatible with the type of piping material you use. Shutoff valves installed in gas systems operating above 0.5 psig must comply with ASME B 16.33. Shutoff valves installed in gas systems operating at 0.5 psig or less must comply with ANSI Z 21.15 or ASME B 16.33. See *IFGC* Section 409.1.1. Figure 16-1 has gas material standards.

Gas valves usually are designed in two patterns for two purposes. If a shutoff valve is outside and accessible to accidental or malicious tampering, it usually has a square nut head that requires a special tool. If a shutoff valve is inside (to control gas appliances) it must also be accessible. It must be manually operated with a lever handle that can be turned off without a special tool. To make inside shutoff valves more convenient to use, they come in two types — a straight pattern and an angle pattern.

The code is strict in requiring an accessible shutoff valve for each piece of gas equipment. You must install an accessible, manually-operated shutoff valve for each interior gas appliance. Install the valve as close to the outlet pipe as possible in the same room as the appliance and not over 6 feet from the appliance. Place the valve upstream from the union, connector or quick-disconnect device the valve serves.

Always make sure any valve is accessible without requiring moving the appliance or portions of the building. See *IFGC* Section 409.5. For a vented decorative appliance or a decorative appliance installed in a vented fireplace, you may locate a shutoff valve in a remote area as long as it's accessible and permanently identified. But the valve can't serve more than the one appliance.

If you ever have to install an equipment shutoff valve in the firebox of a fireplace, be sure you follow the manufacturer's instructions. See *IFGC* Section 409.5.1.

*IFGC* Section 409.4 states that when a gas line requires a pressure regulator valve, you must install a listed shutoff valve directly ahead of the regulator. Any time the line pressure is greater than the designed pressure allowance, you must install an accessible pressure regulator. If the installation is on the building's exterior, protect it from physical damage and make sure it's approved for outdoor use. See *IFGC* Section 410.1.

## MP Regulators

Here are the *IFGC* Section 410.2 requirements for MP regulators:

- Use only approved MP regulators suitable for the inlet and outlet gas pressures for the appliance.
- Be sure an MP regulator can maintain a reduced outlet pressure under a lockup (no-flow) condition.
- The MP regulator capacity must be adequate for all the appliances it serves. Get this information from the manufacturer's published ratings.
- When you install an MP regulator indoors, be sure it's accessible, and either vented to the outside, or equipped with a leak-limiting device. Comply with Section 410.3.
- Install an assembled drip pipe (discussed earlier) between an MP regulator and its upstream shutoff valve. Allow for connection of a pressure-measuring instrument that will also serve as a sediment trap.

- Install another drip pipe at least 10 pipe diameters downstream of an MP regulator outlet. Position it so a pressure-measuring instrument may be connected.

When a pressure regulator requires a vent, *IFGC* Section 410.3 says you must install an independent vent that extends to the outside of the building. Design it so water or foreign objects can't get into it. You don't need an outside vent on a pressure regulator if:

- it has an approved vent-limiting device
- it's properly labeled
- you install it according to the manufacturer's instructions.

## Piping Supports

Generally speaking, if you space your horizontal and vertical gas piping supports the same as for water pipe of like material, you'll be in accord with code requirements. See *IFGC* Section 407 and Chapter 12.

Acceptable materials for pipe supports are pipe hooks, metal pipe straps, bands, brackets, or hangers. Their quality and strength must be adequate for the type and size of pipe being supported. It's important that you anchor the piping so that no undue strain is placed on connected equipment. Never use other piping to support the gas piping. Locate supports at intervals to prevent excessive vibration. Install them so as not to interfere with the free expansion and contraction of the piping.

## Inspecting and Testing Gas Systems

*IFGC* Section 406 tells you about inspecting, testing and purging gas systems. *IFGC* Table 402.4(2) shows the maximum capacity of pipe for low pressure gas (0.5 psig in cubic feet per hour, based on a 0.60 specific gravity gas) for different sizes and lengths of iron or rigid pipe. See also Figure 15-1. The minimum test pressure you can use is $1^1/_2$ times the maximum working pressure you can expect in the system, but never less than 3 psig. See *IFGC* Section 406.4.1. If you're testing for defective material and joint leaks, this will work fine. Most of your work will be with low pressure gas.

The system must remain airtight with no loss of pressure for at least $^1/_2$ hour and definitely until after it's inspected. If you even suspect a pressure loss, test the rough piping for leaks by running a small brush dipped in liquid soap around each joint. If you see any bubbles you'll know you have a repair job to do. Gas piping can't be concealed until it's been pressure-tested and passed by the plumbing inspector. Each outlet must be capped and a pressure gauge installed on one of the outlets.

## Installing Gas Appliances

It's your responsibility to connect the gas from a wall outlet to an appliance, using a rigid pipe or an approved flexible connector. Then the gas supplier will purge the lines of air, check for leaks at the joints that connect the appliance to the gas system, light the pilot lights, and finally adjust each appliance. When you install gas equipment and/or appliances, be sure to follow the conditions of listing, the manufacturer's installation instructions, and the *IFGC* Section 305.

You can install gas appliances such as water heaters and clothes dryers on the floor of residential and public garages under the following code conditions:

- The floor of the garage must be higher than the driveway or adjacent ground. See *IFGC* Section 305.3.
- The combustion chamber must be at least 18 inches above the floor or adjacent ground. See *IFGC* Section 305.3.
- Appliances must be protected from motor vehicle impact. If you can't make sure this won't happen, you must install the appliance according to *IFGC* Section 305.4.

*IFGC* Section 306 says each appliance you install must have enough clearance so work can be done on the appliance without removing any permanent construction or damaging a required fire-resistant-rated assembly. Essentially, if an appliance or piece of equipment was installed correctly, it will be easy to inspect, service, repair or replace it. The architect should design a building to make sure the gas appliances or equipment are properly located. The *IFGC* says that any room where appliances are installed must have a door and an unobstructed passageway. The passageway must be at least 35 inches wide and 80 inches high. If the appliance is installed in an alcove, compartment, basement or some similar space in a dwelling unit, there are some special requirements:

- There must be access to the appliance(s) through an opening or door.
- There must be a passageway at least 24 inches wide or large enough to remove the largest appliance installed.
- There must be a level area on the service side of the appliance at least 30 inches deep and at least as high as the appliance.

If a building's designed for an appliance, such as a furnace in the attic, there must be a way to get the appliance out that's at least 30 inches high, 24 inches wide, and no more than 20 feet long. The passageway must have a solid floor and be at least 24 inches wide. There must be a 30-inch level space at the front or service side of the appliance. However, if the appliance can be serviced or removed through the required opening, a passageway and level space aren't required.

If an appliance has to be installed under the floor of a dwelling unit, there must be an opening and passageway that's large enough to remove the largest component of the appliance. The passageway must be at least 30 inches high, 22 inches wide and no more than 20 feet long. There must be a level space that's 30 inches wide and 30 inches deep at the front or service side of the appliance. However, if the appliance can be serviced or removed through the required opening, you won't have to worry about the level service space or the passageway. If the passageway is 12 inches below the adjoining grade, you must line it with concrete or masonry to 4 inches above grade and make sure it's strong enough to not collapse.

If you have to install an appliance on the roof of a building or an elevated structure, there are certain precautions. If the appliance installation is 16 feet or more above grade or floor level, there must be a permanent approved access for servicing or removing the appliance (usually the architect's job). Such access must not include climbing over any obstructions 30 inches high or walking on any roof whose slope is more than 33 percent.

Other particulars in *IFGC* Section 308 are definitely for the architect and the mechanical trade and should cause no concern for the plumber.

## Appliance Location

*IFGC* Section 303 spells out where you can and can't install gas appliances. Basically, you can't install a gas appliance in a hazardous location unless it's listed and approved for that particular type of installation. You can't install gas appliances in bedrooms, bathrooms, toilet rooms, storage closets and surgical rooms.

You aren't restricted on where you can install a direct-vent appliance that gets its combustion air from the outside. You can install these vented appliances, room heaters, wall furnaces, decorative appliances and unvented decorative appliances in a solid fuel-vented fireplace if the room isn't a confined space and if the building's not too tight in its construction. You need 50 cubic feet of space for each 1,000 Btu per hour input rating of any appliance installed in the space. See *IFGC* Section 304.5.1. The *IFGC* considers any space less than this to be a confined space.

You can install a single wall-mounted unvented gas room heater in a bathroom if:

- its input rating isn't more than 6,000 Btu per hour.
- it's equipped with an oxygen depletion safety shutoff system.
- the bathroom isn't a confined space.

You can install a single wall-mounted unvented room heater in a bedroom if:

- its input rating isn't more than 10,000 Btu per hour.
- it's equipped with an oxygen depletion safety shutoff system.
- the bedroom isn't a confined space.

You can install furnaces and boilers indoors, in closets and alcoves if the manufacturer lists them for such installations. You can install equipment outdoors if the manufacturer lists the equipment for outside installation but you must protect it from outdoor environmental factors.

Any time you install an appliance in a pit or some kind of excavation, be sure it doesn't come in contact with the soil. Locate the appliance at least 12 inches from the sides of the pit or excavation. If the bottom of the excavation is more than 12 inches below the adjoining grade, line it with concrete or masonry that extends 4 inches above adjoining grade. Make certain the walls are strong enough so they won't collapse. Also, figure out an approved way to keep the excavation from flooding.

## Connecting Gas Appliances and Equipment

*IFGC* Section 411 specifies the requirements on connecting an appliance to a piping system. You can use rigid metallic pipe and fittings. Or you can make a connection with soft L-type copper or aluminum alloy tubing and fittings. Flared fittings are acceptable. The distance from wall outlet to appliance connection can't be more than 6 feet and must be in the same room as the appliance. Never use an unprotected knockout opening to connect semi-rigid tubing with a motor-operated appliance.

You can use a listed and labeled manufactured appliance connector. You must locate it in the same room as the appliance and install it according to the manufacturer's instructions. You can use a listed and labeled quick-disconnect device with a listed and labeled appliance connector. Be sure the connectors comply with the standards in Figure 16-1 when you install food service equipment.

Appliance connectors can only be 3 feet long, except for range and clothes dryer connectors that may be up to 6 feet long. See Section 411.1.2. Don't conceal a connector within any wall, or extend it through walls, floors, partitions, ceilings or appliance housing. A connector must be in the same room as the appliance or equipment. The shutoff valve must be the same nominal size as the connector and installed ahead of the connector.

Connect movable appliances or equipment as shown in Figure 16-8, using an approved flexible connector designed and labeled for the application. Install and protect the connector according to the manufacturer's recommendation. The maximum length for a flexible connector is 6 feet.

## Venting Gas Appliances

*IFGC* Sections 304 and 501 have information for all types of vents and chimneys. Most of this information is for other trades. But I've provided here most of the information that will apply on most of your jobs. As a rule, most of your work involving gas vents will be for gas water heaters.

If you install a gas appliance in a room that's a confined space, ventilate it through permanent openings with a total free area of 1 square inch for each 1,000 Btu per hour of input rating (that's at least two openings, 100 square inches each). One vent

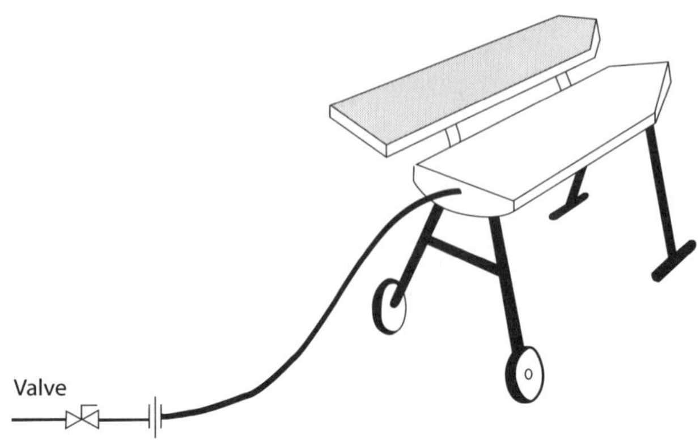


**Figure 16-8**
*Gas ironing equipment*

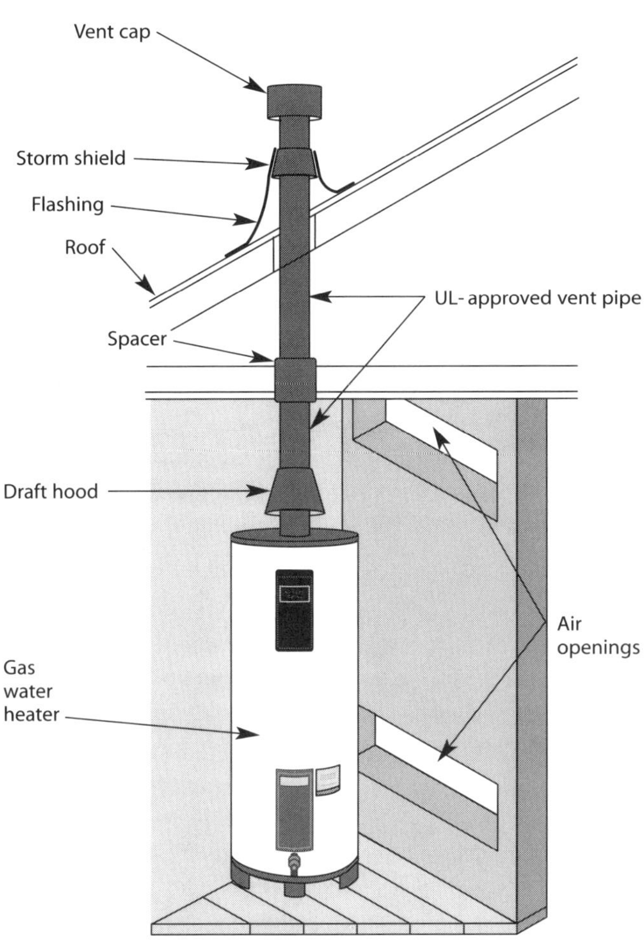


**Figure 16-9**
*Installing a fuel-burning water heater*

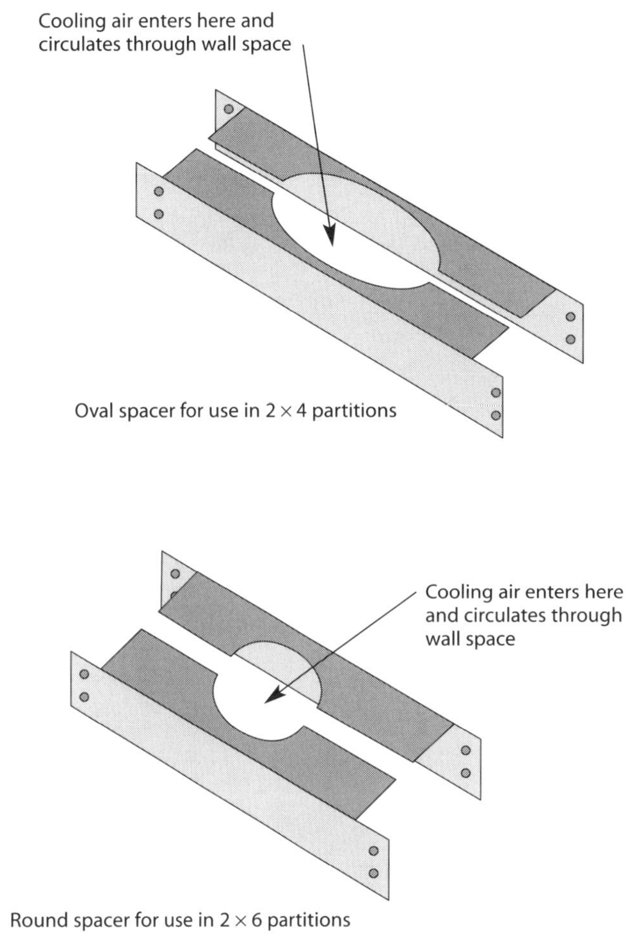


**Figure 16-10**
*Vent pipe spacers*

opening must be at least 12 inches above the floor. The second opening must be at least 12 inches below the ceiling. This provides enough air circulation for the combustion and dilution of flue gases. Air opening dimensions must be at least 3 inches. See *IFGC* Section 304.5.3.1 and Figure 16-9.

Vent an appliance as close as possible to the vent pipe. If a draft hood is required, the vent pipe can't be smaller than the opening of the draft hood. Provide enough clearance between gas appliances and combustible materials to avoid a fire hazard. Install water heaters with insulated jackets at least 2 inches from combustible material, or at least 1 inch from one-hour fire-rated materials.

There are two acceptable types of vent pipe material you'll probably use. The most commonly used is the double-wall metal pipe and fittings. The clearance distance recommended by the manufacturer is stamped in the metal. It's usually at least 1 inch from any combustible material. Metal vent pipes installed in partitions built of combustible materials must have an approved metal spacer that keeps surface temperature below 160° Fahrenheit. See Figure 16-10.

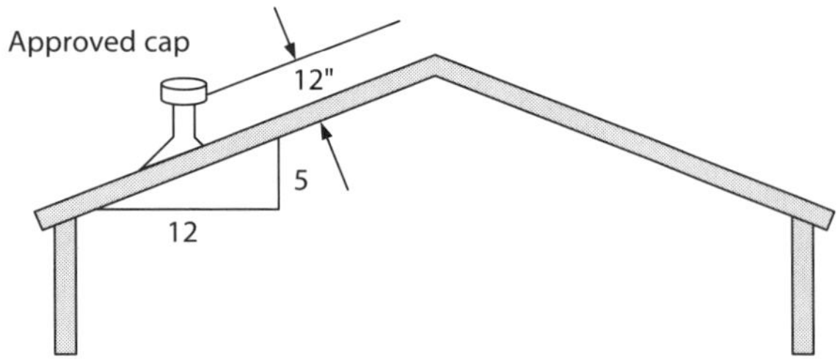


Roof pitch 5'/12' ($22\frac{1}{2}$°) or less (includes flat roofs). Maintain minimum clearance of 12" as illustrated.

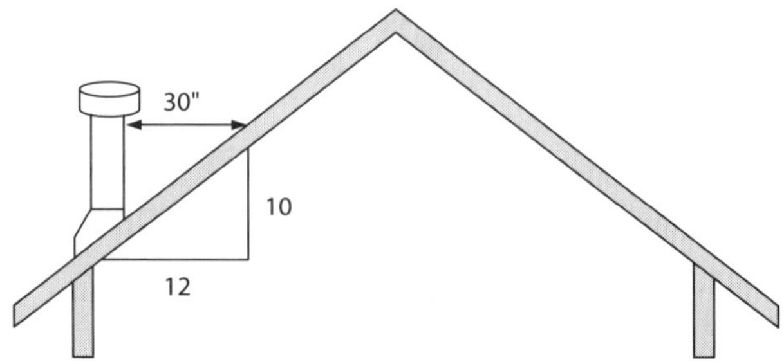


Roof pitch $5\frac{1}{2}$'/12' to 12'/12' (45°). Maintain 30" horizontal distance as illustrated.

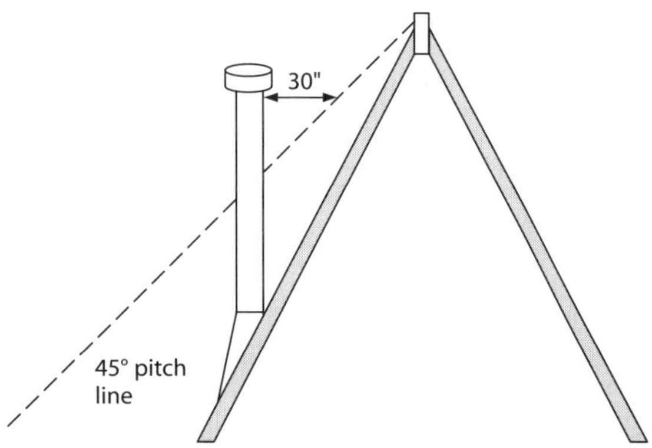


Roof pitch greater than 12'/12' or vertical wall. Maintain 30" horizontal distance from 45° pitch (12'/12') line.

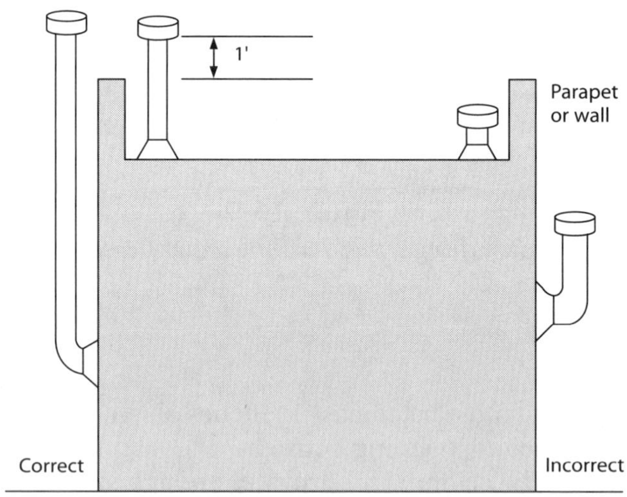


Vent top should be located 1' above parapet or wall when within 30".

**Figure 16-11**

*Acceptable pipe terminations*

The second type of vent pipe material is the single-wall vent pipe. Because special provisions must be taken to avoid the possibility of fire, *IFGC* Section 503.7 has a number of strict installation instructions you must follow:

- Single-wall metal pipe should be made of galvanized sheet steel that's at least 0.0304 thick. Or use other approved, non-combustible, corrosion-resistant material.
- Don't use single-wall uninsulated metal pipe outdoors in cold climate areas.
- Terminate single-wall metal pipe at least 5 feet vertically above the highest connected draft hood or flue collar. Extend it at least 2 feet above the highest point where the pipe passes through a roof. Actually, it must extend at least 2 feet higher than any portion of an adjoining building within a horizontal distance of 10 feet. Be sure it ends with an approved cap or roof assembly.
- Use a non-ventilating thimble to extend a single-wall metal pipe through a roof made of combustible material. Extend the thimble at least 18 inches above the roof and at least 6 inches below the roof. The annular space must be open at the bottom and closed at the top.
- Support all horizontal vent pipes to prevent sagging or misalignment. Some acceptable supporting materials are straps or hangers constructed of at least 20-gauge sheet metal.
- Use a UL-approved cap to terminate all vent pipes above the roof of a building. See Figure 16-11 for examples of acceptable vent pipe terminations outside a building.

# Chapter 16 — Questions and Answers

**16-1 When you use steel or wrought iron pipe in a gas system, the *IFGC* says you must use at least Schedule _____ pipe.**

(A) 20

(B) 30

(C) 40

(D) 80

***Answer:* C**

*Code response: Section 403.4.2*
*This book, page 263*

**16-2 According to the *IFGC*, gas piping with imperfections may be used**

(A) if the imperfections are only surface.

(B) under no circumstances.

(C) if repaired to code standards.

(D) if the inspector approves it.

***Answer:* B**

*Code response: Section 403.7*
*This book, page 263*

**16-3 According to the *IFGC*, a clothes dryer fuel connector shall not exceed ____________ feet.**

(A) 2

(B) 4

(C) 6

(D) 8

***Answer:* C**

*Code response: Section 411.1.2*
*This book, page 274*

**16-4 Cast-iron _______ are always acceptable for installation in a cast-iron gas system.**

(A) flanges

(B) fittings 6 inches or larger

(C) indoor-type fittings 4 inches and larger

(D) bushings

***Answer:* A**

*Code response: Section 403.10.4 (5)*
*This book, page 265*

**16-5 According to the *IFGC*, you can't install aluminum alloy pipe**

(A) in a wooden partition.

(B) in a metal partition.

(C) outside or underground.

(D) Within 4 feet of an electrical conduit.

***Answer:* *C***

*Code response: Section 403.4.4*
*This book, page 265*

**16-6 According to the *IFGC*, plastic pipe and fittings can be used only**

(A) if approved for use in a gas system.

(B) in interior installations.

(C) outside, in locations not subject to freezing.

(D) outside, underground.

***Answer:* *D***

*Code response: Section 403.6*
*This book, page 266*

**16-7 According to the *IFGC*, it's permissible to install _______________ in a concealed gas piping location.**

(A) galvanized unions

(B) tubing joined by brazing

(C) square head bushings

(D) compression couplings

***Answer:* *B***

*Code response: Section 404.3*
*This book, page 267*

**16-8 When you install gas piping outdoors, the *IFGC* always requires you to**

(A) use an approved wrapping.

(B) bury it no less than 12 inches below grade.

(C) encase it in a conduit.

(D) protect it from physical damage.

***Answer:* *D***

*Code response: Section 404.7 and 404.9.1*
*This book, page 267*

**16-9 When you install gas piping outlets through finished walls, you must extend them at least______ beyond the pipe threads.**

(A) $^1/_2$ inch

(B) $^3/_4$ inch

(C) 1 inch

(D) two pipe thicknesses

***Answer: C***

*Code response: Section 404.13*
*This book, page 269*

**16-10 When installing a gas water heater in a garage, the *IFGC* requires that its combustion chamber be at least ______ inches above the garage floor.**

(A) 14

(B) 18

(C) 20

(D) 24

***Answer: B***

*Code response: Section 305.3*
*This book, page 272*

# Abbreviations

*The abbreviations here are often found on blueprints (building plans) and in plumbing reference books (including the code) to identify plumbing fixtures, pipes, valves and nationally-recognized associations.*

**A** air or area

**ACC** access

**AD** area drain

**AFF** above finished floor

**AGA** American Gas Association

**AISI** American Iron and Steel Institute

**AL** aluminum

**ALT** alternate

**ANSI** American National Standards Institute

**AP** access panel

**API** American Petroleum Institute

**APX** approximate

**ARCH** architect

**ARI** Air Conditioning & Refrigeration Institute

**ASA** American Standard Association

**ASB** asbestos

**ASC** above suspended ceiling

**ASCE** American Society of Civil Engineering

**ASHRAE** American Society of Heating, Refrigeration and Air Conditioning Engineers

**ASME** American Society of Mechanical Engineers

**ASSE** American Society of Sanitary Engineering

**ASTM** American Society for Testing and Materials

**AUTO** automatic

**AVG** average

**AWWA** American Water Works Association

**B** bidet

**B&S** bell and spigot

**BBL** barrel

**BEL** below

**BET** between

**BIT** bituminous

**BLDG** building

**BM** bench mark

**BOCA** Building Officials Conference of America (part of ICC)

**BOD** biochemical oxygen demand

**BOT** bottom

**BRZ** bronze

**BS** bar sink

| | |
|---|---|
| **BSMT** | basement |
| **B.T.** | bathtub |
| **BTU** | British thermal unit |
| **C** | Centigrade or condensate |
| **°C** | degrees Centigrade |
| **C to C** | center to center |
| **CB** | catch basin |
| **CFL** | counterflashing |
| **CFM** | cubic feet per minute |
| **CHT** | ceiling height |
| **CI** | cast iron |
| **CIPC** | cast-in-place concrete |
| **CIR** | circle |
| **CIRC** | circumference |
| **CISP** | cast iron soil pipe |
| **CISPI** | Cast Iron Soil Pipe Institute |
| **CK** | caulk |
| **CLG** | ceiling |
| **C.O.** | cleanout |
| **COL** | column |
| **CPR** | copper |
| **CS** | cast steel |
| **CS** | Commercial Standard |
| **CS** | countersunk |
| **CT** | ceramic tile |
| **CTR** | counter |
| **CU FT** | cubic foot |
| **CU IN** | cubic inch |
| **CV** | check valve |
| **CW** | cold water |
| **C.W.M.** | clothes washing machine |
| **D** | drain |
| **DEG or °** | degree |
| **DF** | drinking fountain |
| **DIA** | diameter |
| **DIAG** | diagonal |
| **DIM** | dimension |
| **DL** | dead load |
| **DRB** | drainboard |
| **DS** | downspout |
| **DT** | drain tile |
| **DW** | dishwasher |
| **DWG** | drawing |
| **E** | east |
| **E to C** | end to center |
| **EL** | elevation |
| **ELEC** | electric |
| **ELEV** | elevation |
| **EMER** | emergency |
| **ENC** | enclose |
| **EP** | electric panelboard |
| **EQ** | equal |
| **EQP** | equipment |
| **EST** | estimate |
| **EWC** | electric water cooler |
| **EXCA** | excavate |
| **EXG** | existing |
| **EXH** | exhaust |
| **EXP** | exposed |
| **EXT** | exterior |

| | |
|---|---|
| **°F** | degrees Fahrenheit |
| **F** | Fahrenheit |
| **FA** | fire alarm |
| **FB** | foot bath |
| **FCO** | floor cleanout |
| **FD** | floor drain |
| **FDC** | fire department connection |
| **FE** | fire extinguisher |
| **FEC** | fire extinguisher cabinet |
| **FF** | finish floor |
| **FFE** | finished floor elevation |
| **FG** | finish grade |
| **FGL** | fiberglass |
| **FHC** | fire hose cabinet |
| **FHS** | fire hose station |
| **FL** | fire line |
| **FLCO** | floor cleanout |
| **FLG** | flashing |
| **FLR** | floor |
| **FND** | foundation |
| **FP** | fire plug |
| **FRA** | fresh air |
| **FS** | full size |
| **FSP** | fire standpipe |
| **FTG** | footing |
| **FU** | fixture unit |
| **FUR** | furred |
| **FUT** | future |
| **FWD** | food waste disposal |
| **G** | gas |

| | |
|---|---|
| **GA** | gage or gauge |
| **GAL** | gallons |
| **GALV** | galvanized |
| **GB** | grab bar |
| **GD** | grade |
| **GI** | galvanized iron |
| **GKT** | gasket |
| **GP** | galvanized pipe |
| **GPD** | gallons per day |
| **GPM** | gallons per minute |
| **GS** | glass sink |
| **GV** | gate valve |
| **H** | hydrogen |
| **HB** | hose bib |
| **HD** | head |
| **HDR** | header |
| **HOR** | horizontal |
| **HR** | hour |
| **HT** | height |
| **HTG** | heating |
| **HVAC** | heating/ventilating/air conditioning |
| **HW** | hot water |
| **HWH** | hot water heater |
| **HWR** | hot water return |
| **HWT** | hot water tank |
| **IN** | inch |
| **IAPMO** | International Association of Plumbing and Mechanical Officials |
| **ICBO** | International Conference of Building Officials (part of ICC) |
| **ICC** | International Code Council, Inc. |

| | |
|---|---|
| **ID** | inside diameter |
| **INS** | insulation |
| **INT** | interior |
| **INV** | invert |
| **IPS** | iron pipe size |
| **IW** | indirect waste |
| **J** | joist |
| **JC** | janitor's closet |
| **JT** | joint |
| **KIT** | kitchen |
| **K.S.** | kitchen sink |
| **KW** | kilowatt |
| **L** | length |
| **L. or LAV** | lavatory |
| **LB** | pound |
| **LH** | left hand |
| **LT** | laundry tray |
| **M** | meter |
| **MAS** | masonry |
| **MAX** | maximum |
| **MC** | medicine cabinet |
| **MCAA** | Mechanical Contractors Association of America |
| **MECH** | mechanical |
| **MFR** | manufacturer |
| **MGD** | million gallons per day |
| **MH** | manhole |
| **MI** | malleable iron |
| **MIN** | minimum |
| **MS** | mild steel |
| **N** | north |
| **NAPHCC** | National Association of Plumbing Heating and Cooling Contractors |
| **NBFU** | National Board of Fire Underwriters |
| **NBS** | National Bureau of Standards |
| **NFPA** | National Fire Protection Association |
| **NPS** | nominal pipe size |
| **NTS** | not to scale |
| **O** | oil or oxygen |
| **OD** | outside diameter |
| **OZ** | ounce |
| **P** | pressure |
| **PAR** | parallel |
| **PB** | polybutylene |
| **PCC** | precast concrete |
| **PCF** | pounds per cubic foot |
| **PD** | planter drain |
| **PE** | polyethylene plastic, porcelain enamel |
| **PED** | pedestal |
| **PER** | perimeter |
| **PEX** | polyethylene, cross-linked |
| **PFB** | prefabricated |
| **pH** | hydrogen-ion concentration |
| **PK** | parking |
| **PL** | property line |
| **PLF** | pounds per lineal foot |
| **PP** | pool piping |
| **PPM** | parts per million |
| **PR** | pressure regulator |
| **PSF** | pounds per square foot |
| **PSI** | pounds per square inch |

| | |
|---|---|
| **PVC** | polyvinyl chloride |
| **QT** | quart |
| **R** | riser |
| **RA** | return air |
| **RAD** | radius |
| **RCP** | reinforced concrete pipe |
| **RD** | roof drain or rate of demand |
| **RED** | reducer |
| **RF** | recovery factor |
| **RH** | right hand |
| **RL** | roof leader |
| **RS** | rate of supply |
| **RWC** | rainwater conductor |
| **S** | south |
| **S&W** | solid and waste |
| **SAN** | sanitary |
| **SB** | sitz bath |
| **SBCCI** | Southern Building Code Congress International |
| **SD** | storm drain |
| **SEC** | second |
| **SH.** | shower |
| **SJ** | slip joint |
| **SMLS** | seamless pipe |
| **SP** | swimming pool |
| **SPEC** | specification |
| **SQ** | square |
| **SQ FT** | square foot |
| **SS** | service sink |
| **SST** | stainless steel |
| **ST** | steel |
| **STD** | standard |
| **STG** | seating |
| **STO** | storage |
| **STR** | structural |
| **SV** | service |
| **SW** | service weight |
| **T** | temperature |
| **TBE** | threaded both ends |
| **T&C** | threaded and coupled |
| **TB** | towel bar |
| **TC** | terra cotta |
| **TO** | threads only |
| **TPD** | toilet paper dispenser |
| **TPI** | threads per inch |
| **TPTN** | toilet partition |
| **TSL** | top of slab |
| **U or URN** | urinal |
| **UL** | Underwriters Laboratories |
| **V** | volume, velocity, vent or vacuum cleaner line |
| **VERT** | vertical |
| **VT** | vinyl tile |
| **V.T.R.** | vent through roof |
| **W** | waste |
| **W.C.** | water closet |
| **WH** | water heater, wall hydrant or wall hung |
| **WHP** | water horsepower (head x gpm/3960) |
| **WL** | water level |
| **WT** | weight |
| **XH** | extra heavy |

# Glossary

*The terms included here are found in most plumbing codes. Some words in the code have become so descriptive and specialized that their meaning is different from what a standard dictionary might give.*

## A

***Absorption*** The process of being absorbed in a drainfield absorption area.

***Accessible*** Within physical reach. For example, a valve to control the flow of water to a battery of fixtures may be located in a partition. An access panel may have to be removed to make the valve accessible.

***Accessible, readily*** Implies that the valve is physically accessible without necessitating the removal of an access panel or similar obstruction.

***Acid waste*** Waste which requires special treatment before entry into the conventional drainage system.

***Administrative authority*** The individual plumbing official, board, department, or other agency established and authorized by a state, county, city, or other political subdivision created by law to administer and enforce the provisions of the plumbing code as adopted or amended.

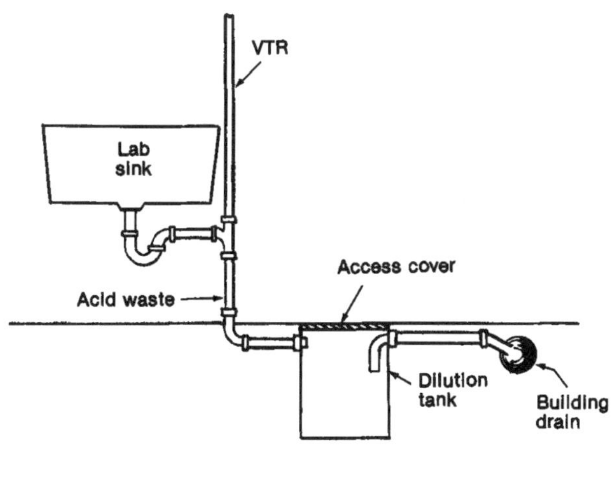


**Acid waste**

***Air admittance valve*** A valve used instead of the traditional vent. It permits air into the pipe to equalize the pressure while maintaining the fixture trap seal.

***Air break (in a drainage system)*** An indirect piping arrangement in which a drain from a fixture, appliance, or other device discharges indirectly into a waste receptor at a point below the flood level rim of the receptor.

***Air gap (in a drainage system)*** The unobstructed vertical distance through the free atmosphere between the lowest outlet of waste pipe and the flood level rim of the receptor into which it is discharging.

***Air gap (in a water distribution system)*** In a water supply system; the unobstructed vertical distance through the free atmosphere between the lowest opening from any pipe or faucet conveying water to a tank, plumbing fixture, or other device, and the flood level rim of the receptacle.

***Air lock*** Air, gas or vapor entrapped between two liquid surfaces in a pipe or liquid container. The flow of liquid may be impeded, or stopped entirely, by the entrapped air, gas or vapor.

***Anaerobic*** Living without free oxygen. For example, anaerobic bacteria function in septic tanks to digest organic matter.

***Angle stop*** A shutoff valve installed in the line to a faucet, usually creating a 90-degree angle from the pipe to the fixture. Its primary purpose is to shut off water to the faucet in case of failure of the faucet, or to permit repair or replacement.

***Approved*** Accepted under an applicable specification or standard stated in the code, or accepted for the proposed use under procedures and powers of the administrative authority.

***Approved testing agency*** A recognized agency established primarily to conduct tests to determine whether materials meet standards established by the administrative authority.

***Area drain*** A receptacle designed to collect surface or rain water from an open area.

***Aspirator*** A special fitting or device which may be supplied by water, or some other fluid under positive pressure, which passes through an integral orifice or constriction, causing a vacuum.

***Auto pilot*** A device on the supply side of a gas appliance that automatically shuts off the gas if the pilot light goes out.

## B

***Back pressure*** Pressure in a pipe that impedes or prevents the flow of the liquid it carries.

***Back siphonage*** The flow of water or other liquids, mixtures or substances into the distributing pipes of a potable supply of water, or into any other fixture, device, or appliance, from any sources other than its intended course, due to a negative pressure in the pipe.

***Back wash*** A method of cleaning a filter by passing water through it in the reverse direction of flow. The waste water is channeled to the waste line.

***Backfill*** That portion of the trench excavation up to the original earth line which is replaced after the sewer or other piping has been laid.

***Backflow*** The flow of water or other liquids, mixtures, or substances into the distributing pipes of a potable water supply from any source other than its intended course.

***Backflow connection*** Any arrangement whereby backflow can occur. Also known as backflow condition.

***Backflow preventer*** A device or means to prevent backflow into the potable water system.

Hose bibb vacuum breaker for frost-proof hydrants

Hose bibb vacuum breaker

**Backflow preventer**

***Backwash piping*** Piping, including the piping connected to the backwash outlet of the filter, that conveys waste water to an approved point of disposal.

***Backwater valve*** A special device installed in a drainage pipe to prevent backflow of liquid waste into a drainage system.

***Barometric loop*** A loop formed from pipe and fittings, rising vertically some 35 feet above the highest fixture it serves to prevent back siphonage into the potable water supply.

***Barrier-free*** Type of construction or installation that allows access to persons in wheelchairs. Some code requirements for barrier-free layout of the more-common plumbing fixtures are shown in Chapter 14 of this book.

***Base*** The lowest point of any vertical pipe.

***Basement*** The floor of a building which is 2 feet or more below outside grade and the ceiling of which is not more than 4 feet 6 inches above outside grade.

***Battery of fixtures*** A group of two or more similar adjacent plumbing fixtures that discharge into a common horizontal waste or soil branch.

***Battery venting*** (A special term used by some codes). A horizontal wet vent system that receives individual fixture drains which are connected horizontally to a branch soil or waste pipe which is circuit or loop vented.

***Boiler blow-off*** An outlet on a boiler that permits the emptying or discharging of water or sediment into a boiler blow-off tank.

***Boiler blow-off tank*** A tank designed to receive the discharge from a boiler blow-off outlet. Its purpose is to cool the water to an acceptable temperature for safe discharge into the drainage system.

***Branch*** Any part of the piping system other than a main, riser or stack.

***Branch drainage pipe*** That portion of the drainage system extending from a trailer park sanitary drainage system to the trailer site, including the terminal end that connects to the trailer drain hose.

***Branch interval*** A length of soil or waste stack (vertical pipe), generally one story high (approximately 9 feet, but not less than 8 feet), which connects the horizontal branches from one floor or story of a building to the stack.

***Branch vent*** A horizontal vent that connects one or more individual fixture vents with a vent stack or a stack vent.

***Branch water pipe*** That portion of the water distribution system extending from a trailer park service main to a trailer site, including the terminal end which connects to the trailer water supply pipe.

***Brazed joint*** The joining of metal parts with alloys having a melting point higher than 800 degrees F, but lower than the metal parts to be joined.

***Building classification*** The arrangement in local codes for the designation of buildings into classes based on use and occupancy. Note: Classification determines the required type and number of plumbing fixtures necessary to meet minimum fixture requirements.

***Building drain, combined*** The main horizontal collection system which conveys sewage, waste liquids and rainwater. Prohibited by most codes.

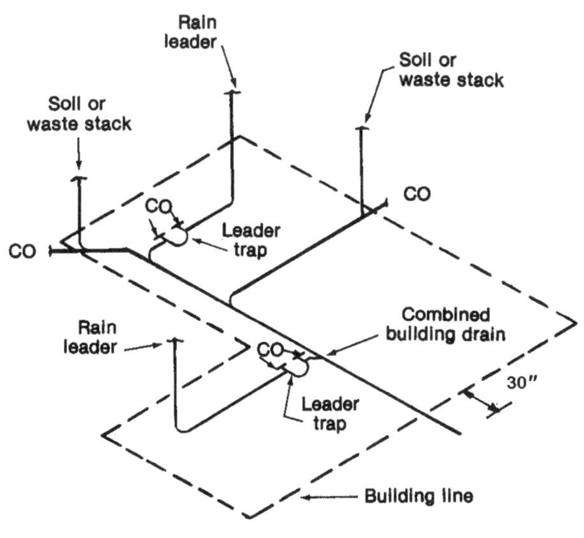


**Building drain, combined**

***Building drain, sanitary*** The main horizontal sanitary collection system, inside the wall line of the building, that conveys sewage and other liquid waste from stacks and other drainage pipes (excluding storm water) to the building sewer, beginning 30 inches outside the building wall. Note: Most codes prohibit the entry of storm water into a building sanitary drainage system.

***Building sewer*** That part of the horizontal piping of a drainage system that connects to the end of the building drain and conveys the contents to a public sewer, private sewer, or individual sewage disposal system or other legal point of disposal.

***Building sewer, combined*** A building sewer pipe which conveys sewage, rainwater or other drainage. Prohibited by most codes.

***Building storm drain*** A drain used to receive and convey rainwater, surface water, ground water, subsurface water and other approved clear water waste, and discharge these waste products into a building storm sewer or a combined building sewer (if approved) 30 inches outside the building wall.

***Building storm sewer*** The pipe that connects to the end of the building storm drain to receive and convey its contents to a public storm sewer, combined sewer (if approved), soakage pit or other approved disposal point.

***Building subdrain*** Any portion of a drainage system which cannot drain by gravity into the building sewer.

***Building supply*** The pipe that carries potable water from the water meter or other approved source (private well) to a building. Also referred to as water service.

***Building trap*** A special fitting or assembly of fittings installed in the building drain to prevent circulation of air between the drainage system of the building and the building sewer. Most codes prohibit the use of a building trap in a drainage system.

## C

***Casing*** A watertight lining, usually of steel, around the inside of a well hole designed to prevent groundwater and other contaminants from entering the well.

***Catch basin*** A manufactured receptacle (sometimes job-constructed) in which rainwater and other approved liquids are retained for a sufficient time, preventing sand and other sediment from entering the storm drainage system.

***Caulking*** Any approved method for rendering a joint water- and gas-tight. For cast iron pipe and fittings with hub joints, the term refers to caulking the joint with lead and oakum.

***Check valve*** A special valve designed to close automatically to prevent the flow of liquid in a reverse direction.

***Circuit vent*** A vent that serves two or more fixture traps. It connects to a horizontal branch drain from the downstream side of the highest (last) fixture connection and then to the vent stack.

***Cleanout*** A plug joined to an opening in a pipe, which can be removed for cleaning purposes.

***Clear water waste*** Principally, the cooling and condensate drainage from refrigeration and air-conditioning equipment, or the cooled condensate from steam heating systems and similar liquid discharge.

***Closet bend*** A curved fitting that connects the closet flange on a toilet to the building drain.

***Code*** Regulations and their subsequent amendments, or any emergency rule or regulation, lawfully adopted by the administrative authority having jurisdiction to control plumbing work.

***Combination Fixture*** A fixture combining a kitchen sink and laundry tray into a single unit. Also a two- or three-compartment sink or laundry tray combined into a single unit.

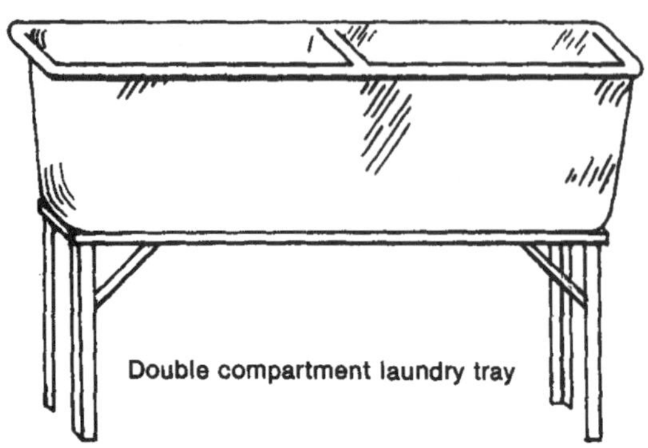


**Combination fixture**

***Combination waste and vent system, horizontal (permitted by some codes)*** A specially-designed waste piping embodying the horizontal wet venting of one or more sinks, floor sinks or floor drains by means of a common waste and vent pipe, adequately sized to provide free movement of air above the flow line of the drain. Generally used in restaurants.

***Combination waste and vent system, vertical (permitted by some codes)*** A specially-designed waste piping embodying the vertical wet venting of special low-rated fixtures (i.e., sinks, lavatories and showers, but no water closets or pressure discharge type fixtures). Fixtures must be located in a direct vertical line on two or more floors, requiring no offset in vertical pipe. Pipes of limited length must be adequately sized to provide free air movement. Generally used in residential buildings.

***Common vent*** The vertical vent portion which serves to vent two fixture drains that are installed at the same level in a vertical stack.

***Compression fitting*** A pipe connection where a nut is tightened on a sleeve, compressing the sleeve to form a tight seal.

***Condensate*** A liquid which separates from a gas, due to a reduction in temperature.

***Continuous vent*** A vertical pipe that is a continuation of the drain pipe to which it connects.

***Continuous waste*** A drain connecting two or more fixtures, or a single fixture with more than one compartment, to a common trap.

***Critical level (in the plumbing code)*** A point on a backflow prevention device or vacuum breaker (usually stamped on the device by the manufacturer) to indicate the minimum elevation the device may be installed above the flood level rim of the fixture or receptacle it serves. If a backflow prevention device is not so stamped, the bottom of the device constitutes the critical level.

***Cross connection*** Any physical connection or arrangement between two separate piping systems, one containing potable water and the other containing water of unknown or questionable safety (such as steam, gas, chemicals, polluted or contaminated water). The code prohibits cross connections.

## D

***Dead end*** A branch leading from a soil, waste or vent pipe, building drain or building sewer that is terminated by a plug or other closed fitting (blind plug) at a developed distance of 2 feet or more. A dead end is also classified as an extension for future connection, or as an extension of a cleanout for accessibility.

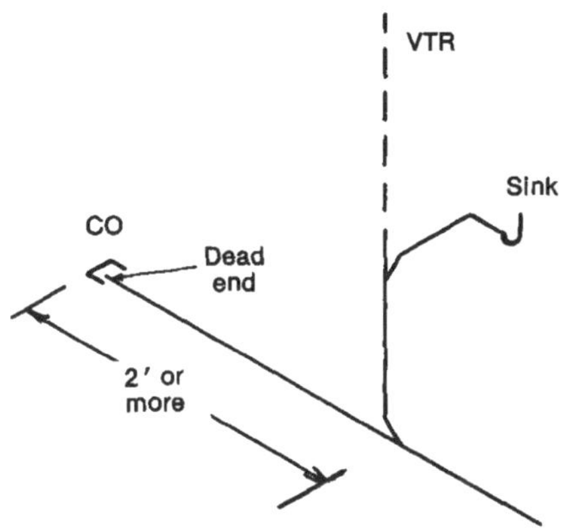


**Dead end**

***Deep-seal trap*** A trap having a water seal of 4 inches or more. Generally refers to interceptors or separator type traps.

***Developed length*** The length of a pipe run measured along the center line of the pipe and fittings.

**Diameter** The nominal diameter (distance straight through the middle) of a pipe or fitting as designated commercially, unless specifically stated otherwise.

***Direct vent*** A vent that draws combustion air from, and exhausts it to, the outside, eliminating the need for a chimney.

***Domestic sewage*** The waterborne wastes derived from ordinary living processes, free from industrial wastes, requiring no special separation or treatment.

***Dosing tank*** A watertight tank connected to the outlet pipe or septic tank, located between the septic tank and drainfield distribution box, equipped with an automatic pump designed to discharge effluent intermittently to a disposal field.

***Double offset*** Two changes in direction installed in succession, or series, in continuous piping.

***Downspout*** See *Leader*.

***Drain*** Any pipe which carries liquid, waste water or other waterborne wastes in a building drainage system to an approved point of disposal.

***Drain hose connection*** An approved flexible hose that is easily detachable and is used to connect a trailer drain to a park's sewer inlet connection.

***Drainage fixture unit*** A design factor used to determine the load-producing effects of different types of plumbing fixtures on the plumbing system. For instance, most codes accept that 1 fixture unit equals up to 7.5 gallons per minute of flow.

***Drainage system*** All the piping within public or private premises that conveys sewage, rainwater, or other liquid wastes to a legal point of disposal.

***Drainage well*** Any drilled, driven or natural cavity which taps the underground water table and into which surface waters or treated waste water, industrial waste or sewage is disposed. Always requires approval from local authority.

***Drum trap*** A trap consisting of a cylinder with a removable cover for cleaning purposes, whose minimum diameter is 4 inches, with a smaller size inlet and outlet. Its use is prohibited or limited by many codes.

***Dry vent*** That portion of a vent system that receives no sewage or waste discharge.

***Durham system*** An all-threaded pipe system of rigid construction, using recessed drainage fittings to correspond to the types of piping being used in drain, waste and vent systems.

## E

***Effective opening*** The minimum cross-sectional area of the diameter of a pipe at the point of water supply discharge.

***Effluent*** The liquid waste discharged from a septic tank outlet pipe into the drainfield.

***Escutcheon*** A decorative flange that covers the surrounding part of the hole in the wall or fixture through which a fixture such as a shower head or faucet stem enters.

## F

***Filter piping*** All the piping, fittings, and valves necessary to connect the filter system of a swimming pool together as a unit.

***Fire lines*** The complete wet standpipe system of a building, including the water service, standpipe, roof manifold, Siamese connections, and pumps.

***Fixture branch*** A drain that serves two or more fixtures and empties to another drain or to a stack.

***Fixture drain*** A drain from the trap of a fixture that connects to any other drain pipe.

***Fixture supply*** A pipe supplying water to a fixture and connected to a branch water supply pipe, or directly to a main water supply pipe.

***Flood level rim*** The top edge of a plumbing fixture or other receptacle from which water or other liquids can overflow, such as the rim of a bathtub.

***Floor drain*** An opening or receptacle located at approximately floor level connected to a trap to receive discharge from indirect waste pipes and provide floor drainage.

***Floor sink*** An opening or receptacle, usually made of enameled cast iron, located at approximately floor level, which is connected to a trap designed to receive the discharge from indirect waste and floor drainage. More sanitary and easier to clean than a regular floor drain, it is usually used in restaurants and hospitals.

**Floor sink**

***Flush tank*** A tank connected to the top of a water closet bowl or above a battery of urinals or similar fixtures designed for the purpose of flushing the useable portion of the fixture.

***Flush valve*** A device located at the bottom of the flush tank, for the purpose of flushing water closets or similar fixtures.

***Flushometer valve*** A device actuated by direct water pressure which discharges a predetermined quantity of water to fixtures for flushing purposes.

***Freeboard*** A term used in plumbing to define the vertical distance (air space) between the maximum liquid level in a septic tank and the top of the septic tank.

***French drain*** A covered ditch filled with rock or other pervious material to hold excess water long enough to be absorbed into the surrounding soil, rather than running off.

## G

***Gate valve*** A valve that shuts off flow through a pipe by means of a gate, usually of brass, that is raised or lowered by means of a screw stem operated by a wheel handle.

***Grade*** The slope or pitch of drainage piping, known as the fall, usually expressed as a fraction of an inch per foot.

***Gray water*** Waste water from any fixture other than toilets.

***Grease interceptor*** A tank of at least 750-gallon capacity, located outside of a building, designed and sized to serve one or more fixtures.

## H

***Hangers*** See *Supports.*

***Heat trap*** A device designed to prevent automatic circulation of heated water (and thus heat loss) in the outlet pipe of a water heater.

***Horizontal branch drain*** A drain pipe extending laterally from a soil or waste stack or building drain. It may or may not have vertical sections or branches.

***Horizontal pipe*** Any pipe or fitting that makes an angle of not more than 45 degrees with the vertical.

***Hot water*** Water heated to a temperature of not less than 120 degrees F.

***Hot water return*** A piping system designed to circulate the heated water.

## I

***Impervious*** Having a smooth surface resistant to absorption. (Plumbing fixtures must not be constructed of materials that will absorb liquids or odors.)

***Indirect waste pipe*** A pipe that conveys liquid wastes (other than body waste) by discharging them into an open plumbing fixture or receptacle such as a floor drain or floor sink. The overflow point of such a fixture or receptacle is at a lower elevation than the item drained.

***Individual vent*** A pipe installed to vent a single fixture drain. It may connect with the building vent system above the fixture served, or terminate outside the building into the open air. Also called a Back vent.

***Industrial waste*** Liquid waste, free of body waste, resulting from the processes used in industrial and commercial establishments.

***Inlet coupling*** The terminal end of a trailer park's water system at each trailer site. The water service connection from the trailer coach is made by a swivel fitting or threaded pipe end.

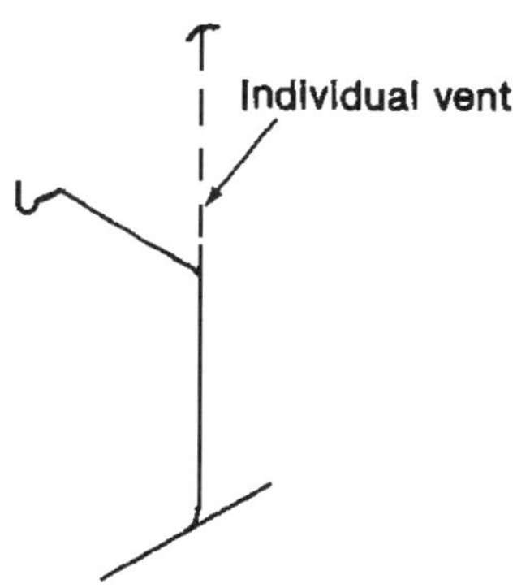


**Individual vent**

***Interceptor*** A device designed and installed to separate and retain deleterious, hazardous, or undesirable matter from normal wastes, and permit normal sewage or liquid wastes to discharge by gravity into the disposal terminal or sewer.

## L

***Labeled*** Equipment or materials bearing the label of a listing agency accepted by the administrative authority.

***Leader*** The vertical water conductor from the roof to the building storm drain, combined building sewer, or other approved means of disposal. Also called a downspout or conductor.

***Liquid waste*** The liquid discharge from any fixture, appliance or appurtenance that connects to a plumbing system that does not receive body waste.

***Load factor*** The percentage of the total connected fixture unit flow rate that is likely to occur at any point, with the probability factor of simultaneous use. It varies with the type of occupancy and the code being used. (See Drain fixture unit.)

***Loop vent*** A vent connecting a horizontal drainage branch or individual fixture drain to a common vent looped back and connected to a stack vent of the originating waste or soil stack.

***Lot*** A single lot or individual parcel or area of land on which is situated a building or buildings, or which is the site of any work regulated by code. It shows the yards, courts, and unoccupied spaces, with their dimensions. It must show the source of water and sewage disposal facilities and other facets relating to any plumbing required or not required by code.

## M

***Main*** The principal artery of any system of continuous piping, to which branches may be connected.

***Main vent*** The principal pipe of the venting system, to which vent branches may be connected.

***May*** A permissive term used in the code, which means allowable or optional, but not required.

***Mezzanine*** An intermediate floor constructed in any story or room. When the total area of a mezzanine floor exceeds $33^1/_3$ percent of the total floor area in that room or story, it's considered by code an additional story, rather than a mezzanine.

## N

***Non-potable water*** Water of unknown or questionable safety, not considered safe for personal or culinary use.

***Nuisance*** Obnoxious, inconvenient, or insanitary. Any public nuisance known at common law or in equity jurisprudence; i.e., whatever building, structure, or premise is not properly ventilated, sewered, drained, cleaned or lighted, and whatever renders the air or human food or drink or water supply unwholesome in violation of the plumbing code.

## O

***Offset*** A combination of elbows or bends which brings one section of pipe out of line but into a parallel line with the other section of pipe.

## P

***P-trap*** A device which maintains the water seal in a drain line to prevent sewer gas from entering the building.

***Pitch*** Grade or slope of the piping. See Grade.

***Plumbing (includes any or all of the following)*** (1) The materials, including pipe, fittings, valves, fixtures and appliances, which are attached to and are a part of a system for the purpose of creating and maintaining sanitary conditions in a building, camp or swimming pool on private property where people live, work, play, assemble or travel. (2) That part of a water supply and sewage and drainage system extending from either the public water supply mains or private water supply to the public sanitary, storm or combined sanitary and storm sewers, or to a private sewage disposal plant, septic tank, disposal field, pit, box filter bed or any other receptacle, or into any natural or artificial body of water or water course upon public or private property. (3) The design, installation or contracting for installation, removal and replacement, repair or remodeling, of all or any part of the materials, appurtenances or devices attached to and forming a part of a plumbing system, including the installation of any fixture, appurtenance or devices used for cooking, washing, drinking, cleaning, fire fighting, mechanical or manufacturing purposes.

***Plumbing appliance*** A special class of plumbing fixture whose operation and/or control may be dependent upon one or more energized components, such as motors, controls, heating elements or pressure- or temperature-sensing elements. It's designed for a specific purpose and not generally indispensable to the operation of the plumbing system. (Example: a dishwasher.)

***Plumbing appurtenance*** A manufactured device, or a prefabricated assembly of component parts, which is an adjunct to the basic plumbing system. Requires no additional water supply and contributes no additional load to the drainage system. It presumably serves some useful function in the operation, maintenance, or safety of the plumbing system.

***Plumbing contractor*** A person who has been qualified as a master plumber by an examining board, who contracts on predetermined terms to provide labor, materials and knowledge, and to be responsible for work installed under his supervision to comply with established practices and specifications.

***Plumbing fixtures*** Receptacles, devices, or appliances that are supplied with water or that receive or discharge liquids or liquid-borne waste, with or without discharge, into the drainage system to which they may be directly or indirectly connected.

***Plumbing inspector*** A trained person qualified to pass judgment on plumbing installations.

***Plumbing official*** The chief administrative officer charged with the administration, enforcement and application of the plumbing code and all amendments thereto.

***Plumbing system*** The drainage system, water supply, water supply distribution pipes, plumbing fixtures, traps, soil pipes, waste pipes, vent pipes, building drains, building sewers, building storm drain, building storm sewer, liquid waste piping, water treating equipment, water using equipment, sewerage treatment, sewerage treatment equipment, fire standpipes, fire sprinklers, and relative appliances and appurtenances, including their respective connections and devices, within the private property lines of a premise.

***Potable water*** Water which is satisfactory for drinking, culinary and domestic purposes and that meets the requirements of the health authority having jurisdiction. Potable water is considered purified, having been treated by one or several processes as required by its original untreated condition.

***Private property*** For the purposes of the code, private property includes all property except streets or roads dedicated to the public, and easements (excluding easements between private parties).

***Private or private use fixtures*** The plumbing fixtures in residences and apartments, and in private bathrooms of hotels and similar installations, where the fixtures are intended for the use of a family or an individual.

***Private sewage disposal system*** A septic tank discharging the effluent into a subsurface disposal field, into one or more seepage pits, or into a combination of both.

***Private sewer*** A sewer privately owned and not directly controlled by a public authority.

***Private water supply*** Any approved water supply other than a public water supply that serves one or more buildings.

***Public or public use fixtures*** Plumbing fixtures in commercial and industrial establishments, restaurants, bars, public buildings, comfort stations, schools, gymnasiums, railroad stations or places to which the public is invited or which are frequented by the public without special permission or special invitation, and other installations (whether paid or free) where a number of fixtures are installed so that their unrestricted use is available to the public.

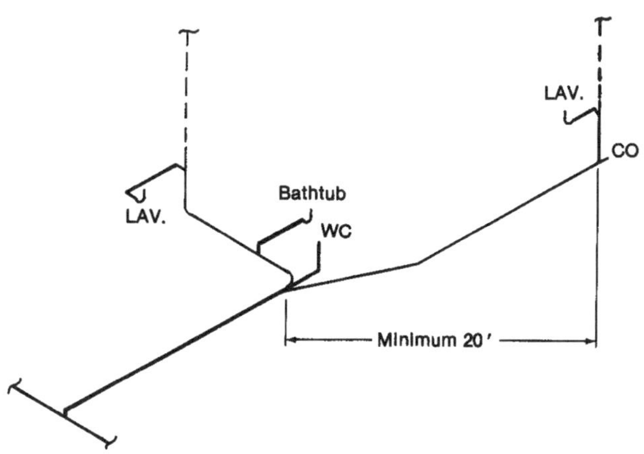


**Remote fixture**

***Public sewer*** A common sewer directly controlled by public authority.

***Public swimming pool*** A pool, together with its buildings and appurtenances, where the public is allowed to bathe or which is open to the public for bathing purposes by consent of the owner, whether or not a fee is charged. It may be operated by an owner, lessee, operator, licensee, or concessionaire.

## R

***Receptor*** An approved plumbing fixture or other device adequately designed to receive the discharge from indirect waste pipes.

***Recirculating piping (also called return piping or pool inlet piping)*** The piping connected to the discharge side of the pool pump. It returns water to the pool after filtering.

***Reducer*** A fitting that adapts one size of pipe to another, in either direction.

***Refuse*** All solid wastes, including garbage, rubbish and ashes, but excluding body waste.

***Relief vent*** A vent, the primary function of which is to provide air circulation between drainage and vent systems.

***Remote fixture*** A single fixture installed more than 20 feet from the upper end of the branch line.

***Rim*** An unobstructed open edge at the overflow point of a fixture.

***Riser*** A water supply pipe extending vertically one full story or more to convey water to branch piping or plumbing fixtures.

***Rock drainfield*** Rock generally used in a disposal field, consisting of 3/4-inch rock, 100% passing a 1-inch screen, a maximum of 10% passing a 1/2-inch screen.

***Roof area*** The area on the building roof bounded by barriers, thus creating a drainage surface.

***Roof drain*** An outlet fitted with a strainer to receive water that collects on the surface of a roof and that discharges that water into a leader or downspout.

***Roughing-in*** The installation of all the parts of a plumbing system that can be completed prior to the installation of the plumbing fixtures. It includes drainage, water supply, and vent piping, and the necessary fixture supports.

***Running trap*** A fitting constructed of cast iron, 4 inches or larger, in which the inlet and outlet are a horizontal straight line; however, between these two points the waterway is depressed below the bottom side tangent of either the inlet or outlet, forming a water seal trap in the building drain or sewer. Prohibited in many codes.

## S

***Saddle fitting*** A fitting attached with clamps to the outside of a pipe, sealed to the pipe with a gasket, located over an opening drilled or cut through the pipe wall to create a change in direction, or to create a branch. Prohibited in many codes.

***Safe pan*** A collector (receptacle) placed beneath a water heater located above the first floor level to catch leakage and dispense it safely to an approved point of disposal. Also used to waterproof shower stalls. (Also referred to as a drain pan or shower pan.)

***Sand interceptor*** A device installed in a drainage pipe to prevent sand and other gritty material from entering the building drainage system.

***Sanitary fitting*** A fitting that joins pipes in waste, drain, and vent systems through which solids may pass without clogging the system.

***Sanitary sewer*** A pipe which carries sewage, excluding storm, surface and ground water.

***Schedule*** A numerical designation given to pipes, based on their wall thickness, such as Schedule 40.

***Scupper*** A drain on a roof or deck to allow drainage of rainwater or melting snow.

***Seamless pipe*** A pipe without welds.

***Second hand*** A term applied to material or plumbing equipment which has been installed and used, or removed from its first installation.

***Seepage pit*** A covered and lined pit with open-jointed walls and bottom through which the effluent from a septic tank may seep into the surrounding soil.

***Separator*** A device that accumulates and recovers objectionable substances from liquid waste.

***Septic tank*** A watertight receptacle which receives the discharge of a drainage system or part thereof, so designed and constructed as to separate solids from liquid, digest organic matter through a period of detention, and allow the liquids (effluent) to discharge into the soil outside the tank through a subsurface system of open-joint or perforated piping, or other approved methods.

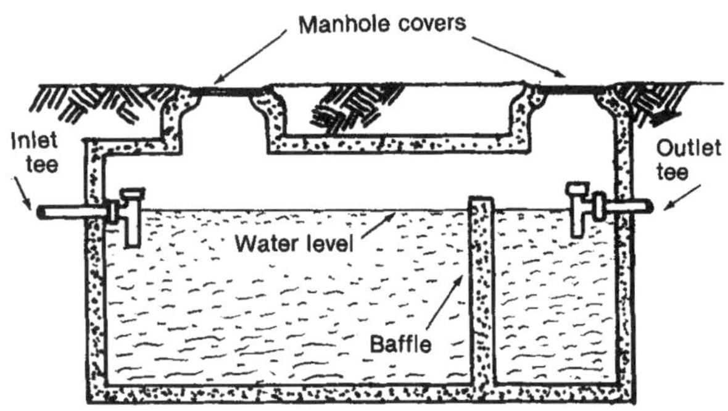


**Septic tank**

***Sewage*** Any liquid waste containing animal, mineral or vegetable matter in suspension or solution. May include liquids containing chemicals in solution; i.e., the used-water supply of a building.

***Sewage ejector*** A device that operates automatically, designed for lifting and discharging sewage.

***Shall*** A mandatory term. When used in the code, it requires compliance without deviation.

***Side vent*** A vent which connects to a horizontal drain pipe through a fitting at an angle no greater than 45 degrees to the vertical.

***Slip joint*** An adjustable tubing connection in which the joint is made tight with a compression nut and compression ring, or a gasket, washer, or O-ring.

***Slope*** See *Grade.*

***Soil pipe*** Any pipe which conveys the discharge of water closets or fixtures having similar functions, with or without the discharge from other fixtures, to the building drain or building sewer.

***Soil stack*** A vertical stack which receives water closet discharge and is connected to the building drain.

***Special wastes*** Wastes which require special handling before entry into the building drainage system or to a legal point of disposal, including but not limited to: oil, sand, grease, glass and storm water.

***Stack*** The vertical pipe of a soil, waste or vent piping system.

***Stack vent*** The extension of a soil or waste stack above the highest horizontal drain connected to the stack.

***Stack venting*** A method of venting fixtures through the soil or waste stack.

***Standpipe system*** A system of piping installed for fire protection purposes which has a primary water supply constantly or automatically available at each hose outlet.

***Storm sewer*** A pipe located outside of the building wall line for conveying rainwater and/or surface water.

***Subsoil drain*** A drain which collects only subsurface or seepage water and conveys it to a place of disposal.

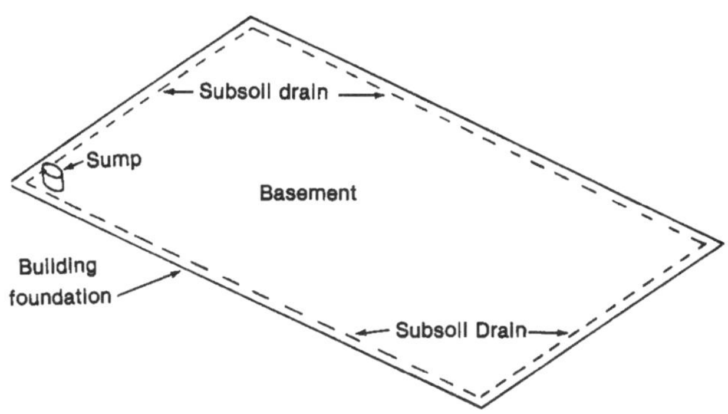


**Subsoil drain**

***Subsurface drain*** A drain that receives only subsurface or seepage water and conveys it to a place of disposal.

***Sump*** A tank or pit, located below the normal grade of the gravity system, which receives sewage or liquid waste and which must be emptied by mechanical means, such as a pump.

***Supply well*** Any artificial opening in the ground designed to conduct water from a source bed through the surface when water from such well is used for public, semipublic or private use.

***Supports*** Devices for supporting and securing pipe and fixtures to walls, ceilings, floors or structural members. Also known as hangers or anchors.

## T

***Trap*** A fitting or device designed and constructed to provide a liquid seal that prevents the back passage of air and sewer gases without materially affecting the flow of sewage or water.

***Trap arm*** The drain from the trap of a fixture to the junction of that drain with the vent pipe. See Fixture drain.

***Trap primer*** A device to automatically maintain a water seal in a trap. (Example: a floor drain.)

***Trap seal*** The maximum vertical depth of liquid that a trap will retain, measured between the crown weir and the top of the dip of the trap.

## U

***Unacceptable substances*** Substances that are considered harmful to drainage systems, sewers, or treatment plants. These include grease, flammable waste, sand, plaster, cinders, oil, etc.

***Union*** A three-piece fitting, generally used with steel pipe, that joins two sections of pipe.

## V

***Vacuum breaker*** A device which prevents contaminated water from backflowing into the water supply.

***Vacuum fitting*** A device which connects the pool sweeper hose to the side of the pool for cleaning purposes. Every pool must have a vacuum fitting, located a maximum of 10 inches below the water surface inside the pool, in an accessible location.

***Valve*** A mechanical device that starts, stops or regulates the liquid which flows through it.

***Valve seat*** The part of the valve which stops the flow when it contacts with the valve's movable part.

***Vent stack*** A vertical vent pipe installed primarily for the purpose of providing circulation of air to and from any part of the drainage system.

***Vent system*** A pipe or pipes installed to provide a flow of air to or from a drainage system, or to provide a circulation of air within such system which prevents back pressure and siphonage from breaking water trap seals serving the fixtures on the system.

***Vertical pipe*** Any pipe or fitting installed in a vertical position or that makes an angle of not more than 45 degrees with the vertical.

***Vitreous*** A material with a high glaze and low porosity. Because of the presence of glass in its manufacture, it usually has translucent qualities.

***Venturi*** A short tube with a tapered constriction in the middle, causing an increase in flow velocity and corresponding decrease in pressure.

## W

***Waste*** See *Liquid waste.*

***Waste arm*** An extension pipe generally used to extend a sink drain.

***Waste & Overflow*** In a bathtub, a drain with a top outlet which eliminates any overflow during filling, and a bottom drain which removes any remaining wastewater.

***Waste pipe*** Any pipe which receives the discharge of any fixture, except water closets or fixtures having similar functions, and conveys it to the building drain or to the soil or waste stack.

***Water closet*** Toilet.

***Water-distributing pipe*** A pipe within a building which conveys water from the water service pipe to the plumbing fixtures, appliances and other water outlets.

***Water main*** A water supply pipe for public or community use. The public water distribution system is located in the street, alley or a dedicated easement adjacent to individual parcels of land.

***Water outlet*** As used in connection with the water-distributing system, the discharge opening for the water (1) to the fixture, (2) to atmospheric pressure (except into an open tank which is part of the water supply system), (3) to a boiler or heating system, (4) to any water-operated device or equipment requiring water to operate, but not part of the plumbing system.

***Water service*** The supply of potable water from the water meter or other approved source (private well) to a building. Also referred to as a building supply.

***Water service pipe*** The pipe from the water main or other source of water supply to the building served.

***Water supply system*** A piping system consisting of the water service pipe, the water-distributing pipes, the standpipe system and the necessary connecting pipes, fittings, control valves and all appurtenances related to the water supply in or on private property.

***Water table*** The upper limit of the portion of the ground completely saturated with water.

***Weir*** A raised section in the path of a fluid, such as inside a diaphragm valve.

***Well casing*** Piping in a well which protects the potable water from contamination by the surrounding material.

***Wet vent*** A waste pipe designed to vent and convey waste from fixtures other than water closets or similar fixtures.

***Working pressure*** The maximum pressure allowed in a system.

***Wye*** A fitting with three openings, in the shape of a "Y", to join pipes at a 45-degree angle.

## Y

***Yoke vent*** A pipe connecting upward from a soil or waste stack for the purpose of preventing pressure changes in the stacks.

# Index

## D



## E



## F

## G



## H



## I

## J



## K



## L

## R



## S

## Y

# Practical References for Builders

## Basic Engineering for Builders

If you've ever been stumped by an engineering problem on the job, yet wanted to avoid the expense of hiring a qualified engineer, you should have this book. Here you'll find engineering principles explained in non-technical language and practical methods for applying them on the job. With the help of this book you'll be able to understand engineering functions in the plans and how to meet the requirements, how to get permits issued without the help of an engineer, and anticipate requirements for concrete, steel, wood and masonry. See why you sometimes have to hire an engineer and what you can undertake yourself: surveying, concrete, lumber loads and stresses, steel, masonry, plumbing, and HVAC systems. This book is designed to help the builder save money by understanding engineering principles that you can incorporate into the jobs you bid.
**400 pages, 8½ x 11, $36.50**

## National Plumbing & HVAC Estimator

Manhours, labor and material costs for all common plumbing and HVAC work in residential, commercial, and industrial buildings. You can quickly work up a reliable estimate based on the pipe, fittings and equipment required. Every plumbing and HVAC estimator can use the cost estimates in this practical manual. Sample estimating and bidding forms and contracts also included. Explains how to handle change orders, letters of intent, and warranties. Describes the right way to process submittals, deal with suppliers and subcontract specialty work. Includes a CD-ROM with an electronic version of the book with *National Estimator*, a stand-alone *Windows*™ estimating program, plus an interactive multimedia video that shows how to use the disk to compile construction cost estimates.
**400 pages, 8½ x 11, $48.25. Revised annually**

## Rough Framing Carpentry

If you'd like to make good money working outdoors as a framer, this is the book for you. Here you'll find shortcuts to laying out studs; speed cutting blocks, trimmers and plates by eye; quickly building and blocking rake walls; installing ceiling backing, ceiling joists, and truss joists; cutting and assembling hip trusses and California fills; arches and drop ceilings — all with production line procedures that save you time and help you make more money. Over 100 on-the-job photos of how to do it right and what can go wrong. **304 pages, 8½ x 11, $26.50**

## 2003 *International Building Code*

Updated means of egress and interior finish requirements, comprehensive roof provisions, seismic engineering provisions, innovative construction technology, revamped structural provisions, reorganized occupancy classifications and the latest industry standards in material design.
**768 pages, 8½ x 11, $78.00**

## Wiring a House

A master electrician gives you the tips and shortcuts he's learned in over 30 years of wiring houses to meet code and to provide years of reliable service. Here you'll learn what materials work best in what application, which tools get the job done in the fastest time, and how to figure out and design the right wiring layout for any residential job, whether the entire house, or just a room. You also see how to install main service panels, ensure that the house is properly grounded, and install receptacles, GFCIs, switches, fixtures and appliances per *NEC* requirements. Dozens of diagrams and full-color illustrations show you exactly how the work goes together.
**248 pages, 8½ x 11, $24.95**

## Precision Framing

Here you'll find over 200 color photos and 40 detailed drawings along with the practical how-to information you need to frame a house. You'll find how to level a mudsill, plan and install center beams, planning and laying out exterior bearing walls, blocking, strapping, fire stops, radius openings and more. **218 pages, 8½ x 11, $24.95**

## National Repair & Remodeling Estimator

The complete pricing guide for dwelling reconstruction costs. Reliable, specific data you can apply on every repair and remodeling job. Up-to-date material costs and labor figures based on thousands of jobs across the country. Provides recommended crew sizes; average production rates; exact material, equipment, and labor costs; a total unit cost and a total price including overhead and profit. Separate listings for high- and low-volume builders, so prices shown are specific for any size business. Estimating tips specific to repair and remodeling work to make your bids complete, realistic, and profitable. Includes a CD-ROM with an electronic version of the book with *National Estimator*, a stand-alone *Windows*™ estimating program, plus an interactive multimedia video that shows how to use the disk to compile construction cost estimates.
**296 pages, 8½ x 11, $48.50. Revised annually**

## Wood-Frame House Construction

Step-by-step construction details, from the layout of the outer walls, excavation and formwork, to finish carpentry and painting. Packed with clear illustrations and explanations updated for modern construction methods. Everything you need to know about framing, roofing, siding, interior finishings, floor covering and stairs — your complete book of wood-frame homebuilding. **320 pages, 8½ x 11, $25.50. Revised edition**

## The Contractor's Legal Kit

Stop "eating" the costs of bad designs, hidden conditions, and job surprises. Set ground rules that assign those costs to the rightful party ahead of time. And it's all in plain English, not "legalese." For less than the cost of an hour with a lawyer you'll learn the exclusions to put in your agreements, why your insurance company may pay for your legal defense, how to avoid liability for injuries to your sub and his employees or damages they cause, how to collect on lawsuits you win, and much more. It also includes a FREE computer disk with contracts and forms you can customize for your own use. **352 pages, 8½ x 11, $59.95**

## Contractor's Survival Manual

How to survive hard times and succeed during the up cycles. Shows what to do when the bills can't be paid, finding money and buying time, transferring debt, and all the alternatives to bankruptcy. Explains how to build profits, avoid problems in zoning and permits, taxes, time-keeping, and payroll. Unconventional advice on how to invest in inflation, get high appraisals, trade and postpone income, and stay hip-deep in profitable work.
**160 pages, 8½ x 11, $25.00**

## National Renovation & Insurance Repair Estimator

Current prices in dollars and cents for hard-to-find items needed on most insurance, repair, remodeling, and renovation jobs. All price items include labor, material, and equipment breakouts, plus special charts that tell you exactly how these costs are calculated. Includes a CD-ROM with an electronic version of the book with *National Estimator*, a stand-alone *Windows*™ estimating program, plus an interactive multimedia video that shows how to use the disk to compile construction cost estimates.
**568 pages, 8½ x 11, $49.50. Revised annually**

## Markup & Profit: A Contractor's Guide

In order to succeed in a construction business, you have to be able to price your jobs to cover all labor, material and overhead expenses, and make a decent profit. The problem is knowing what markup to use. You don't want to lose jobs because you charge too much, and you don't want to work for free because you've charged too little. If you know how to calculate markup, you can apply it to your job costs to find the right sales price for your work. This book gives you tried and tested formulas, with step-by-step instructions and easy-to-follow examples, so you can easily figure the markup that's right for your business. Includes a CD-ROM with forms and checklists for your use. **320 pages, 8½ x 11, $32.50**

## Profits in Buying & Renovating Homes

Step-by-step instructions for selecting, repairing, improving, and selling highly profitable "fixer-uppers." Shows which price ranges offer the highest profit-to-investment ratios, which neighborhoods offer the best return, practical directions for repairs, and tips on dealing with buyers, sellers, and real estate agents. Shows you how to determine your profit before you buy, what "bargains" to avoid, and how to make simple, profitable, inexpensive upgrades. **304 pages, 8½ x 11, $22.75**

## Plumber's Handbook Revised

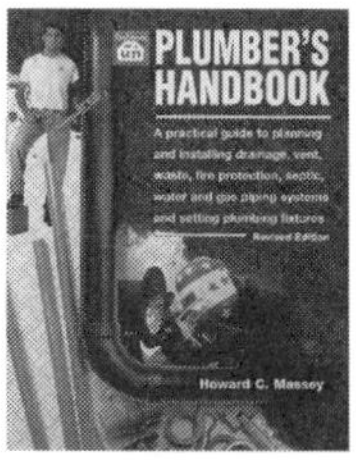


This new edition shows what will and won't pass inspection in drainage, vent, and waste piping, septic tanks, water supply, graywater recycling systems, pools and spas, fire protection, and gas piping systems. All tables, standards, and specifications are completely up-to-date with recent plumbing code changes. Covers common layouts for residential work, how to size piping, select and hang fixtures, practical recommendations, and trade tips. It's the approved reference for the plumbing contractor's exam in many states. Includes an extensive set of multiple choice questions after each chapter, and in the back of the book, the answers and explanations. Also in the back of the book, a full sample plumber's exam. **352 pages, 8½ x 11, $32.00**

## 2003 *International Residential Code*

Replacing the CABO *One- and Two-Family Dwelling Code*, this book has the latest technological advances in building design and construction. Among the changes are provisions for steel framing and energy savings. Also contains mechanical, fuel gas and plumbing provisions that coordinate with the *International Mechanical Code* and *International Plumbing Code*. **578 pages, 8½ x 11, $53.00**

## Contractor's Guide to the Building Code Revised

This new edition was written in collaboration with the International Conference of Building Officials, writers of the code. It explains in plain English exactly what the latest edition of the *Uniform Building Code* requires. Based on the 1997 code, it explains the changes and what they mean for the builder. Also covers the *Uniform Mechanical Code* and the *Uniform Plumbing Code*. Shows how to design and construct residential and light commercial buildings that'll pass inspection the first time. Suggests how to work with an inspector to minimize construction costs, what common building shortcuts are likely to be cited, and where exceptions may be granted. **320 pages, 8½ x 11, $39.00**

## Roof Framing

Shows how to frame any type of roof in common use today, even if you've never framed a roof before. Includes using a pocket calculator to figure any common, hip, valley, or jack rafter length in seconds. Over 400 illustrations cover every measurement and every cut on each type of roof: gable, hip, Dutch, Tudor, gambrel, shed, gazebo, and more. **480 pages, 5½ x 8½, $22.00**

## Plumber's Exam Preparation Guide

Hundreds of questions and answers to help you pass the apprentice, journeyman, or master plumber's exam. Questions are in the style of the actual exam. Gives answers for both the Standard and Uniform plumbing codes. Includes tips on studying for the exam and the best way to prepare yourself for examination day. **320 pages, 8½ x 11, $34.00**

## Construction Surveying & Layout

A practical guide to simplified construction surveying. How to divide land, use a transit and tape to find a known point, draw an accurate survey map from your field notes, use topographic surveys, and the right way to level and set grade. You'll learn how to make a survey for any residential or commercial lot, driveway, road, or bridge — including how to figure cuts and fills and calculate excavation quantities. Use this guide to make your own surveys, or just read and verify the accuracy of surveys made by others. **256 pages, 8½ x 11, $49.95**

## Construction Estimating Reference Data

Provides the 300 most useful manhour tables for practically every item of construction. Labor requirements are listed for sitework, concrete work, masonry, steel, carpentry, thermal and moisture protection, doors and windows, finishes, mechanical and electrical. Each section details the work being estimated and gives appropriate crew size and equipment needed. Includes a CD-ROM with an electronic version of the book with *National Estimator*, a stand-alone *Windows*™ estimating program, plus an interactive multimedia video that shows how to use the disk to compile construction cost estimates. **384 pages, 11 x 8½, $39.50**

## Residential Wiring to the 2002 *NEC*

Shows how to install rough and finish wiring in new construction, alterations, and additions. Complete instructions on troubleshooting and repairs. Every subject is referenced to the most recent *National Electrical Code*, and there's 22 pages of the most-needed *NEC* tables to help make your wiring pass inspection — the first time. **352 pages, 5½ x 8½, $29.00**

## Renovating & Restyling Older Homes

Any builder can turn a run-down old house into a showcase of perfection — if the customer has unlimited funds to spend. Unfortunately, most customers are on a tight budget. They usually want more improvements than they can afford — and they expect you to deliver. This book shows how to add economical improvements that can increase the property value by two, five or even ten times the cost of the remodel. Sound impossible? Here you'll find the secrets of a builder who has been putting these techniques to work on Victorian and Craftsman-style houses for twenty years. You'll see what to repair, what to replace and what to leave, so you can remodel or restyle older homes for the least amount of money and the greatest increase in value. **416 pages, 8½ x 11, $33.50**

## Planning Drain, Waste & Vent Systems

How to design plumbing systems in residential, commercial, and industrial buildings. Covers designing systems that meet code requirements for homes, commercial buildings, private sewage disposal systems, and even mobile home parks. Includes relevant code sections and many illustrations to guide you though what the code requires in designing drainage, waste, and vent systems. **192 pages, 8½ x 11, $29.95**

## National Construction Estimator

Current building costs for residential, commercial, and industrial construction. Estimated prices for every common building material. Provides manhours, recommended crew, and gives the labor cost for installation. Includes a CD-ROM with an electronic version of the book with *National Estimator*, a stand-alone *Windows*™ estimating program, plus an interactive multimedia video that shows how to use the disk to compile construction cost estimates. **616 pages, 8½ x 11, $47.50. Revised annually**

## Estimating Home Building Costs

Estimate every phase of residential construction from site costs to the profit margin you include in your bid. Shows how to keep track of manhours and make accurate labor cost estimates for footings, foundations, framing and sheathing finishes, electrical, plumbing, and more. Provides and explains sample cost estimate worksheets with complete instructions for each job phase. **320 pages, 5½ x 8½, $17.00**

## Construction Forms & Contracts

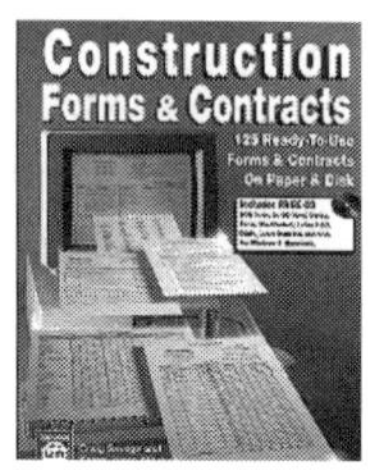


125 forms you can copy and use — or load into your computer (from the FREE disk enclosed). Then you can customize the forms to fit your company, fill them out, and print. Loads into Word for *Windows*™, *Lotus 1-2-3, WordPerfect, Works,* or *Excel* programs. You'll find forms covering accounting, estimating, fieldwork, contracts, and general office. Each form comes with complete instructions on when to use it and how to fill it out. These forms were designed, tested and used by contractors, and will help keep your business organized, profitable and out of legal, accounting and collection troubles. Includes a CD-ROM for *Windows*™ and Mac. **432 pages, 8½ x 11, $41.75**

## Moving to Commercial Construction

In commercial work, a single job can keep you and your crews busy for a year or more. The profit percentages are higher, but so is the risk involved. This book takes you step-by-step through the process of setting up a successful commercial business; finding work, estimating and bidding, value engineering, getting through the submittal and shop drawing process, keeping a stable work force, controlling costs, and promoting your business. Explains the design/build and partnering business concepts and their advantage over the competitive bid process. Includes sample letters, contracts, checklists and forms that you can use in your business, plus a CD-ROM with blank copies in several word-processing formats for both Mac and PC computers. **256 pages, 8½ x 11, $42.00**

## Contractor's Guide to the Plumbing Code

This new publication by ICBO explains the 2000 *International Plumbing Code* in plain English. It's filled with illustrations and figures to help contractors meet code on all plumbing work. Covers history of the code, conventional DWV systems, cost factors that may affect plumbing design and much more. **112-pages, 8½ x 11, $25.95**

## Plumbing & HVAC Manhour Estimates

Hundreds of tested and proven manhours for installing just about any plumbing and HVAC component you're likely to use in residential, commercial, and industrial work. You'll find manhours for installing piping systems, specialties, fixtures and accessories, ducting systems, and HVAC equipment. If you estimate the price of plumbing, you shouldn't be without the reliable, proven manhours in this unique book.
**224 pages, 5½ x 8½, $28.25**

## Contractor's Plain-English Legal Guide

For today's contractors, legal problems are like snakes in the swamp — you might not see them, but you know they're there. This book tells you where the snakes are hiding and directs you to the safe path. With the directions in this easy-to-read handbook you're less likely to need a $200-an-hour lawyer. Includes simple directions for starting your business, writing contracts that cover just about any eventuality, collecting what's owed you, filing liens, protecting yourself from unethical subcontractors, and more. For about the price of 15 minutes in a lawyer's office, you'll have a guide that will make many of those visits unnecessary. Includes a CD-ROM with blank copies of all the forms and contracts in the book.
**272 pages, 8½ x 11, $49.50**

## Basic Lumber Engineering for Builders

Beam and lumber requirements for many jobs aren't always clear, especially with changing building codes and lumber products. Most of the time you rely on your own "rules of thumb" when figuring spans or lumber engineering. This book can help you fill the gap between what you can find in the building code span tables and what you need to pay a certified engineer to do. With its large, clear illustrations and examples, this book shows you how to figure stresses for pre-engineered wood or wood structural members, how to calculate loads, and how to design your own girders, joists and beams. Included FREE with the book — an easy-to-use limited version of NorthBridge Software's *Wood Beam Sizing* program.
**272 pages, 8½ x 11, $38.00**

## Residential Plumbing Inspection

You never know what you'll find in a plumbing inspection. Corrosion between pipes, illegal traps, and even unsafe shut-off valves can be found in crawl spaces and attics of homes that appear on the surface to be built perfectly to code. This unique book is filled with illustrations and photographs of typical plumbing code violations you're likely to encounter during inspections. Then it explains what should be done to bring them to code. It covers all the areas: drainage systems, vents and venting, fixtures, water distribution, fuel gas piping, water heaters, vents and combustion, private wells and pumps, solar heating systems, and more.
**90 pages, 8½ x 11, $49.95**

## Inspecting a House

Here you'll find a rational sequence of home inspection so that you won't miss anything in heating, electrical, plumbing, and drainage systems. Whether it's an overloaded electrical panel, a driveway that sends water into the basement when it rains, or a gas furnace that's improperly vented, you'll be shown how to spot and evaluate problem areas.
**266 pages, 8 x 10, $24.95**

## Finish Carpentry: Efficient Techniques for Custom Interiors

Professional finish carpentry demands expert skills, precise tools, and a solid understanding of how to do the work. This new book explains how to install moldings, paneled walls and ceilings, and just about every aspect of interior trim - including doors and windows. Covers built-in bookshelves, coffered ceilings, and skylight wells and soffits, including paneled ceilings with decorative beams. **288 pages, 8½ x 11, $34.95**

## Estimating With Microsoft *Excel*

Most builders estimate with *Excel* because it's easy to learn, quick to use, and can be customized to your style of estimating. Here you'll find step-by-step instructions on how to create your own customized automated spreadsheet estimating program for use with *Excel*. You'll learn how to use the magic of *Excel* to create detail sheets, cost breakdown summaries, and links. You'll put this all to use in estimating concrete, rebar, permit fees, and roofing. You can even create your own macros. Includes a CD-ROM that illustrates examples in the book and provides you with templates you can use to set up your own estimating system. **148 pages, 7 x 9, $39.95**

## CD Estimator

If your computer has *Windows*™ and a CD-ROM drive, CD Estimator puts at your fingertips 85,000 construction costs for new construction, remodeling, renovation & insurance repair, electrical, plumbing, HVAC and painting. Quarterly cost updates are available at no charge on the Internet. You'll also have the *National Estimator* program — a stand-alone estimating program for *Windows*™ that *Remodeling* magazine called a "computer wiz," and Job Cost Wizard, a program that lets you export your estimates to QuickBooks Pro for actual job costing. A 60-minute interactive video teaches you how to use this CD-ROM to estimate construction costs. And to top it off, to help you create professional-looking estimates, the disk includes over 40 construction estimating and bidding forms in a format that's perfect for nearly any *Windows*™ word processing or spreadsheet program.
**CD Estimator is $68.50**

## Vest Pocket Guide to HVAC Electricity

This handy guide will be a constant source of useful information for anyone working with electrical systems for heating, ventilating, refrigeration, and air conditioning. Includes essential tables and diagrams for calculating and installing electrical systems for powering and controlling motors, fans, heating elements, compressors, transformers and every electrical part of an HVAC system. **304 pages, 3½x 5½, $18.00**

## Roofing Construction & Estimating

Installation, repair and estimating for nearly every type of roof covering available today in residential and commercial structures: asphalt shingles, roll roofing, wood shingles and shakes, clay tile, slate, metal, built-up, and elastomeric. Covers sheathing and underlayment techniques, as well as secrets for installing leakproof valleys. Many estimating tips help you minimize waste, as well as insure a profit on every job. Troubleshooting techniques help you identify the true source of most leaks. Over 300 large, clear illustrations help you find the answer to just about all your roofing questions. **432 pages, 8½ x 11, $38.00**

## Builder's Guide to Room Additions

How to tackle problems that are unique to additions, such as requirements for basement conversions, reinforcing ceiling joists for second-story conversions, handling problems in attic conversions, what's required for footings, foundations, and slabs, how to design the best bathroom for the space, and much more. Besides actual construction, you'll even find help in designing, planning, and estimating your room addition jobs.
**352 pages, 8½ x 11, $34.95**

**Craftsman Book Company**
6058 Corte del Cedro
P.O. Box 6500
Carlsbad, CA 92018

☎ 24 hour order line
**1-800-829-8123**
Fax (760) 438-0398

## In A Hurry?

We accept phone orders charged to your
❍ Visa, ❍ MasterCard, ❍ Discover or ❍ American Express

Card#________________________

Exp. date____________Initials____________

**Tax Deductible:** Treasury regulations make these references tax deductible when used in your work. Save the canceled check or charge card statement as your receipt.

**Order online** http://www.craftsman-book.com
**Free on the Internet!** Download any of Craftsman's estimating costbooks for a 30-day free trial! www.craftsman-book.com/downloads

Name
e-mail address (for order tracking and special offers)
Company
Address
City/State/Zip ❍ This is a residence
Total enclosed______________(In California add 7.25% tax)
*We pay shipping when your check covers your order in full.*

**10-Day Money Back Guarantee**

- ❍ 69.95 ADA Accessibility Guidelines for Buildings and Facilities With ADA Technical Assistance Manuals
- ❍ 36.50 Basic Engineering for Builders
- ❍ 38.00 Basic Lumber Engineering for Builders
- ❍ 14.75 Blueprint Reading for the Building Trades
- ❍ 34.95 Builder's Guide to Room Additions
- ❍ 59.95 California Accessibility Reference Manual (CARM) 2000
- ❍ 68.50 CD Estimator
- ❍ 39.50 Construction Estimating Reference Data with FREE *National Estimator* on a CD-ROM.
- ❍ 41.75 Construction Forms & Contracts with a CD-ROM for *Windows*™ and Macintosh.
- ❍ 49.95 Construction Surveying & Layout
- ❍ 39.00 Contractor's Guide to the Building Code Revised
- ❍ 25.95 Contractor's Guide to the Plumbing Code
- ❍ 59.95 Contractor's Legal Kitc
- ❍ 49.50 Contractor's Plain-English Legal Guide
- ❍ 25.00 Contractor's Survival Manual
- ❍ 36.00 Craftsman's Illustrated Dictionary of ConstructionTerms
- ❍ 17.00 Estimating Home Building Costs
- ❍ 39.95 Estimating with Microsoft *Excel*
- ❍ 34.95 Finish Carpentry: Efficient Techniques for Custom Interiors
- ❍ 24.95 Inspecting a House
- ❍ 78.00 2003 *International Building Code*
- ❍ 53.00 2003 *International Residential Code*
- ❍ 32.50 Markup & Profit: A Contractor's Guide
- ❍ 42.00 Moving to Commercial Construction
- ❍ 47.50 National Construction Estimator with FREE *National Estimator* on a CD-ROM.
- ❍ 48.25 National Plumbing & HVAC Estimator with FREE *National Estimator* on a CD-ROM.
- ❍ 49.50 National Renovation & Insurance Repair Estimator with FREE *National Estimator* on a CD-ROM.
- ❍ 48.50 National Repair & Remodeling Estimator with FREE *National Estimator* on a CD-ROM.
- ❍ 29.95 Planning Drain, Waste & Vent Systems
- ❍ 34.00 Plumber's Exam Preparation Guide
- ❍ 32.00 Plumber's Handbook Revised
- ❍ 28.25 Plumbing & HVAC Manhour Estimates
- ❍ 24.95 Precision Framing
- ❍ 22.75 Profits in Buying & Renovating Homes
- ❍ 33.50 Renovating & Restyling Older Homes
- ❍ 49.95 Residential Plumbing Inspection
- ❍ 29.00 Residential Wiring to the 2002 *NEC*
- ❍ 32.50 A Roof Cutter's Secrets to Custom House
- ❍ 22.00 Roof Framing
- ❍ 38.00 Roofing Construction & Estimating
- ❍ 26.50 Rough Framing Carpentry
- ❍ 39.75 Steel-Frame House Construction
- ❍ 18.00 Vest Pocket Guide to HVAC Electricity
- ❍ 24.95 Wiring a House
- ❍ 25.50 Wood-Frame House Construction
- ❍ 37.00 Illustrated Guide to the *IPC & IFGC*
- ❍ FREE Full Color Catalog

**Prices subject to change without notice**

NO POSTAGE NECESSARY IF MAILED IN THE UNITED STATES

**BUSINESS REPLY MAIL**

FIRST CLASS MAIL PERMIT NO. 271 CARLSBAD, CA

POSTAGE WILL BE PAID BY ADDRESSEE

**Craftsman Book Company**
6058 Corte del Cedro
P.O. Box 6500
Carlsbad, CA 92018-9974

NO POSTAGE NECESSARY IF MAILED IN THE UNITED STATES

**BUSINESS REPLY MAIL**

FIRST CLASS MAIL PERMIT NO. 271 CARLSBAD, CA

POSTAGE WILL BE PAID BY ADDRESSEE

**Craftsman Book Company**
6058 Corte del Cedro
P.O. Box 6500
Carlsbad, CA 92018-9974

NO POSTAGE NECESSARY IF MAILED IN THE UNITED STATES

**BUSINESS REPLY MAIL**

FIRST CLASS MAIL PERMIT NO. 271 CARLSBAD, CA

POSTAGE WILL BE PAID BY ADDRESSEE

**Craftsman Book Company**
6058 Corte del Cedro
P.O. Box 6500
Carlsbad, CA 92018-9974